BLACK ROUTES TO ISLAM

THE CRITICAL BLACK STUDIES SERIES

INSTITUTE FOR RESEARCH IN AFRICAN-AMERICAN STUDIES

Edited by Manning Marable

The Critical Black Studies Series features readers and anthologies examining challenging topics within the contemporary black experience in the United States, the Caribbean, Africa, and across the African diaspora. All readers include scholarly articles originally published in the acclaimed quarterly interdisciplinary journal *Souls*, published by the Institute for Research in African-American Studies at Columbia University. Under the general editorial supervision of Manning Marable, the readers in the series are designed both for college and university course adoption as well as for general readers and researchers. The Critical Black Studies Series seeks to provoke intellectual debate and exchange over the most critical issues confronting the political, socioeconomic, and cultural reality of black life in the United States and beyond.

Titles in this series published by Palgrave Macmillan:

Racializing Justice, Disenfranchising Lives: The Racism, Criminal Justice, and Law Reader (2007)

> Edited by Manning Marable, Keesha Middlemass, and Ian Steinberg

Seeking Higher Ground: The Hurricane Katrina Crisis, Race, and Public Policy Reader (2008)

> Edited by Manning Marable and Kristen Clarke

Transnational Blackness: Navigating the Global Color Line (2008)

> Edited by Manning Marable and Vanessa Agard-Jones

Black Routes to Islam

> Edited by Manning Marable and Hishaam D. Aidi

Barack Obama and African-American Empowerment: The Rise of Black America's New Leadership

> Edited by Manning Marable and Kristin Clarke

New Social Movements in the African Diaspora: Challenging Global Apartheid

> Edited by Leith Mullings

The New Black History: The African-American Experience since 1945 Reader

> Edited by Manning Marable and Peniel E. Joseph

Beyond Race: New Social Movements in the African Diaspora

> Edited by Manning Marable

The Black Women, Gender, and Sexuality Reader

> Edited by Manning Marable

Black Intellectuals: The Race, Ideology, and Power Reader

> Edited by Manning Marable

BLACK ROUTES TO ISLAM

Edited by Manning Marable and Hishaam D. Aidi

palgrave
macmillan

First published in 2009 by PALGRAVE MACMILLAN® in the United States—a division of St. Martin's Press LLC, 175 Fifth Avenue, New York, NY 10010.

Where this book is distributed in the UK, Europe and the rest of the world, this is by Palgrave Macmillan, a division of Macmillan Publishers Limited, registered in England, company number 785998, of Houndmills, Basingstoke, Hampshire RG21 6XS.

Palgrave Macmillan is the global academic imprint of the above companies and has companies and representatives throughout the world.

Palgrave® and Macmillan® are registered trademarks in the United States, the United Kingdom, Europe and other countries.

Hardcover ISBN: 978-1-4039-8400-5
Paperback ISBN: 978-1-4039-7781-6

Library of Congress Cataloging-in-Publication Data

Black routes to Islam / edited by Manning Marable and Hishaam D Aidi.
    p. cm.
  Includes bibliographical references and index.
   ISBN 978-1-4039-7781-6—ISBN 978-1-4039-8400-5 1. African American Muslims—History. 2. African Americans—Religion—History. 3. Islam—United States—History. 4. African Americans—Travel—Islamic countries. I. Marable, Manning, 1950– II. Aidi, Hishaam D.
  BP67.U6B53 2009
  297.8'7—dc22   2009016317

A catalogue record of the book is available from the British Library.

Design by Scribe Inc.

First edition: August 2009

10 9 8 7 6 5 4 3 2 1

Tranferred to Digital Printing 2015

This volume includes articles that originally appeared in *Souls*, published by the Institute for Research in African-American Studies, Columbia University.

CONTENTS

THE EARLY MUSLIM PRESENCE AND ITS SIGNIFICANCE

HISHAAM D. AIDI AND MANNING MARABLE

In the name of God The Compassionate . . . I am not able to write my life. I have forgotten much of the language of the Arabs. I read not the grammatical, but very little of the common dialect. I ask thee, O brother, to reproach me not, for my eyes are weak, and my body also.

—Omar ibn Said, Muslim slave in North Carolina,
Letter to a friend in West Africa (1836)

A LITTLE KNOWN FACT THAT CONTINUES TO INSPIRE INCREDULITY IS THAT America's first Muslims arrived chained in the hulls of slave ships. In 1977, when ABC television broadcast the miniseries *Roots* to an unprecedented 130 million viewers, this skepticism came into full view. The miniseries based on Alex Haley's eponymous novel would spark a great interest in African cultural retentions and ethnic and racial genealogy, but also disbelief—particularly over the scenes showing how Kunta would avoid eating pork and kneel in prayer facing east. A year after the novel's publication, critic James Michener took issue with a scene that described Africans praying to Allah in the interior of a ship making the transatlantic crossing: "To have Kunta Kinte, or one of his fellows praying to Allah while chained in the bottom of a Christian ship is an unjustified sop to contemporary developments rather than a true reflection of the past."[1] Critics suspected that Haley had inserted the scenes of Kunta Kinte praying simply to lend some historical precedent and legitimacy to the Black Muslim and Black Power movements of the 1960s and 1970s.[2] This skepticism would obviously increase as various genealogists contested Haley's claims that Kunta was his seventh-generation ancestor taken from the village of Juffure in the Gambia in 1767 and sold into slavery in Maryland. It was long considered improbable that the African slaves in America may have been Muslim, and even more unlikely that they continued to practice their faith and transmitted their culture to their children in the New World. As one popular textbook put it, "what Muslim faith they [the African slaves] brought with them was quickly absorbed into their new Christian milieu and disappeared."[3]

But scholarship over the last two decades, inspired in part by the debate over *Roots*, has revealed a "subtle Muslim presence" in America since the early 1500s. Historians have unearthed texts written by Muslim slaves in English and Arabic, shedding light on a far-flung population of Muslim Africans enslaved throughout the New World, many of whom were distinguished by their literacy, and who struggled to maintain their faith through rituals and naming practices, by reading the Koran and writing Arabic, sometimes even launching jihads against their overlords. While historians will continue to debate the number of Muslims enslaved in the New World,[4] it has become increasingly clear that Muslim slaves in Anglo-America were treated comparatively better than their counterparts in Latin America, and their presence would shape not only racial categories and stratification in the United States but also inform early American views of the "Orient." The impact of the early Muslim presence on American racial discourse and representations of the Islamic world is critical to understanding the different Islamic and quasi-Islamic movements that emerged in early twentieth-century African America.

TEXTS AND RETENTIONS

The historical research on African Muslim slaves in the New World reveals a difference between the treatment of Muslim slaves in Spanish territories and those in French and English territories. Michael Gomez notes that whereas the experience of Muslim slaves in Latin America, the Spanish-speaking Caribbean, and Brazil was characterized by "severe political repression," since these communities were often seen as threatening, the "Muslim communities in the United Sates were comparatively quiet and compliant." In addition, throughout the French- and English-speaking Caribbean and North America, Muslim slaves—who were often Mande, Fulbe, and from Senegambia—would be elevated above other slaves and given less arduous work. "Patterns of privileging Muslim individuals would develop all over Anglophone America," writes Gomez, "and they stand out in sharp relief against the anti-Muslim mania of the Spanish and Portuguese domains." Why this difference? A central reason is that these "new" encounters were an extension of centuries-old interactions in the Old World, and the differing treatments of Muslim slaves reflect the European states' disparate relationships to North Africa and the wider Muslim world. The Spanish and Portuguese states, who were battling Islam within their borders, were obsessed with the Moorish threat and would take great measures to prevent the importation of Muslim Africans to the New World, even persecuting those slaves suspected of being Muslim. If the relationship between Africans and Spanish and Portuguese in Latin America was a continuation of a history of violence that had existed on the Iberian Peninsula for centuries, "England had no such tradition and therefore had no reason to anticipate religious hostility in any a priori fashion."[5] One indicator of this comparative tolerance—even preference—for Muslim slaves that existed in the United States was the considerable writing (journalistic, scholarly, and even fiction) that emerged about Muslim slaves and their narratives.

Slave narratives are a precious resource for understanding the experiences of Muslim slaves in America and how they were treated and represented. The Muslim slaves who captured the American imagination in the eighteenth and nineteenth centuries were all distinguished by their leadership qualities and learning. Some of these individuals would become local legends spawning a spate of literature trying to explain their literacy and supposed "Oriental" origins. The firsthand accounts of contemporaries who encountered Muslim slaves have turned out to be critical in shedding light on the lives and habits of Muslim slaves. Consider the reference in Georgia Conrad's memoir, *Reminiscences of a Southern Woman*, to a Muslim family she met on Sapelo Island, off the coast of Georgia, in the mid-1850s: "On Sapelo Island near Darcen, I used to know a family of Negroes who worshipped Mahomet. They were tall and well-formed, with good features. They conversed with us in English, but in talking among themselves they used a foreign tongue that no one else understood. The head of the tribe was a very old man named Bi-la-li. He always wore a cap that resembled a Turkish fez. These negroes held themselves aloof from others as if they were conscious of their own superiority."[6] Conrad was referring to Bilali, a Muslim slave who would gain notoriety for his literacy and valor. Wylly Spaulding, the grandson of Thomas Spaulding, who was Bilali's master, also writes about his grandfather's slaves of "Moorish or Arabian descent, devout Mussulmans, who prayed to Allah . . . morning, noon, and evening."[7]

The said Bilali would make a lasting mark in the American cultural and literary imagination, becoming the subject of two children's books by Joel Chandler Harris—*The Story of Aaron (So Named) the Son of Ben Ali* (1896) and *Aaron in the Wildwoods* (1897); he would be invoked as a Muslim ancestor in Tony Morrison's novel *Song of Solomon* (1977), and mentioned in Julie Dash's 1991 film *Daughters of the Dust*. Bilali gained notoriety after the War of 1812, during which he and eighty slaves successfully fought and prevented the British from invading Sapelo. He is also remembered for an Arabic text that he wrote and had placed in his coffin along with his Quran and his prayer rug. The text—a collection of pieces from the Maliki legal text *ar-Risala*—attempted to reconcile the law of Islam with leading a principled life and showed how the author was struggling to maintain his *iman* in the land of America.[8] Because of his outstanding qualities, Bilali would be appointed the manager of his master's plantation, overseeing approximately five hundred slaves.

Ethnographic work in the 1930s about the Muslim community of Sapelo and St. Simon's Island during the antebellum period—by linguist Lorenzo Dow Turner and the Georgia Writers' Project—showed that the islands' relative isolation allowed for a preservation of Muslim traditions. This decades-old research on the Georgia Sea Islands has proven to be an invaluable source on Muslim slave life (and further added to Bilali's reputation since a number of the individuals interviewed during the 1930s were Bilali's descendents).[9] Particularly intriguing was Lorenzo Turner's hypothesis that the "Ring Shout," a Southern Baptist ritual in which worshippers move in a circle around an altar at the center of the church clapping, shuffling their feet, and praying aloud, may derive from the Muslim ritual of circumambulating the Kaaba. Lorenzo Turner noted that "Shout = a religious ring dance in which the participants continue to perform until they are exhausted" and proposed that the origin of the word "Shout" is the Arabic *shaut*, which means "to move around the Kaaba (the small stone building at Mecca which is the chief object of the pilgrimage of the Mohamedans) until exhausted."[10] Another possible Muslim retention that still absorbs scholars involves the fact that the Christian congregation on Sapelo Island prays toward the east and the church itself is built facing that direction. Worshippers are taught to pray facing east because "the devil is in the other corner"; and in this congregation, even the dead are buried facing east. While "the east" in this case might simply mean Africa and not necessarily Mecca, historians continue to ponder if this peculiarity—along with other Shouters' prayer rituals—including being barefoot at church, kneeling on a piece of fabric, performing a ritual handshake followed by the touching of the left breast by the right hand—are "Islamic traits."[11] Without more evidence to substantiate these claims of Muslim retentions in African American culture, these hypotheses remain speculative; but these claims of Muslim "survivals" and "continuities" would become critical to the twentieth-century social movements discussed in this volume that would use these alleged retentions as raw material for their ideological and cultural repertoires.

DRIVERS AND OVERSEERS

While the cultural influence of the early Muslim presence remains a matter of speculation, the impact of the Muslim slave population on the hierarchy of the American South is less disputed. As Gomez has argued, "the most lasting [and] . . . most salient impact" of Islam in colonial and antebellum America "was its role in the process of social stratification within the larger African American society," adding that "the early Muslim community contributed significantly to the development of African American identity."[12] Many of the Muslim slaves were of Fulbe, Mande, and Senegambian background, whose features were thought to be closer to those of Europeans than of Africans, and American slave owners invariably saw Muslim slaves "as more intelligent, more reasonable, more physically attractive, more dignified people."[13] The "European" features and intelligence were seen as obviously linked. This bias reflected the influence of the notorious Hamitic thesis—the argument, popular in nineteenth-century European colonial thought, that North Africans and Central Africans with "thin" features were actually Africans of Caucasian origin who had migrated from the West to Africa; thus, Senegambians, the Fulbe, and Tutsis, for instance, were said to be not of African origin but of European, Egyptian, or Arab descent. The Hamitic thesis and the belief in the alleged superiority of the "Mohammedans" would make its way across the Atlantic and be invoked to explain the literacy of Muslim slaves in the New World. In his *America Negro Slavery*, Ulrich Phillips observes that plantation owners often thought the Senegalese were the most intelligent because they "had a strong Arabic strain in their ancestry."[14]

Because of this thinking, Muslims slaves in the United States were often placed in positions of power over other non-Muslim slaves. In his book *Prince Among Slaves*—about the life of another prominent Muslim slave, Ibrahima Abdal Rahman—Terry Alford notes that Muslim slaves were used as "drivers, overseers and confidential servants with a frequency their numbers did not justify."[15] Muslim overseers informed on disobedient slaves and crushed slave uprisings, earning the distrust of other Africans. In a treatise he wrote calling for the "benign" treatment of slaves, Zephaniah Kingsley noted how Muslim drivers on Sapelo Island jointly crushed a slave revolt. Referring to events along the Georgia coast during the Anglo-American War of 1812, Kingsley alludes to "two instances, to the southward, where gangs of negroes were prevented from deserting to the enemy [England] by drivers, or influential negroes, whose integrity to their masters and influence over the slaves prevented it; and what is still more remarkable, in both instances the influential negroes were Africans; and professors of the Mahommedan religion."[16] Muslim slaves often took the side of the slave owner, and this—along with the preferential treatment they received—created a rift between Muslim and non-Muslim slaves.

LITERACY AND GENEALOGY

The impact of the Muslim presence on American racial discourse is also evident in how scholars and journalists of the day sought to explain the origins of Muslim slaves. A great deal of energy was expended by antebellum writers and owners trying to ascertain the genealogy of the Muslim slaves and to find a way to classify them so that their Arabic literacy would not threaten the ideological underpinnings of racial slavery. The case of Omar ibn Said is illustrative: because of his Arabic literacy, he was identified as a Moor. "He is an Arab by birth of royal blood" who had been enslaved by Africans, "whom he had always hated," explained *The Providence Journal* in 1846.[17] But it is the tales spun around Omar's origins that are most dubious—and revealing. On January 22, 1847, *The Wilmington Chronicle* accurately described Omar as a "Pulo" (Fulbe), but the writer notes, "the Foulahs, or Falatas, are known as descendents of the Arabian Mohamedans who migrated to Western Africa in the seventh century. They carried with them the literature of Arabia, as well as the religion of their great Prophet, and have ever retained both. The Foulahs stand in the scale of civilizations at the head of all African tribes."[18] This tendency to depict African Muslim slaves as Moorish, Arab, or of Oriental origin was widespread. The aforementioned Bilali of Sapelo Island would be similarly "Orientalized," portrayed as "an Arab—man of the desert—slave hunter" in Joel Chandler Harris's novel *The Story of Aaron (So Named) the Son of Ben Ali*.

The literacy and education of the Muslim slaves was rarely seen as a result of their exposure to Islam or Arabic education in West Africa but was attributed to their Arab, Berber, or Moorish origin. Muslim slaves were thus separated from other Africans, not only physically—in the work and positions they were given—but also ideologically and epistemologically. And these Muslim slaves—many of whom distanced themselves from their non-Muslim counterparts—would accept these labels and often eagerly claim Moorish, Arab, or Berber origin themselves. (For instance, Ibrahima Abd ar-Rahman, a Fulbe, is reported to have insisted that he had no "Negro" blood and "placed the negro in a scale of being infinitely below the Moor.") This air of superiority would fuel the resentment of other slaves, with those of Muslim, Moorish, and Arab identity being suspected of being proslavery and antiblack, while also luring other slaves to claim Moorish or Muslim identity.

But why the need for these utterly fictitious genealogies? Why the effort to "de-Negroize" the Muslim slave? As Henry Louis Gates has observed, the Enlightenment discourse of this era viewed writing as "the *visible* sign of reason,"[19] and, as such, the literacy of Muslim slaves posed a challenge. Diouf argues that accepting these Muslim slaves as Africans would have posed an ideological threat to the American racial order: "It was more acceptable to deny any Africanness to the distinguished Muslims than to recognize that a 'true' African could be intelligent and cultured but enslaved nonetheless. So, gradually, the African Muslims were seen as owing their perceived superiority not to their own 'genes,' not even to their culture or proximity to the Arab world, but to foreign 'blood.'"[20] This reasoning grew out of a peculiarly American race regime, which, unlike its counterparts in Latin America, did not recognize the intermediate strata of mulattos, creoles, or *mestizo*, with different meanings of intelligence and privilege attached to these categories. As Diouf observes, in the more rigid American racial order with its Manichean categories, an "intelligent black" was categorized not as black but as close to white—a Moor or Arab; blacks were thus still seen as inferior, but those of alleged Moorish or Arab origins were seen as suited for higher positions where they could monitor the majority of blacks and enforce the rules of the slave society.

If Muslim slaves in the United States had to be categorized and represented in a way that would not unsettle the country's racial order, their narratives—the slave narrative—would emerge as a literary genre in the United States in a way unseen in Latin America and pose a challenge to reigning assumptions about human races, modernity, and the Enlightenment. As one literary critic asked, "What happened when a significant number of literate enslaved Muslims gave the incontestable lie to the notion that blacks were incapable of reason—possessing neither books nor history nor scriptural/revealed religion—were, in short, a cultural tabula rasa?"[21] One way to deal with this challenge was by presenting the Muslim slave's life story as a tale of uplift from heathenism and an espousal of Christianity and American values. The popular narrative that emerged around Omar ibn Said, for example, tells of an Arab Muslim's difficult journey to freedom and enlightenment: after enduring slavery in South Carolina, he ran away before he was taken in by John Owen, the former governor of North Carolina who was impressed by his literacy and helped him convert to Christianity.

In these accounts, the Muslim slave's conversion to Christianity was also often accompanied by an express desire to return to Africa to spread the gospel and American civilization. The first Muslim slave known to have gained freedom because of his literacy was Ayyub bin Suleiman (also known as Job the Son of Solomon), a Fulbe from a prominent family who was enslaved in Maryland, where white children "would mock him and throw dirt in his face" as he attempted to recite his prayers in the woods. His literacy and piety would draw the attention of an English minister who would begin to press for his manumission. Henry Louis Gates identifies the 1731 letter of Ayyub bin Suleiman Diallo to his father—asking that he ransom him out of slavery—as the first instance in modernity when an African literally wrote his way out of slavery.[22] The London-based Royal African Company would eventually arrange for Ayyub's return to Bondu, where he would assist the Royal African Company in their trade of rum, gold, and slaves. Literacy would in fact help a number of African Muslim slaves earn their liberty, as they offered their services to American and English colonial projects in West Africa.

The widespread interest in the stories of the Muslim slaves must be viewed in historical context. Of the hundreds of narratives written in North America, very few were written by African-born slaves and even fewer—the most notable exception being Olaudah Equiano's autobiography published in 1789—would gain a wide readership. Abolitionists who promoted slave narratives generally did not seek the texts of African-born slaves since they thought these would be seen as too alien to gain a wide readership and aid the abolitionists' cause.[23] Moreover, unlike most slave narratives that detailed the horrors of life under slavery and crackled with tales of escape and adventure, the Muslim slaves' narratives were not so suspenseful, often simply giving descriptions of Islamic teachings and portrayals of daily life in Africa. Yet despite their few numbers, the Muslim slave narratives would draw considerable attention because of the Haitian Revolution (1791–1803) and because of the Barbary wars (1801–5), which made the figure of a disobedient slave or educated Muslim slave rather unsettling. As African American slave narratives were gaining a readership, the captivity narratives of white Americans held in North Africa were also becoming popular, and, jointly, these texts would engage the American popular imagination and disturb the racial status quo.

WHITE SLAVES, AFRICAN MASTERS

The attacks on American vessels off the Barbary Coast, and the enslavement of scores of white Americans by Africans in a distant Muslim land since the mid-eighteenth century, produced a range of reactions from American observers and touched on myriad cultural and political issues. Since the 1700s, American observers would view their country's racial predicament through the prism of Barbary and the wider Orient.[24]

Critics would point to North African slavery and despotism to tell America that it had to live up to its principles and abolish bondage—that it could not descend to such barbarous behavior. As early as 1700, the renowned jurist Samuel Sewell wrote *The Selling of Joseph*, the first antislavery pamphlet condemning bondage in America and highlighting the hypocrisy of those who decried slavery in North Africa: "Methinks, when we are bemoaning the barbarous Usage of our Friends and Kinsfolk in *Africa*: it might not be unreasonable to enquire whether we are not culpable in forcing the Africans to become *Slaves* amongst our selves."[25] In 1776, Samuel Hopkins, a Congregationalist minister and early abolitionist, excoriated his fellow Northerners who seemed appalled by North African slavery but silent about bondage at home: "If many thousands of our children were slaves in Algiers or any part of the Turkish dominions . . . how would the attention of the country be turned to it! Would it not become the chief topic of conversation? Would any cost or labor be spared . . . in order to obtain their freedom? . . . And why are we not as much affected with the slavery of the thousands of blacks among ourselves whose miserable state is before our eyes." He continued, "The reason is obvious . . . It is because they are Negroes and fit for nothing but slaves. 'Tis because they are negroes, and we have been used to look on them in a mean, contemptible light."[26] Hopkins would go on to ban slave owners from his congregation and found the American Colonization Society to return blacks to Africa.

The "Barbary problem" emerged as a contentious issue at the Constitutional Convention of 1787, with the Federalists calling for the establishment of a navy to meet the corsair threat and the anti-Federalists resisting the expansion of the federal state's power. At the convention in Philadelphia, representative Hugh Williamson of North Carolina would ask publicly, "What is there to prevent an Algerine pirate from landing on your coast, and carrying your citizens into slavery? You have not a single sloop of war?"[27] After the Barbary wars, the regencies of North Africa would cease to pose a threat to American commercial interests, but the narratives written by Americans enslaved in North Africa would gain a new significance in the antebellum years, as these texts, like the Muslim slave narratives, became part of the debate over slavery. Abolitionists and white supremacists alike would use the narratives of released captives to protest or defend slavery on American soil.

The Barbary captivity narratives had existed since the 1600s, but only became popular in the first half of the nineteenth century as the number of slaves in the United States increased rapidly—from one million in 1807 to two million by 1837[28]—at the same time as African American slave narratives rose in popularity. As the dispute over slavery intensified, so did the interest in the writings of ex-slaves—whites held captive in Barbary but also the accounts of free blacks—from Omar ibn Said's autobiography (1831) to Frederick Douglas's best-selling narrative (1845) to Harriet Jacobs's *Incidents in the Life of A Slave Girl* (1861)—which were used to bolster the abolitionist case. In his *White Slavery in the Barbary States* (1853), the abolitionist Charles Sumner observed that the interest in Barbary captives fed interest in slavery at home: "The interest awakened for the slave in

Algiers embraced also the slave at home. Sometimes they were said to be alike in condition; sometimes, indeed it was openly declared that the horrors of our American slavery surpassed that of Algiers."[29]

The Barbary captivity narrative depicted a trajectory of religious redemption that was the reverse of the Muslim slave's account, which portrayed morally redeemed figures crossing the Atlantic to return to their African homelands, but both texts depicted individuals set on helping America live up to its principles. In the African American Muslim narrative, the slave would suffer in bondage before a kind master would discover his Muslim ("Oriental") origins and manumit him; the grateful African would then embrace Christianity and pledge to serve America upon returning to his homeland, enlightened and intent on spreading American ideals. In the Barbary narrative, by contrast, the white slave would endure brutality and resist conversion to Islam before defeating his Moorish master, who would then embrace Christianity and American antislavery liberalism, and the captive would then return to America vowing to abolish slavery. Despite their predictable "plotline," both types of narratives resonated deeply, challenging American racial hierarchies and the young republic's moral authority and self-image.

The Barbary narratives helped shape, and were shaped by, the debate over slavery that was dividing the country. Tales of white bodies possessed and violated by African masters presented a distorted mirror image of the situation in America, causing considerable anxiety and challenging racial slavery's black-white divide and its foundational assumptions about white supremacy and African primitivism. North African slavery posed a similarity and a parallel between America and the Orient that was profoundly disturbing: not only were *they* capable of enslaving white Americans, but how could *we* denounce their slavery while maintaining slavery at home? This contradiction produced varied reactions from those who would compare the two systems of slavery and denounce the South for stooping to Oriental levels of cruelty to those who were sickened by North African "barbarousness" and concluded that the blacks in America could never be ready for emancipation.

The most popular, and possibly the most influential, Barbary narrative to be sold in the United States was Captain James Riley's 1817 account of his capture by "wandering Arabs" in northwest Africa. His book, *Suffering in Africa*, which drew attention to the country's moral blind spot of slavery while speaking of America's providential role, sold nearly a million copies and was adapted into an illustrated children's storybook. The narrative told the story of how Captain Riley and his crew endured slavery in the desert after their ship *Commerce* was shipwrecked off the coast of modern-day Mauritania. When finally liberated by a compassionate Muslim traveler, Riley vows to devote the rest of his life to ending slavery in America. Recalling black slaves he had seen on the auction block in New Orleans, Riley's narrative ends with a plea to Americans to chop down the "cursed tree of slavery" and obliterate "the rod of oppression."[30] When he returns to Washington, Riley meets President Monroe, who encourages him to publish his story. Riley's *Suffering in Africa* came out in at least twenty-eight editions and is still in print today. It is perhaps the most influential Barbary captivity narrative since it is said to have been one of Abraham Lincoln's all-time favorite books—along with *The Bible* and John Bunyan's *Pilgrim's Progress*—and is believed to have shaped the future president's views on slavery.[31] These narratives would shape early American representations of Africa and the Orient, unsettling American racial formations and showing how such racial constructs were often projected onto the Muslim world. From the mid-eighteenth century onward, in part because of these narratives, many Americans would view the Islamic world through the prism of race and slavery.

MUSLIM SLAVES AND THE COLONIZATION OF LIBERIA

A number of the abolitionists who pointed to North Africa to highlight America's bane of slavery came to believe that returning blacks to West Africa was the best solution to the country's racial discord. Thus, the ardent abolitionist Samuel Hopkins thought that equality was not possible for blacks in America because of widespread racism and white fears of uprisings by freed blacks, and came to believe in "colonization" or the repatriation of freed blacks to Africa. The reverend underlined the economic gains of planting a colony of African Americans in Africa. "Such a settlement, promoted by the Americans," he enthusiastically wrote a fellow minister in 1793, "would not only tend to the good of the Africans, but would, in time, be a source of profitable trade to America, instead of the West India trade, which will probably fail more and more."[32]

The American Colonization Society's attitude toward Muslim slaves is particularly interesting. For the repatriation movement, Muslim slaves, with their Arabic literacy, were seen as a valuable tool for opening up West Africa to American economic and political interests. The Muslim slaves were seen as natural intermediaries with the Muslims and pagans of the West African Coast; they had, after all, embraced Christianity, were indebted to America for their newfound freedom and faith, and would help spread American civilization in Africa. As the repatriation movement grew in influence, Muslim slaves and their narratives gained greater political significance, and individuals like Ibrahima Abdul Rahman would—along with other Muslim slaves—become central to the debate about Liberia, and, often feigning conversion to Christianity, would volunteer to return to Africa to spread the Gospel's good news. Muslim slaves would work with the American Colonization Society to gain passage to Africa. According to *The African Repository*, the mouthpiece of the American Colonization Society, in 1835, Lamine Kaba ("Paul"), who had served for almost thirty years as a slave in Georgia and South Carolina, addressed a group of white philanthropists about his sudden conversion to Christianity following his manumission and his desire to pursue missionary work in Africa, where he would join his family. Traveling under the name of Paul A. Mandingo, Kaba would reach Liberia in August 1835 and would settle in Sierra Leone, but he would remain in contact with Muslim slaves that he had befriended in America. (The letter quoted in the beginning of this chapter was written by Omar Ibn Said to Kaba in 1836, showing that literate Muslim slaves maintained ties across the Atlantic.[33]) The Muslim slaves' eagerness to get involved in the repatriation effort—as a way to gain passage home—undoubtedly incensed black abolitionists opposed to the Liberia project, which they saw as a ruse to rid America of its African population.

African American Muslim slave narratives continue to be discovered: Omar bin Said's Arabic text turned up in an old trunk in Virginia in 1995; the narrative of Shaykh Sana See of Panama was discovered in 1999; and the 1873 autobiography of Nicolas Said of Alabama was stumbled upon in a library annex in 2000.[34] As more of these texts come to light, we will undoubtedly learn more about the lives of Muslim slaves and ex-slaves. The foregoing discussion of the early Muslim presence on the American plantation, the literacy of Muslim slaves, the reactions to Muslim Arabic writing, and the simultaneous appeal of narratives by and about Muslim slaves and the Barbary captivity narratives are all critical to understanding early (black and white) American attitudes toward Islam, the rise of modern Islamic movements, and contemporary representations of Islam and the Orient. Since the American Revolution, Islam has been contentiously associated with race and slavery—as revealed by the politics surrounding the Barbary and Muslim slave narratives—and with American interests in Africa and the Orient.

By the early twentieth century, there would be a clear association in the African American community between Arabic and erudition, and between Islam, liberation, and black empowerment. The experiences and narratives of Muslim slaves—combined with the writings of Edward Blyden, who, while in Liberia, was deeply impressed by the civilization and political autonomy of Islamic states in West Africa—laid the intellectual groundwork for the Black Muslim movements of the early and mid-twentieth century. Yet, at the same time, there was a growing resentment of Muslims as arrogant "house negroes" who denied their blackness and preferred to pass for "Arab" or "Oriental" or even as race traitors all too willing to defend the institution of slavery and partake in pernicious colonization projects. The preferential treatment afforded Muslim slaves by the American racial order encouraged other blacks to claim Moorish or Arab descent and even "incentivized" conversion to Islam.[35] The association of Islam with preferential treatment and "passing" would grate the nerves of prominent black leaders. Reflecting this view, Booker T. Washington, in his autobiography *Up From Slavery*, scoffs at how a "dark-skinned man . . . a citizen of Morocco" is allowed into a "local hotel" from which he, "an American Negro," is banned.[36] The belief that Muslim identity and conversion to Islam constituted a form of "passing" would echo deep into the twentieth century and would only be reinforced by the fact that many African Americans who embraced Islam were not "Jim Crowed." As Dizzy Gillespie famously pointed out in the 1950s, many jazz musicians who took on Muslim names could enter "whites only" restaurants and even had "white" stamped on their union cards.[37]

In short, decades before the post-1945 era when the United States began to establish a presence in the Middle East, and before the advent of immigrants from North Africa, the Middle East, and South Asia, the labels "Arab," "Moorish," and "Muslim" already referred to loaded, polarizing categories. By the mid-nineteenth century, a discourse had evolved that separated the ("sophisticated") Arab/Moor from the African, and split Northern Africa ("The Orient") from Black Africa ("Africa"). These efforts to separate Arab from African identity, seen in the Orientalizing ("de-Negroizing") of Muslim slaves, is still evident in contemporary discourses on Northern Africa, specifically Sudan, where Arabic-speaking blacks are habitually described as "settlers" or "nonindigenous." The emergence of various Islamic and quasi-Islamic movements in African America in the early twentieth century, which would overlap with the rise of the northern ghetto and American global power, would resourcefully adapt and contest these historic categories and representations. The postwar ascent of the United States and the country's subsequent expansion into the Muslim world—particularly the Middle East—combined with the influx of immigrants from these regions into the United States, would all increase the appeal of Islam in the country's urban core, producing myriad trends and movements—secular and religious—that would claim solidarity or a connection to some region or people in the Islamic world.

The chapters in this volume address various aspects of this historic and evolving relationship between Islam and black America. The authors included in this volume explore different dimensions of the more than century-long interaction between black America and Islam. Starting with the nineteenth-century narratives of African American travelers to the Holy Land, the following chapters probe Islam's role in urban social movements, music, and popular culture, gender dynamics, relations between African Americans and Muslim immigrants, and the racial politics of American Islam with the ongoing war in Iraq and deepening U.S. involvement in the Orient.

NOTES

1. James A Michener, "*Roots*, Unique In Its Time," *New York Times Book Review*, February 26, 1977, 41.

2. Alex Haley, *Roots: The Saga of an American Family* (Garden City, NY: Doubleday, 1976). The novel contains various scenes highlighting Kunta's Islamic faith and practices: "As the sun began to set, Kunta turned his face toward the East, and by the time he had finished his silent evening prayer to Allah, dusk was gathering" (*Haley*, 173). Elsewhere, Haley writes, "Through the night, he lay drifting into and out of sleep and wondering about these black ones who sounded like Africans but ate pig. It meant that they were all strangers—or traitors—to Allah. Silently he begged Allah's forgiveness in advance if his lips would even touch any swine without his realizing it, or even if he ever ate from any plate that any swine meat had ever been on" (*Haley*, 178).

3. Caesar E. Farah, *Islam: Beliefs and Observances*, 7th ed. (New York: Baron's Educational Services, 2003), 323.

4. Allan Austin estimates that 7 to 8 percent of the Africans enslaved in America between the seventeenth and nineteenth centuries were Muslim. Based on a 10 percent estimate of all

West Africans introduced between 1711 and 1808, Austin has also suggested that there was a total of 29,695 Muslim slaves in the Americas. Allan D. Austin, *African Muslims in Antebellum America: A Sourcebook* (New York: Routledge, 1984). Michael Gomez thinks "thousands if not tens of thousands" of Muslim slaves were brought to the United States. Michael Gomez, "Muslims in Early America," *Journal of Southern History* 60 (November 4, 1994): 682.

5. Michael Gomez, *Black Crescent: The Experience and Legacy of African Muslims in the Americas* (London: Cambridge University Press, 2005), 58.

6. Georgia Bryan Conrad, "Reminiscences of a Southern Woman," from *Southern Workman*, 1901, cited in Lydia Parrish, *Slave Songs of the Georgia Sea Islands* (Athens: University of Georgia Press), 28.

7. Charles Spaulding Wylly, *The Seed That Was Sown in Georgia* (New York: Neale, 1910).

8. For more on Bilali's Arabic manuscript, see Joseph H Greenberg, "The Decipherment of the 'Ben Ali Diary,' a Preliminary Statement," *Journal of Negro History* 25 (July 1940): 372–75.

9. Katie Brown, one of Bilali's great-granddaughters—and, at the time of the WPA interview, described as "one of the oldest inhabitants" of Sapelo Island—told her interviewer that Bilali's seven daughters were named "Margret, Bentoo, Chaalut, medina, Yaruba, Fatima, and Hestuh." Katie went on to describe her grandmother Margret's Islamic headdress, which differed from the scarf that she (Katie) wore, and to explain how her grandmother would make rice cakes (*saraka*) for the children on religious holidays, and how Bilali and his wife Phoebe used to "pray on duh bead," a reference to the Islamic rosary (*tasbih*): "Margaret an uh daughter Cotto use tuh say dat Bilali an he wife Phoebe pray on duh bead. Dey wuz bery puhticluh bout duh time dey pray and dey bery regluh bout duh hour. Wen duh sun come up, when it straight obuh head and wen it set, das duh time dey pray. Dey bow tuh duh sun an hab lill mat tuh kneel on. Duh beads is on a long string. Belali he pull bead an he say, 'Belambi, Hakabara, Mahamadu.' Phoebe she say, 'Ameen, Ameen.'" See *Drums and Shadows: Survival Studies among the Georgia Coastal Negroes*, by the Georgia Writers' Project Savannah Unit (Athens: University of Georgia Press, 1940), 158–61.

10. Lorenzo Dow Turner, *Africanisms in the Gullah Dialect* (1949; New York: Arno, 1969).

11. Sylviane A. Diouf, *Servants of Allah: African Muslims Enslaved in the Americas* (New York: New York University Press, 1998), 190. Washington Creel notes that the Gullah people buried their dead so that their body faced east. See Margaret Washington Creel, "A Peculiar People": Slave Religion and Community-Culture among the Gullahs," *Journal of Southern History* 56, no. 2 (May 1990): 332–33.

12. Michael Gomez, *Exchanging Our Country Marks: The Transformation of African Identities in the Colonial and Antebellum South* (Chapel Hill: University of North Carolina Press, 1998), 60.

13. See Newbell N. Puckett, *Folk Beliefs of the Southern Negro* (Chapel Hill: University of North Carolina Press, 1926), 528–29.

14. Ulrich Bonnell Phillips, *American Negro Slavery* (New York: D. Appleton, 1918), 42. The influence of the Hamitic thesis in the American South is evident in the writings of scholar-diplomat William Brown Hodgson. In his *Notes on Northern Africa, the Sahara and the Sudan* (New York, 1844), he states emphatically that "the Foulahs are *not* Negroes. They differ essentially from the Negro race, in all the charactertistics that are marked by physical anthropology. They may be said to occupy the intermediate space between the Arab and the Negro. All travelers concur in representing them as a distinct race, in moral as in physical traits . . . They concur also in the report, that the Foulahs of every region

represent themselves to be white men, and proudly assert their superiority to the black tribes, among whom they live."

15. Terry Alford, *Prince Among Slaves* (New York: Oxford University Press, 1986).

16. Zephaniah Kingsley, *A Treatise on the Patriarchal or Cooperative System of Society as it Exists in Some Governments and Colonies in America, and in the United States, Under the Name of Slavery, with its Necessity and Advantages*, 2nd ed. (Freeport, NY: Books for Libraries Press, 1971), 13–14.

17. Austin, *African Muslims in Antebellum America*, 474.

18. Ibid., 473.

19. Henry Louis Gates, *The Signifying Monkey: A Theory of Afro-American Literary Criticism* (New York: Oxford University Press, 1988), 13.

20. Diouf, *Servants of Allah*, 102.

21. Keith Cartwright, *Reading Africa into American Literature: Epics, Fables, and Gothic Tales* (University Press of Kentucky, 2004), 158.

22. Henry Louis Gates, *Figures in Black* (1987), 12–13; Gates, *The Signifying Monkey*, 148, 163.

23. See Diouf, *Servants of Allah*, 140.

24. Lotfi Ben Rajab, "America's Captive Freemen In North Africa: The Comparative Method In Abolitionist Persuasion," *Slavery & Abolition* 9, no. 1 (1988): 57–71.

25. Samuel Sewell, *The Selling of Joseph: A Memorial*, ed. Sidney Kaplan (Amherst: University of Massachusetts Press, 1969), 11.

26. Stanley K. Schultz, "The Making of a Reformer: The Reverend Samuel Hopkins as an Eighteenth-Century Abolitionist," *Proceedings of the American Philosophical Society* 115, no. 5 (October 15, 1971).

27. Hugh Williamson, Speech at Edenton, North Carolina, *New York Daily Advertiser*, February 26, 1788, in *The Documentary History of the Ratification of the Constitution* 25, 206.

28. Paul Baepler, ed., *White Slaves, African Masters: An Anthology of American Barbary Captivity Narratives (Chicago*: University of Chicago Press), 25.

29. Charles Sumner, *White Slavery in the Barbary States (1853)* (Ayer, 1947), 83.

30. James Riley, *Sufferings in Africa: An Authentic Narrative* (New York: Potter, 1965), 445–47.

31. Gerald R. McMurtry, "The Influence of Riley's Narrative Upon Abraham Lincoln," *Indiana Magazine of History* 30 (1934): 133–38.

32. Samuel Hopkins, *Discourse* (Boston: Doctrinal Tract and Book Society, 1852), 611, 608–12; the *Dialogue* (1785; reprinted, Arno Press, 1970), 584.

33. George E. Post, "Arabic-Speaking Negro Mohammedans in Africa," *The African Repository* (May 1869): 130–31.

34. Moustafa Bayoumi, "Moving Beliefs: The Panama Manuscript of Sheikh Sana See and African Diasporic Islam," *Interventions* 5, no. 1 (April 2003): 58–81; Precious Rasheeda Muhammad, "The Autobiography of Nicholas Said: A Native of Bornou, Eastern Soudan, Central Africa," *Journal of Islam in America* (2001).

35. While Ibrahima's case was in the headlines, just before his return to Africa, others began claiming Muslim descent and Oriental ancestry, including "Abdullah Mohammed," who claimed to have been abducted from his native Syria by pirates, and "Almourad Ali," who raised $1500 for his passage back home to Turkey before it was discovered that he was from Albany, New York. See Alford, *Prince Among Slaves*, 137.

36. Booker T. Washington, *Up From Slavery* (New York: Carol, 1993), 103.

37. Dizzy Gillespie and Al Frazer, *To Be or Not to Bop* (New York: Da Capo, 1979), 293. As Gillespie put it, "'Man, if you join the Muslim faith, you ain't colored no more, you'll be white,' they'd say. You get a new name and you don't have to be a nigger no more."

PART I

GEOGRAPHIES AND THE POLITICAL IMAGINATION

CHAPTER 1

LOCATING PALESTINE IN PRE-1948

BLACK INTERNATIONALISM

ALEX LUBIN

AT THE 1955 ASIAN-AFRICAN CONFERENCE HELD IN BANDUNG, INDONESIA, Pan-Africanists and anticolonial leaders from Asia and North Africa met to discuss the politics of nonalignment. The conference addressed many themes, including the possibilities for Afro-Asian solidarity, the complexities of anticolonial struggle within Cold War hegemony, and the enduring colonial occupations of North Africa, parts of Southeast Asia, and Palestine.

Written as part orientalist travel narrative and part journalistic report, Richard Wright's *The Color Curtain: A Report on the Bandung Conference* (1956) rendered the conference themes to American audiences by locating it within the contexts of the Cold War and Pan-African politics.[1] While Wright's tone was generally celebratory, *Curtain* was firmly rooted in a Western aesthetic that treated the nonaligned geographies as exotic and irrational. For example, Wright's reportage lacked serious engagement with the North African and Arab contingents at Bandung, and his depictions of the conference as a place where "black Africans mingled with swarthy Arabs" delimited his view of Bandung's potential (176). Thus, while *Curtain* illustrates anticolonial politics in Palestine and North Africa, Wright's report on Bandung illustrates the ways that Palestine, in particular, was an ambivalent geography in Wright's imaginary.

During his flight to Bandung, Wright engaged North African delegates who insisted on making the question of Palestine central to conference's agenda: "[I was shown] photos of Arab refugees driven by the Jews out of their homes. I leafed through the bundle of photos; they were authentic, grim, showing long lines of men, women and children marching barefooted and half-naked over desert sands, depicting babies sleeping without shelter, revealing human beings living like animals" (76). Shocked by scenes of Palestinian subjugation, Wright was none-theless reluctant to support Arab anticolonial politics. Instead, he suggested that the question of Palestine was primarily one of religious fanaticism and irrational-ity, not colonialism: "It was strange how, the moment I left the dry, impersonal, abstract world of the West, I encountered at once: *religion* . . . And it was passion-ate, unyielding religion, feeding on itself, sufficient unto itself. And the Jews had been spurred by religious dreams to build a state in Palestine . . . Irrationalism meeting irrationalism . . . Though the conversation about the alleged aggression of the Jews in Palestine raged up and down the aisles of the plane, I could hear but little of it" (77–78). For Wright, the Arab-Israeli conflict, even as it engaged issues of exile and homeland that were central to his own cosmopolitanism, was illegible as a question of colonialism.

As an exile from the United States writing with the support, in part, of the Central Intelligence Agency, Wright conveys a doubleness in his prose, writing both as a Western observer of the nonaligned world and a critic of Western forms of colonialism and racism.[2] In this way, he not only suggests the sort of double consciousness defined by W. E. B. Du Bois but also a doubleness related to his sta-tus within and outside of the West. For Du Bois, double consciousness described how black people lived within and beyond the world of color, "a world which yields him no true self-consciousness, but only lets him see himself through the revelation of the other world."[3] While Wright expresses double consciousness in a Du Boisian sense, he is similarly "double" regarding his status as a Western subject who, seeking freedom outside of Western geographies and forms of racial belong-ing, was nevertheless fully part of the West. In this way, Wright viewed Palestine in ways similar to many orientalists, while also viewing it as an important site for anticolonial politics.

I am using Edward Said's term "orientalism" with full recognition that African American relationships to orientalism raise certain problematics. For Said, "orien-talism" is the epistemological basis of imperialism—it is a textual formation that divides the world into two uneven entities called "occident" and "orient." The occident defines itself, its notions of progress, and modernity through compari-son to the orient's supposed primitiveness and exoticism. Said imagined oriental-ism in terms of dominant cultural productions; he therefore did not account for the ways subaltern populations within the occident, who lacked access to imperial authority, might relate to orientalism. The concept of orientalism thus obscures the liminality of African American thought about the orient, which was, I con-tend, simultaneously orientalist and critical of Western modernity's reliance on black slaves and antiblack racism.[4] African American orientalism thus represents a variation on Said's definition, suggesting ways that imperial regimes shaped African American modes of viewing the Middle East, even while those regimes excluded black subjectivity.

Black political thought concerning Palestine is often understood within a con-temporary political context that assumes an inevitable Israeli conclusion to Pal-estinian history, while also assuming global Jewish belonging to Israel. Moreover, this framework locates black politics surrounding Palestine within the context of black/Jewish relations in the United States in ways that links black support for Palestinians to ethnic and racial antipathy between black and Jewish Americans.[5] This framework elides important routes of black internationalist politics that extend to the Arab/Islamic world, while also forgetting the pre-1948 history of Palestine.

This chapter is concerned with charting routes linking African America to pre-1948 Palestine in order to trouble some common assumptions about African American relationships to Palestine and Israel. For many African American radicals, Palestine has been a generative site for articulating anticolonial and antiracist politics. Because of the parallels between anti-Jewish anti-Semitism and antiblack racism within the West, Jewish and African diasporic politics have often overlapped. Many African Americans have viewed Jewish Zionism as an anticolonial and antiracist movement; when the state of Israel was created in 1948, African American radicals like W. E. B. Du Bois, Paul Robeson, and many others firmly supported the creation of the Jewish state. Yet, over time, some African Americans who had supported the creation of the modern state of Israel began to identify Zionism as a colonial discourse, such as when Du Bois criticized Israeli aggression in his poem "Suez." Locating Palestine in African American politics prior to the creation of the state of Israel generates an opportunity to examine when antiracist and anticolonial politics become rooted in territory and Western forms of governmentality and the ways that internationalist politics work within and against the governing logics of empire.[6]

It should be noted that different groups of African Americans employed orientalism to different ends. For Pan-Africanists, Palestine was a primitive geography that spoke to the needs of a new modernity in the orient. In this way, orientalism helped Pan-Africanists build a case for settler colonialism in Africa. For civil rights activists, Zionist settlements represented progress to the primitive orient and exemplified proof that peoples exiled from the West could find freedom by looking to the nation-state as the most appropriate rubric for redress. Hence, in the Zionist desire for a Jewish state, African American civil rights leaders saw a model for addressing racial exclusion and terror in the United States. For African American Christian Zionists, the orient was in need of Christian restoration, and this could only be accomplished through Jewish return to historic Israel followed by Jewish conversion to Christianity. This group saw the Zionist movement not in terms of redress for Jewish anti-Semitism but instead in terms of biblical prophecy. Each of these views of the orient rested on a form of orientalism but often for different ends. Moreover, each group of African American travelers employed Zionism in order to better understand how exiled peoples could imagine home and politically mobilize to reach a "promised land."

During the second half of the nineteenth century, African Americans, like many other Americans, began to document their travel to "the orient." The travel narrative was a medium for African Americans to narrate as Western travelers and thereby gain subjectivity and recognition as Western subjects. This writing took place during a ninety-year period that was framed by the end of slavery in the United States, the colonial occupation of "the Levant" in the wake of World War I, and the genocidal Holocaust taking place in Europe. Pan-Africanists, African Americans seeking rights within the United States, and African American Christian Zionists each rendered Palestine through orientalism; yet because they were themselves displaced within the West, Palestine was also viewed with a great deal of identification and longing because of its status as the site of exodus.[7] Thus, the travel narratives discussed here document the presence of pre-1948 Palestine in African American intellectual and political thought; but, more importantly, they illustrate the complexities of black internationalist politics when framed by orientalism and Zionism.

The first African American Holy Land travel narrative was written by David Dorr, who was a slave brought to Europe and the orient by his master. Born in 1827 or 1828 in New Orleans, Dorr traveled through Europe and the Ottoman-controlled orient with his master between 1851 and 1854. Upon returning to the United States, Dorr fled his master, who had failed to guarantee his freedom, and in 1858, he published *A Colored Man 'Round the World*. As Malini Johar Schuller has suggested, Dorr's narrative is marked by a gentlemanly tone that establishes the authority of a slave to speak and define for American readers the non-Western world.[8] In this way, Dorr participates in orientalism through his construction of knowledge about the orient and through his geographic and cultural descriptions of the orient in comparison to the occident.

For example, after describing a bucolic experience traveling through Europe, Dorr relates his journey from Paris to Egypt by juxtaposing Parisian civility with oriental savagery: "If you have, see me alike, pulling away from the festal abode of Paris' comfort, and loosening the tie of familiar smiles, for a hard journey over a rough see, dead lands, and a treacherous people" (63–64). In Egypt, Dorr's accommodations were "sickening," and he was continually appalled by what he considered Arab laziness. While traveling down the Nile, Dorr described "some places, when the boat was shoving out, some great, fat and lazy Arab would come blowing and panting to the edge of the Nile with one single egg, that he had been waiting for the hen to lay . . . To believe what an Arab says when trying to sell anything, would be a sublime display of the most profound ignorance a man could be guilty of" (172–73). Dorr's frequent allusions to the backward orient gained him entrance into a Western authority to gaze east.

Yet Dorr's narrative is not merely orientalist; it also makes subtle allusions to slavery and, in this way, is critical of Western modernity's slavery and racial terror. The book is dedicated "to my slave mother." In the dedication to the book, Dorr writes the inscription, "Mother! Wherever thou art, whether in Heaven or a lesser world; or whether around the freedom Base of a Bunker Hill, or only at the lowest savannah of American Slavery, thou art the same to me, and I dedicate this token of my knowledge to thee mother, Oh, my own mother, Your David." This dedication intervenes in the teleology of occidental modernity made possible through the travel narrative. It underscores that *this* purveyor of the Western gaze is himself the product of an enduring slavery and that travel in the orient signifies a level of freedom unknown to the author in the United States.

Furthermore, Dorr represents the orient as a place of African self-rule when he visits the pyramids of Egypt. There he notes the similarities in Turkish rule in the Ottoman Empire to the rule of Egyptian monarchs, "though black." In this way, Dorr exposes the black origins of Egyptian civilizations, thereby locating blacks within both the orient and occident. This doubleness allows Dorr to assume an authority not only over the orient and its indigenous populations but also over the Christian sites of the Holy Land. Because he claims parts of the orient as his ancestral homeland, Dorr is critical of Christian tourists who cannot "know" what he can. For Dorr, it was the Western tourist and not the black slave who was a foreigner. In Jericho, for example, Dorr scoffs at a guide who explains that the source of a spring is "because the jawbone that Sampson fought so bravely with was buried here." Since he imagines a privileged place for himself in the biblical landscape of exodus, Dorr "was not inclined to believe anything I heard from the people about here, because I knew as much as they did about it. I came to Jerusalem with a submissive heart, but when I heard all the absurdities of these ignorant people, I was more included to ridicule right over these sacred dead bodies, and spots, than pay homage" (186). Dorr makes clear his belief that members of the African diaspora are not foreigners to Palestine when he compares the consistency of the waters of the Jordan to that of the Mississippi; both rivers share in the black imaginary as sites of emancipation and the comparison links Dorr to the orient.

By engaging the legacy of slavery in the context of an orientalist travel narrative, Dorr's story takes on the political significance of a slave narrative, but one routed not South to North but internationally, West to East. As Schuller has noted, Dorr challenges slavery through cultures of taste; his ability to assume Western respectability and to narrate the orient enables him an authority to challenge slavery. Ultimately, however, Dorr's travel narrative is a call for a new modernity in the United States, one that locates exodus and the acknowledgment of black civilization at the center of the West rather than the orient. In this way, Dorr's internationalism is rooted in the West even while it is staged in the orient.

Subsequent African American travel writers would employ orientalist travel accounts to similar ends; but those published after 1890 had only to look to the growing Zionist movement as a touchstone around which criticism of slavery could be elaborated. In 1896, Theodore Herzl, responding to anti-Jewish anti-Semitism in France, published his treatise on Jewish Zionism called *The Jewish State*. Although Herzl set eyes on Palestine as the only homeland for the Jewish state, some Zionists were less committed to Palestine as homeland and would settle for any territory where Jews could establish self-rule.

Arguably, the most sophisticated articulation of the territorial Zionist position was not by a Jewish intellectual but by an African intellectual from the Americas named Edward Wilmot Blyden. Blyden had also witnessed the Dreyfus affair (as had W. E. B. Du Bois), had grown up among Jews in the Danish colony of St. Thomas, and became one of the founding fathers of Pan- Africanism and a supporter of Liberian colonization. Blyden's travel narrative suggests ways that black travel to the orient could deploy orientalism in order to posit a new modernity. As a Pan-Africanist, Blyden was interested in articulating an internationalist politics for all blacks as well as in advocating settler movements in Africa.[9]

Blyden's travel narrative *From West Africa to Palestine* (1873), along with his essays "Mohammedism and the Negro Race" (1877) and "The Jewish Question" (1898), convey the complex alchemy of orientalism, Zionism, and Pan-Africanism in Blyden's writing. *From West Africa to Palestine*, like Dorr's travel narrative, is orientalist in the ways it establishes a Western authority to gaze at the orient. For example, the travel narrative is filled with geographic descriptions of the orients' barrenness as well as its exotica. Blyden is "struck with the bareness of the mountains of all forest trees of natural growth" while also taken by the "sublimity of scenery—the overpowering charms of the *tout ensemble* of a summer-evening view from the summits of Lebanon."[10] These geographic observations helped Blyden establish a hierarchy of the orient's inhabitants, with native Arabs on the bottom and recent settlers, including missionaries at the American University of Beirut, at the top. The orient was, for Blyden, disorganized and poorly ruled: "There seems to be no law or order to regulate the tumultuous and boisterous crowds which overwhelm the new comer to these Oriental ports" (154).

Because Blyden viewed Arabs as irrational and Ottoman rule as insufficient, he advocated Western imperial administration of Palestine:

> When one . . . visits [Palestine] and perceives how, under the misrule of the Turks—a misrule rather of negligence and omission than of elaborate design—everything lies waste and desolate—how the land is infested with thieves and robbers—how some of the most interesting localities cannot be visited without a strong and expensive guard—when he sees sacred places under the surveillance of Turkish solders who have no respect for that which the Christian venerates—he wonders why it is that the land has not passed long ago into the hands of one of the Great Christian Powers . . . the land is desolate and overthrown by strangers. (192–93)

Blyden's orientalism rests on his assumptions about the beneficent role of occidental imperial administration.

Yet Blyden also sought to challenge how Western cultures had assumed blacks' inability to be fully modern and to assume self-rule. Thus, he supported Zionism and embraced Pan-Islamism in order to show how, without the racial slavery and racial terror of the West, blacks were fully capable of being modern and establishing self-rule. To this end, Blyden was especially interested in identifying the African origins of Egyptian civilization. In Egypt, Blyden finds himself most capable of expressing the possibilities for a new and African modernity: "I felt lifted out of the commonplace grandeur of modern times; and, could my voice have reached every African in the world, I would have earnestly addressed him in the language of Hilary Teage—'Retake your fame'" (105).[11]

Blyden sought to illustrate how Egyptian civilization was central to Western civilization and then to show how, because blacks were of the West, they were capable of ruling Liberia: "Now that the slave-holding of Africans in Protestant countries has come to an end . . . it is to be hoped that a large-hearted philosophy and an honest interpretation of the facts of history, sacred and secular, will do them the justice to admit their [black Egyptians'] participation in, if not origination of, the great works of ancient civilization" (106). If blacks originated "the great works of ancient civilization," Blyden argued, they were certainly capable of participating in the civilizing authority of colonial administrations.

Blyden's case for self-rule in Liberia thus rested on the racial logic of colonialism and orientalism. His case for Liberia was made through an orientalist travel narrative touting the benefits of Western colonialism in the Levant. Moreover, Blyden looked to settler-colonialism in North America as a useful model for African American colonization of Liberia: "While the American Indians, who were, without doubt, an old a worn-out people, could not survive the introduction of the new phases of life brought among them from Europe, but sunk beneath the unaccustomed aspect which their country assumed under the vigorous hand of the fresh and youthful Anglo-Saxon and Teutonic races, the Guinea Negro, in an entirely new and distant country, has entirely delighted in the change of climate and circumstances, and has prospered, physically, on all that great continent and its islands, from Canada to cape horn" (109–10). African Americans, unlike American Indians, had prospered in North America and other slaveholding regions; thus, according to Blyden, they were fit to rule.

From 1885 through 1919, the British and French empires acquired vast areas of Liberia's original 1821 boundaries. In Zionism, Blyden likely saw a useful model for black self-determination during a time when Liberian independence was in question. In order to develop his case for Liberian independence, Blyden turned to Jewish diasporic history. In his 1898 essay, "The Jewish Question," he links the Jewish and black diasporas as having "a history almost identical of sorrow and oppression." In Zionism, Blyden saw a movement of suffering people to a national homeland, and, given his interest in bringing African Americans to Liberia, Zionism seemed to him a "marvelous movement": "The question [of Zionism] is similar to that which at this moment agitates thousands of the descendants of Africa in America, anxious to return to the land of their fathers . . . And as the history of the African race—their enslavement, persecution, proscription, and sufferings—closely resembles that of the Jews, I have been led also by a natural process of thought and a fellow feeling to study the great question now uppermost in the minds of thousands, if not millions of Jews."[12] Although Blyden believed "[all] recognize the claim and right of the Jew to the Holy Land," he did not believe that the Zionist movement required settlement of Palestine. Importantly, he therefore did not support imperial intervention on behalf of Jews: "The 'ideals of Zion' can be carried out only by the people of Zion. Imperial races can not do the work of spiritual races."

For Blyden, Zionism was not an imperialist, but a panhumanist, movement that could "bring about the practical brotherhood of humanity by establishing, or rather propagating, the international religion in whose cult men of all races, climes, and countries will call upon the Lord under one name." Blyden was not merely a Zionist, but also part of a black Atlantic tradition that looked beyond the nation-state in order to find freedom.[13] Therefore, while he saw the nation-state as the most appropriate rubric for black self-rule, Blyden also believed that the Zionist movement did not need to be rooted in a particular geography; indeed, he advocated African settlement for Jewish Zionists: "If what I have here written should have no other effect than to attract the attention of thinking and enlightened Jews to the great continent of Africa . . . I should feel amply rewarded."[14]

Blyden embraced Zionism as an internationalist movement operating within and against the West; hence, for him, Zionism was not incompatible with advocating Islam as a humanistic faith in Africa. Blyden recognized that Arab colonial powers for whom Islam was the main religion created relative equality among their colonized subjects. In his 1877 essay, "Mohammedism and the Negro Race," Blyden challenged the Christian teleology of empire embodied in the Crusades and British and French empires and wrote to a Methodist audience about Islam as a humanistic faith. Moreover, he argued that colonial powers guided by Islam were far less oppressive, at least in Africa, than were those guided by Christianity: "Wherever the Negro is found in Christian lands, his leading trait is not docility as has often been alleged, but servility. He is slow and unprogressive . . . there is no Christian community of Negroes anywhere which is self reliant and independent."[15]

On the other hand, Blyden viewed Arab colonial powers, and Islam, as less degrading to Africans: "If the Mohammedan Negro had at any time to choose between the Koran and the sword, when he chose the former he was allowed to wield the latter as the equal of any other Moslem; but no amount of allegiance to the Gospel relieved the Christian Negro from the denigration of wearing the chain which he received with it, or rescued him from the political and, in a measure, ecclesiastical proscription which he still undergoes in all the countries of his exile" (115). For Blyden, who learned Arabic but never converted to Islam, Arab-Islam was less violent in its administration of colonial rule. Moreover, even for Blyden, the Christian minister, embracing aspects of Islam and Zionism were not contradictory—each discourse spoke to Western forms of racial violence and terror.

Although, for Blyden, Zionism could be embraced as a diasporic movement that spoke to African American settler projects in Africa, to Christian Zionists, Jewish Zionism could be embraced as the first stage in Christian restoration in the Holy Land. African American Christian Zionists' eschatological view of the Holy Land employed orientalism and Zionism to different ends than Blyden and Dorr. Indeed, one year prior to Herzl's inauguration of the Zionist movement, an African Methodist Episcopalian minister named Daniel P. Seaton published a travel account espousing Jewish return to Palestine following Jewish conversion to Christianity. Seaton's *The Land of Promise: The Bible Land and Its Revelation* (1895) offered readers a history of the Bible's geography. Seaton, who had traveled to Palestine on at least two occasions, sought to render biblical stories through geographic descriptions of Palestine; yet Seaton also sought to locate African American Christians as the beneficiaries of Zionism. In doing so, Seaton narrated as an orientalist by describing Palestine and its people as primitive, exotic, and in need of Western intervention. Seaton's disdain for Palestine's inhabitants included the Jews he encountered there. His advocacy of Jewish return to Palestine was directed at European Jews who had received the benefits of Western civilization; moreover, Jewish resettlement of Palestine was merely the first stage in Jewish conversion to Christianity.

Palestinians were, according to Seaton, primitive: "These farmers have lived too greatly isolated from the modern people and so far behind the march of civilization they would not know how to use the farming implements used in modern times."[16] Seaton believed the natives required colonial intervention in order to improve their lot: "[In Joppa] the stranger finds himself in a most repulsively filthy place, with a wild looking people, of all complexions, among whom ignorance is dominant, excepting those who have settled there from countries of progressive civilization, and you can find but few natives who have been taught to appreciate a higher state of manhood" (48). Only in the American colony did Seaton find any relief from the primitive landscape: "It should be stated concerning the American Colony at Joppa they are doing well, and have done much to change the habits of many of the natives, who, at the time they landed, were not far above the average heathen: they have built a commodious little village to themselves in the most healthy section of town, and have organized a church and school, which has done an incalculable amount of good" (46).

Although the Zionist movement had not yet formed, Seaton drew on a history of Christian Zionism that viewed Jewish return to Palestine as the precursor to the return of the Messiah. Thus, while he regarded the Jews he encountered in the same contemptuous way he did the Arab Muslims, Seaton nevertheless saw them as redeemable, especially as more European Jews settled in Palestine:

> [Jews] have a hopeful future; the time is coming when they will fully accept Christ, whom their fathers nailed to the cross, and reverently come before Him in devout worship, return to their own land, and pay Him their tribute on the very summit where the pathetic prayer was offered by the Lord Jesus, in their behalf, while the arrows of death were piercing His soul . . . If we have noticed the predictions concerning the future of this people, we cannot be otherwise then inclined to the opinion, that a restoration of the Jews will take place . . . What a glorious time, what a blessed period when the people, once dispersed and unsettled, shall again "sing the Lord's song" in their own land! (140–43)

Seaton sought to participate in a Christian Zionist movement that could return "the people" to "their own land." In this way, *Promise's* engagement with Zionism enabled Seaton to imagine a future promised land for all Christians, even as that future was based on Western colonialism and Jewish conversion to Christianity.

For the Reverend W. L. Jones, Palestine was not merely a metaphor for African American diasporic longing; it was the actual scene of African American restoration.[17]

Jones began his 1907 travel account, *The Travel in Egypt and Scenes of Jerusalem* with the following: "For a long time, yes fifteen years, I have had a desire to visit the old world. I first felt that it was my calling to Africa, and for several years I was troubled with that thought; afterwards my mind was disabused of that idea, for a new one, that of Jerusalem. And for more than ten years I have had a restless desire for the Holy Land and especially for Jerusalem" (5). Jones left for Jerusalem in 1897. Like Seaton's *Promise*, Jones's travel account represents Palestine through orientalist tropes: "There is nothing beautiful about the little city Joppa. The streets are narrow and not as clean as they ought to be, and full of Arabs, Turks, Bedouins, donkeys, and camels" (71). The conflation of the native with the natural—people with animals—was a staple of orientalist travel literature.

Orientalist descriptions of geography also shaped Methodist Episcopalian Bishop William Sampson Brooks's 1915 travel narrative *Footprints of a Black Man*.[18] Brooks was a Pan-Africanist born in Maryland. During the first half of the 1920s, Brooks served as the bishop of West Africa who contributed to the construction of Monrovia Normal and Industrial College in Liberia. By the time of Brooks's travel narrative, Jewish Zionists had begun establishing colonies in Palestine. Brooks observes these colonies and applauded the settler-colonial movement, especially the German colony he encountered in Haifa: "It owes its progressiveness and beauty to the indefatigable industry and thrift of a small Germany colony nestling at the foot of the mountain." According to Brooks, the inhabitants of the colony were German Americans who combined "their American ideas, methods and tools with the incompatible German spirit for progress, and they have accomplished the salvation of the city. They have revolutionized the city from the filth and squalor of the Turks to its present condition" (123).

Brooks also supported European Jewish Zionists who were creating colonies in Jerusalem and other parts of Palestine. He distinguished between the "native" Jews who had lived in Jerusalem for centuries and the new settlers from Europe: "The [Jewish] quarter reeks with filth. About ten thousand men, women, and children live in its wretched tenements in the most abject squalor and wretchedness" (185). This native Jewish neighborhood, however, was described in sharp contrast to the Zionist settlement outside the old city walls. There was, by the time of Brooks's travel, the "'Zion Suburb,' a new settlement of Jews who live in comfortable homes, and commodious tenements, and enjoy real cleanliness and sanitation." Whereas the Jewish settlers were "making themselves respected in the business and commercial life of Palestine, in spite of the great obstacles and restrictions the Turkish Government place in their path . . . The Moslems of Jerusalem are among the most fanatic and rapacious in Palestine, and derive great profit from brisk traffic in souvenirs of the Holy Land and in showing tourists places of interest connect with the life of Christ and the days of the kings" (186). To Brooks, the Arabs he encountered were a people without history, merely there to provide a service industry for Christian travelers. While he looked forward to Jewish conversion to Christianity, he also believed that European Jewish settlement of Palestine would improve the land and make it more hospitable for Christian tourists such as him. Thus as the Zionist movement became a settler movement in Palestine, some African Americans saw it as an opportunity to create a new modernism in a primitive landscape. As an ideal, the new modernism of Palestine, or, in the case of Blyden and Brooks, Liberia, would replicate forms of colonial governmentality yet would challenge Western forms of racial belonging and violence. However, as facts on the ground challenged this hopeful vision and as racial politics in the United States changed, African American relationships to Zionism and modernism were transformed.

By the 1920s, the context for African American travel narratives had dramatically changed. In the wake of the imperial World War I, the British and French empires expanded their reach in the Middle East. Moreover, the Zionist movement gained momentum by the outcomes of the war, as many Zionist Jews fought on behalf of the British army in Palestine. In 1914, the Jewish population in Palestine was estimated at 7.5 percent. In 1922, the year of the British Mandate over Palestine, the Jewish population was at 11.1 percent. This percentage grew steadily to nearly 30 percent in 1941.[19] While these changing demographics meant many different things to different groups in the region, to the West's travelers, it appeared that Palestine was becoming more Western and, by extension, more modern. Many African American travelers saw Jewish return to Palestine, especially in the wake of World War II, as the just solution to racial terror. Yet even with the belief the Jewish return to historic Israel was a just solution, African American writing about Zionism and anti-Semitism was saturated by orientalist tropes and diasporic longings that had structured African American writing about the region prior to a large Western Jewish population in Palestine.

In the wake of the World War I, Zionism changed from what Blyden identified as a humanist antinationalist movement to one operating within the imperial logics of the West. As Israeli historian Ilan Pappe has noted, Zionists joined the imperial movement in the Middle East and would adopt a Western framework, the settler colonial nation-state, as the rubric for achieving its goals.[20] Zionism was less about exile and panhumanism and more about the possibilities of the nation-state as a rubric for redressing anti-Jewish racism. Moreover, as Zionism became tied to the geopolitical question of Palestine, there was growing international concern about the Nazi party's massacre of Jews and many others. These and many other factors transformed how African Americans took up the question of Zionism.

Similarly, major changes were shaping black politics across the Atlantic in the United States. The 1920s and 1930s witnessed the development of a variety of internationalist, diasporic, and anticolonial movements, including the rise of the negritude movement routed through Paris and Marcus Garvey's black-nationalist movement in the United States. Many African Americans, especially those participating in what would be known as the Harlem Renaissance, organized politics that were not only based on inclusion and equality within the U.S. nation-state but were also diasporic and spoke to the shared experience of blacks across the

globe.[21] Within this context of black internationalist politics, the question of Palestine, now framed as a solidly "Jewish question," took on new importance in
African American intellectual thought.

In her 1928 book, *My Trip Through Egypt and the Holy Land*, the African
American writer Carolyn Bagley illustrates the growing presence of Western amenities in the Holy Land. While traveling across Palestine, she encountered a fellow
passenger who shares his personal story: "I learned that he was a Jew living in the
Jewish city of Tel Aviv near Jaffa . . . Here the business was not so remunerative as
before but he felt free and a man of an equal chance with others . . . Passing along
a ridge overlooking large plains below, we passed several villages containing many
stone houses with pretty red tops, surrounded by a background of green hills and
fertile, well-kept farms. All these belonged to the new Zionists districts, which
America has done so much to promote."[22] Bagley viewed Jewish settlements as
European transplants in a primitive landscape, yet she also believed that Jews "felt
free" and had "equal chance with others" in Palestine. Like many African American travelers to British-controlled Palestine, Bagley saw stark differences between
the primitive and the modern; moreover, Zionism, while replicating Western
forms of colonialism, also promised freedom for those excluded and terrorized
within Western metropoles.

It was within a context of black internationalist politics and civil rights activism that the famous pastor of Harlem's Abyssinian Baptist Church, Adam Clayton Powell, Sr., traveled to Palestine, a country then under the British mandate
and wrote a travel narrative. Contained in a two-book volume titled *Palestine
and Saints in Caesar's Household*, the narrative is framed as a challenge to the
growing trend of anti-Semitism sweeping the European continent. Powell was
especially interested in understanding how Zionist settlements in Palestine had
created spaces of freedom for Jewish victims of anti-Semitism. He believed that
the Balfour declaration clearly established a Jewish national home in Palestine, yet
he was critical of Western imperial rule in Palestine and believed it had created
animosity between Jews and Arabs that had not been in force prior to World War
I. He thus looked to Zionism in order to criticize Western racism and imagine
possible routes of black American liberation in the United States.

The preface of *Palestine* frames the travel narrative in orientalist terms as Powell
attempts to dispel for his readers any romantic notions they may have had about
the Holy Land: "Before he had spent a week in the Holy Land, he had met people
characterized by all the bad qualities possessed by the worst in New York and in
the mountains of Kentucky. That little strip of land between the Jordan and the
Mediterranean produced more holy characters and more holy literature than any
one of the five continents, but the men and their literature have had more influence for good upon the citizens for Chicago than upon the natives."[23]

If Powell was unwilling to write a romantic story of the orient it was because,
to him, the orient was geography of liberation and not merely an historical
landscape. The orient's significance was in its sacred history as the site of exodus and thus as a counterexample to Western racism. Powell punctuated his

travel narrative with references to anti-Semitism and the regular abuse of African Americans in the United States, and he framed his travel narrative as an intervention into the West's disregard for Jewish victims European racism: "The second reason for writing . . . this book is to help stem the world's rising tide of fierce, ungodly anti-Semitism . . . The colored people should be the last, even by their silence, to give consent to the brutal persecution of the Jews. For if this campaign of inhuman cruelty against the Jews should succeed, show knows by what the same evil forces would next attempt to put the colored people on the rack" (viii). Powell develops authority as an orientalist narrator while also locating racism and violence at the center of Western imperialism. He therefore embraced Zionism as representative of a new modernism within a primitive landscape. As he describes his interaction with "the Zion movement" in Tel-Aviv, for example, he represents Jewish settlements in terms of modernity—the neighborhoods' technology, cleanliness, and civility: "The Zion movement, one of the most significant in the world today, is made up of Jews in all parts of the world, some of whom are moving back to Palestine to live the remainder of their lives. These settlers, who have met with such bitter antagonism, are more prosperous in the Jaffa section than in any other part of Palestine" (24).

Because he understood the Zionists movement as an antiracist movement, and one that could be emulated by African Americans, Powell was unable to understand Arab protest as anything but anti-Jewish. He witnessed daily violence between settlers and indigenous Arabs, yet he interpreted Arab animosity toward Jews only as anti-Semitic without recognizing how the process of settler-colonialism in the region shaped Arab responses to Jewish colonization: "Arab after Arab said to me, 'Before we will let the Jews come back here and rule the Holy Land they desecrated, every one of us will die with our shoes on.' They say this with a look of cruel murder on their faces and the hiss of serpent in their voices" (29).

Yet Powell also understood that prior to the British invasion of Palestine "the Jews and Arabs lived side by side, in Palestine and other countries, without experiencing any serious trouble" (29). Moreover, Powell noted that "for 450 years under the powerful reign of Arab princes in Spain, the Jews experienced the happiest and most prosperous era of their racial existence" (16). Here Powell suggests that the orient, and the Arab world in general, had not created the sort of racial horrors one could claim were central to the occident. Embracing Zionism converged, for Powell, with his commitment to illuminating the acts of racial terror underpinning occidental modernity. For example, *Palestine's* conclusion firmly establishes Powell's political project of describing Palestine as a means critique antiblack racism in the United States: "As I stood there [in the Holy Land], I could not help but recall that both Moses and I represented an enslaved, persecuted and despised race. Moses was born a slave; he tramped and traveled and sacrificed for forty years to reach Canaan, but died without attaining the overmastering ambition of his life. I was born in a one-room log cabin in Virginia, twenty-six days after the chains of slavery were broken from the black man's wrist and the white man's conscience" (91–92). To a black American writer struggling with the daily abuse to black subjects, Zionism seemed an intervention into Western notions of progress and modernity; in fact, what Powell sought was a new form of modernism, one distinct from "the orient" but also distinct from the West.

There is a relationship between African Americans and Palestine that is at once orientalist and internationalist. Prior to 1948, African American travel writers represented Palestine as an exotic destination in need of colonial intervention, yet they were also attached to the region as geography of liberation. For these writers, Zionism, orientalism, and even Islamic humanism were narrative tropes as well as political movements that spoke to the needs for African Americans to engage antiblack politics globally. Palestine was an ambivalent space in African American travel writing because it was at once non-Western yet was also the scene of exodus and, for some, an extension of Northern Africa.

The pre-1948 travel narratives not only help give nuance to Richard Wright's ambivalent writing about the region, but they also reveal some of the complexities of African American engagement with the Arab/Israeli conflict after 1948. African Americans who embraced the creation of the state of Israel in the wake of World War II—W. E. B. Du Bois and Paul Robeson among them—were likely enthusiastic supporters of the promise of diasporic politics culminating in the formation of a homeland. In order to understand why most African American radicals embraced the creation of the state of Israel, one must consider how African American political struggles have been rooted within and against notions of the national and international. As Nikihl Singh shows in *Black is a Country*, black Americans have waged struggles that have been shaped by the desire for international and diasporic movements and by the desire for redress within the framework of the nation-state. When the modern state of Israel was created in 1948, African American radicals were committed to a civil rights strategy in the United States that looked to the nation-state's logic of inclusion as a rubric for the movement; this may have led some African Americans to see the formation of a Jewish state as the most appropriate means to challenge anti-Jewish anti-Semitism. Ralph Bunche, for example, a communist internationalist during the 1930s, was the United Nation's representative in charge of administering the partition of Palestine and the creation of Israel. Bunche's role in the creation of Israel illustrates the complexities of African American anticolonial politics that were themselves operating against and within the logic of empire.[24]

As scholars contemplate what Earl Lewis has called "overlapping discourses of Diaspora," they will need to attend to the complexities of Palestine in the African American global imaginary.[25] A complex alchemy of orientalism, Zionism and Pan-Africanism structured African American internationalist politics centered in Palestine. The Middle East is therefore a generative region for examining the possibilities and limits of black internationalism. Moreover, these complexities speak to the critical need to locating the Arab/Islamic world and the question of Palestine in African American political and intellectual thought.

NOTES

1. Richard Wright, *The Color Curtain: A Report on the Bandung Conference* (Cleveland: World Publishing Company, 1956).

2. Wright's trip to Bandung was partially funded by the American Congress for Cultural Freedom, an organization later shown to be a front for the CIA and State Department. See Bill Mullen, *Afro-Orientalism* (Minneapolis: University of Minnesota Press, 1995), 66.

3. W. E. B. Du Bois, *The Souls of Black Folk* (New York: Dover, 1994).

4. Edward Said, *Orientalism* (New York: Vintage, 1979).

5. See, for example, Robert Weisbord and Richard Kazarian, *Israel in the Black American Imagination* (New York: Greenwood Press, 1985). Also see Michael Lerner and Cornel West, *Jews and Blacks: A Dialogue on Race, Religion and Culture in America* (New York: Plume, 1996).

6. But for a couple notable examples, such as Cedric Robinson's classic, *Black Marxism: The Making of the Black Radical Tradition* (Chapel Hill: University of North Carolina Press, 2000); parts of Melanie McAlister's wonderful *Epic Encounters: Culture Media, and U.S. Interests in the Middle East, 1945–2000* (Berkeley: University of California Press, 2001); and Scott Trafton's recent *Egypt Land: Race and Nineteenth Century Egyptomania* (Durham, NC: Duke University Press, 2004), there remains very little consideration of the Middle East in African American political and intellectual thought.

7. These categories are somewhat arbitrary, and they are in no way exclusive. Pan-Africanists, for example, could also be Christian Zionists (as in the case of William Sampson Brooks). I use the category in order to show how the travel writing was shaped by a particular set of political and religious motives.

8. David Dorr, *A Colored Man 'Round the World*, ed. Malini Johar Schueller (1858; Ann Arbor: University of Michigan Press, 1999).

9. For an excellent essay on Blyden's fascination with Palestine, see Hilton Obenzinger, *American Palestine: Melville, Twain, and the Holy Land Mania* (Princeton, NJ: Princeton University Press, 1999).

10. Edward Wilmot Blyden, *From West Africa to Palestine* (London: T. J. Sawyer, 1873), 136, 141.

11. Teage was one of the founding fathers of Liberian independence that penned that country's declaration of independence.

12. Blyden, *The Jewish Question* (Liverpool: Lionel Hart, 1898).

13. Ibid. For a discussion of postnationalist and internationalist African American politics, see Nikihl Singh, *Black Is a Country: Race and the Unfinished Struggle for Democracy* (Cambridge: Harvard University Press, 2005). Also see Gilroy, *The Black Atlantic: Modernity and Double Consciousness* (Cambridge: Harvard University Press, 1993).

14. Blyden, *The Jewish Question*.

15. Edward Wilmot Blyden, "Mohammedism and the Negro Race," *Methodist Quarterly Review* (January 1877): 111.

16. Daniel P. Seaton, *The Land of Promise: Or, the Bible Land and its Revelation* (Philadelphia, African Methodist Episcopalean Church, 1895), 15.

17. W. L. Jones, *The Travel in Egypt and Scenes of Jerusalem* (Atlanta: Converse and Wing, 1908).

18. William Sampson Brooks, *Footprints of a Black Man* (St. Louis: Eden, 1915).

19. Israeli Pro-Con.org, "What are the solutions to the Israeli-Palestine conflict," http://www.israelipalestinianprocon.org/populationpalestine.html#sources1.

20. See Ilan Pappe, *A History of Modern Palestine: One Land, Two Peoples* (Cambridge, UK: Cambridge University Press, 2004).

21. See, for example, Nikihl Singh, *Black is a Country*. For a story of the Negritude movement, see Brent Hayes Edwards, *The Practice of Diaspora* (Cambridge: Harvard University Press, 2003). Also see Penny Von Eschen, *Race Against Empire: Black Americans and Anti-Colonialism, 1937–1957* (Ithaca: Cornell University Press, 1997) and Brenda Gayle Plummer, *Rising Wind: Black Americans and U.S. Foreign Affairs, 1935–1960* (Chapel Hill: University of North Carolina Press, 1996).

22. Carolyn Bagley, *My Trip Through Egypt and the Holy Land* (New York: Grafton, 1928), 187.

23. Adam Clayton Powell, Sr. *Palestine and Saints in Caesar's Household* (New York: R. R. Smith, 1939), vii.

24. See Nikihl Singh's *Black is a Country*. For a more general discussion of how the Cold War changed African American internationalist politics, see Von Eschen's *Race Against Empire* and Plummer's *A Rising Wind*; Gerald Horne's *Black and Red: W. E. B. Du Bois and the Afro-American Response to the Cold War, 1944–1963* (Albany, NY: SUNY Press, 1986). Although the Cold War changed the context of black internationalism, it is also important to note that many African Americans continued to embrace anticolonial politics during the 1950s and 1960s.

25. Earl Lewis, "To Turn as on a Pivot: Writing African Americans into a History of Overlapping Diasporas," *American Historical Review* 100, no. 3 (1995): 765–87.

CHAPTER 2

BLACK ORIENTALISM

ITS GENESIS, AIMS, AND SIGNIFICANCE FOR AMERICAN ISLAM

SHERMAN A. JACKSON

IN 1978, EDWARD SAID PUBLISHED HIS NOW-FAMOUS *ORIENTALISM*.[1] A CHRISTIAN Palestinian, Said devoted *Orientalism* to exposing the manner in which the prejudices and power of Europe, and later, the United States, created both a geographical entity called "the Orient" and a scholarly tradition of speaking and writing about it. This was not the Orient of Japan or China; this was the "Near East" and "Middle East." While Jews, Christians, and others contributed to the cultures and history of this region, Islam and Muslims were the primary, if not exclusive, targets of this new discourse. As the incubator and projector of Western fears, repressions, and prejudices, occidental discourse about the Orient normalized a whole series of self-serving, condescending stereotypes about Arab and Muslim "Orientals." These, in turn, justified the propriety and explained the inevitability of Western domination and privilege. This self-referencing, power-driven psychological predisposition, deeply rooted and often consciously indulged, constituted what Said meant by Orientalism.

Said noted that Orientalism was not a purely political affair, something that only Western governments and armies used against Oriental despots and their cowering subjects. Western intellectuals and academicians also played a role in this enterprise. Even when British, French, or American scholars approached the Orient with no conscious political aims, they could neither transcend nor disengage themselves from the social, historical, and institutional forces that shaped their mental schemas. Indeed, the Western scholar, wrote Said, "c[a]me up against the Orient as a European or American first, as an individual second."[2] As an individual, she or he might look *across* the Atlantic or Mediterranean *to* the Orient; but as a Westerner, she or he could only look *down* from a self-described superior civilization, a perspective destined to shape the Orient into a projection of the most deeply ingrained Western fears, prejudices, and obsessions.

If white Westerners approached the Orient as white Europeans and Americans, one would only expect Blackamerican thinkers and scholars to approach it as Blackamericans.[3] The meaning and implications of this would depend, of course, first, on where Blackamericans were in their own existential struggle and, second, on what influence the Orient was perceived as exerting on their lives. Prior to the 1970s, what little role the Orient played in Blackamerican consciousness was almost invariably positive. From the 1970s on, however, a palpable change begins to emerge. This change coincided with the coming of large numbers of Muslims from the Middle East and Asia to the United States, a development that produced major shifts in the priorities, sensibilities, and image of Islam in America.

Prior to the 1970s, Islam in the United States—and I include here the "proto-Islamic movements such as the Nation of Islam—had been dominated by a black presence and thus a black, American agenda. From the 1970s on, however, "real" Islam increasingly came to be perceived as the religion of Arabs and foreigners, who were neither knowledgeable about nor genuinely interested in the realities of Blackamericans. With this development, Blackamericans who identified with Islam, especially Sunnis, came under increasing criticism as "cultural apostates," "racial heretics" and self-hating "wannabee A-rabs" who had simply moved from the back of the bus to the back of the camel. Given that Blackamerican converts to Islam had all defected either from the Black Church or some other Blackamerican movement, certain elements within the Blackamerican community had always perceived Islam's gains as their own loss. The synergy between their negative predisposition and the cultural and ideological dislocations now dogging Blackamerican Islam would ultimately contribute to the rise of the phenomenon of Black Orientalism.

Unlike Said's "White" Orientalism, the aim of Black Orientalism had nothing to do with a desire to control or dominate the Orient. Like Said's Orientalism, however, its target was emphatically Islam. At bottom, Black Orientalism is a reaction to the newly developed relationship between Islam, Blackamericans, and the Muslim world. It comes in the aftermath of the shift in the basis of religious authority in Blackamerican Islam, from Black Religion to the intellectual tradition of historical Islam. Its ultimate aim is to challenge, if not undermine, the esteem enjoyed by Islam in the Blackamerican community by projecting onto the Muslim world a set of images, perceptions, resentments, and stereotypes that are far more the product of the black experience in the United States than they are of any direct relationship with or knowledge of Islam or the Muslim world. By highlighting, in other words, the purported historical race-prejudice of the Muslim world and, in some instances, alleged responses to it, Black Orientalism seeks to impugn the propriety of the new relationship between Islam, Blackamericans and the Muslim world and ultimately to call into question Blackamerican Muslims' status as "authentic," "orthodox" Blackamericans.

In this chapter, I shall trace the rise, nature, and significance of Black Oriental-ism. This will include a brief examination of the development of Islam among Blackamericans in order to place Black Orientalism in a more meaningful histori-cal context. It will be followed by a more detailed look at the shifts and disloca-tions in Blackamerican Islam engendered by the influx of Muslim immigrants to the United States following the changes in immigration quotas in 1965. I will then sharpen my definition of Black Orientalism, highlighting the distinction between it and valid criticisms of Arabs and/or Muslims. That will be followed by a brief, synecdochic response to one particular expression of Black Orientalism, what I refer to as nationalist Black Orientalism. I will conclude with a word about the significance of Black Orientalism for the present and future of Islam in the United States.

RELIGION, IDENTITY, AND THE SPREAD
OF ISLAM AMONG BLACKAMERICANS

In tracing the history of Islam among Blackamericans, it is important to begin with the fact that the United States is unique among the Western democracies in that a significant proportion of its Muslim population was born in this country. The spread of Islam among Blackamericans did not follow any of the patterns familiar to Islam in other parts of the world: it was not the result of immigra-tion, conquest, or the efforts of traveling Sufis. The rise of Islam among Blacka-mericans owes its impetus to a masterful feat of appropriation via the vehicle of Black Religion. The early Blackamerican "Islamizers," for example, Noble Drew Ali and the Honorable Elijah Muhammad, enlisted Islam not only as a strictly religious expression but as a basis for developing an alternative modality of Ameri-can Blackness. Blackamericans at large came to see in this religion not only a path to spiritual salvation but also to a more authentic Blackamerican self. Elijah Muhammad campaigned not only against Christianity as a theology but against those "finger-poppin', chitlin'-eatin', yes sa bossin' Negroes" whose modality of blackness he perceived as aiding and abetting the enemy. Islam, in other words, as "the Black Man's Religion," was as much about identity-formation and what E. E. Curtis IV refers to as "cultural nationalism" as it was about religion in the restricted (Western) sense.[4]

This very practical dimension of being enlisted as the basis of an alternative modality of American Blackness is crucial to understanding the rise of Islam among Blackamericans. It is also critical, however, to a proper understanding of the rise of Black Orientalism. For on the one hand, it was primarily this dimen-sion of Blackamerican Islam that thrust it into competition with the Black Church and other Blackamerican movements. At the same time, it was this dimension of Blackamerican Islam that was neither understood nor appreciated by the masses of immigrant Muslims who came to monopolize the authority to define a prop-erly constituted Islamic life in the United States post-1965. In the face of this new, immigrant authority, ostensibly grounded in the supertradition of historical

Islam, Blackamerican Muslims found themselves unable to address their realities in ways that were likely to prove effective in an American context or be recognized as Islamic in a Muslim one. In addition, this new ideological dependency left them unable to insulate the positive features of their Blackamerican culture and legacy from the hostile reflexes of an immigrant Islam still reeling in reaction to its nemesis: the modern West. All of this would leave Blackamerican Muslims open to the charge of being followers of a religion that countenanced, where it did not actually endorse, the devaluation, marginalization, and subjugation of blacks.

FROM BLACK RELIGION TO HISTORICAL ISLAM

The history of Islam among Blackamericans begins, for all intents and purposes, in the early twentieth century, with the marriage of Islam and Black Religion. Black Religion, however, should not be understood to constitute a distinct religion *per se* but rather as a religious *orientation*. It has no theology or orthodoxy, no institutionalized ecclesiastical order and no public or private liturgy; it has no foundational documents, like the Bible or the Baghavad Ghita, and no founding figures like Buddha or Zoroaster. The God of Black Religion is neither specifically Jesus, Yahweh, or Allah. It is, rather, an abstract category into which any and all of these can be fit. In a word, Black Religion might be described as the deism or natural religion of Blackamericans, a spontaneous folk orientation grounded in the belief in a supernatural power located outside human history yet uniquely focused on that power's manifesting itself in the form of interventions in the crucible of American race relations. Black Religion is essentially a holy protest against white supremacy and its material effects. As C. Eric Lincoln put it, its point of departure was American slavery, and had it not been for slavery, there would have been no Black Religion.[5]

The Black Church ultimately emerged out of the marriage between Black Religion and Protestantism in the eighteenth to nineteenth century and conferred a palpably religious dimension upon the black struggle in America. Indeed, the Black Church remained the dominant host of Black Religion until the beginning of the twentieth century. In the closing decades of the nineteenth century and the opening years of the twentieth, Blackamericans began to migrate en masse from the South to Northern metropolises, where the relationship between the Black Church and Black Religion was ruptured and the latter was forced to look for new accommodations. Joseph R. Washington, Jr., has described this alienation from the Black Church as follows: "Since the 1920s, black religion, the religion of the folk, has been dysfunctional. From this period on the once subordinate and latent stream of white Protestant evangelicalism has been dominant and manifest, relegating the uniqueness of black religion to verbal expression from the pulpit in such a way that action was stifled."[6]

The early "Islamizers," Noble Drew Ali and the Honorable Elijah Muhammad, emerged in the context of this ruptured relationship between the Black Church and Black Religion, offering asylum to Black Religion in what they presented as

Islam. By using Black Religion as a vehicle for appropriating Islam and making it meaningful and valuable to Blackamericans, these early Islamizers were able to popularize the religion and render it the cultural property of Blackamericans as a whole. This establishment of a sense of ownership was critical to the rising rate of Blackamerican conversion. It was also an important factor in the Islamizer's ability to influence Blackamerican culture at large. One sees signs of this in the newly developed disdain for pork or in the spread of Arabic names, in both their proper and bastardized forms (i.e., those of the "a-ee-a" pattern, such as Lakeesha, Tamika, Shameeka). In short, this historical feat of appropriation marked the true beginning of the history of communal conversion to Islam among Blackamericans and gave the religion itself bona fide roots in American soil. Indeed, without this historical achievement, it is doubtful that Islam would have come to enjoy the success that it has come to enjoy among Blackamericans.

If only by default, Black Religion remained the primary means by which Blackamerican proto- and Sunni Islam validated itself prior to 1965. For the Islamizers, and even many Sunni Muslims, it was not primarily the Qur'an, *Sunnah*, or books of law and exegesis that authenticated a view as Islamic. It was, rather, the extent to which a view was perceived as contributing to the throwing off the yoke of white domination or to conforming to the dictates of the new black cultural orthodoxy. As long as this remained the case, however, Black Orientalism existed only as a cry on the margins of Blackamerica. With the repeal of the National Origins Act in 1965, however, and the subsequent massive influx of Muslims from the Middle East and Asia, a new basis of religious authority was introduced into American Islam. The primary authenticators of Islam were no longer Black Religion nor Black Americans. They were now immigrants and the traditional Islamic religious sciences in whose name they ostensibly spoke. Beyond its impact on Blackamerican Muslims, this shift in the basis of Islamic religious authority coincided with a fundamental change in the attitude of Blackamerican non-Muslims toward the Arab/Muslim world.

Prior to the shift from Black Religion to historical Islam, the Arab and Muslim worlds were invariably included as constituents of an idealized Third World, a regiment of Franz Fanon's *Wretched of the Earth*, grinding out the universal ground offensive against white supremacy and Western imperialism.[7] After this shift and the establishment of critical masses of immigrant Muslims in the United States, there was a growing number of Blackamerican scholars who denied the Arab and Muslim worlds this status and began to portray it instead as a precursor, partner, or imitator of the West in its denigration and subjugation of blacks. Several works by Blackamerican writers from the early 1970s reflected this development: C. Williams, *The Destruction of Black Civilization*; S. Maglangbayan, *Garvey, Lumumba, and Malcolm: Black National-Separatists*; Y. Ben-Jochannan, *African Origins of Major Western Religions*; and H. Madhubati (Don L. Lee), *Enemies: The Clash of Races*.[8] This was the beginning of Black Orientalism, a trend that has continued into the new millennium.

BLACK ORIENTALISM AND WHAT IT IS NOT

Not every criticism of the stereotypes, prejudices, and practices of Muslim Orientals is an expression of Black Orientalism. Valid criticism, however, is distinct from ideologically driven projection. The former is based on direct experience or knowledge of verifiable facts; the latter, on imagination, ideology, and a will to denigrate. When Blackamericans condemn the exploitative activities of (Muslim!) Arab liquor store magnates in greater Detroit or Chicago, this is no more an exercise in anti-Muslim Black Orientalism than earlier critiques of Jewish slumlords were of anti-Semitism. Similarly, if the old antimiscegenation laws prove how deeply ingrained antiblack racism was among American whites, de facto antimiscegenation sentiment among Muslim Orientals cannot be written off as a benign "cultural preference." In short, if the association between Islam, Blackamericans, and the Muslim world should not be taken as a license for wild and unwarranted projections, neither should it require turning a blind eye to real offenses experienced firsthand.

Nor must Blackamerican criticism of Muslim Orientals be limited to contemporary facts or experience. The premodern Islamic legacy remains the repository of the greatest authority for contemporary Muslims and it continues to inform the thought and sensibilities of Islam in the United States. When we turn to this legacy, we find that Muslim legal, historical, exegetical, and belle-lettristic literature are replete with antiblack sentiment. It is neither Black Orientalism nor a manifestation of anti-Muslim bias to criticize and analyze such works. On the contrary, such criticism and analysis is necessary for the establishment of a standard that can be fairly and consistently applied across the board.

Consider the following example. In his famous *Prolegomenon*, Ibn Khaldûn (808/1406) says that blacks in the southern portion of Africa "are not to be numbered among humans."[9] The early Meccan jurist, Tâ'ûs, refused to attend weddings between a 'black' and 'white' because, given his understanding of the Qur'ânic verse about the Satanic impulse to "change God's creation" (*taghyîr khalq Allâh* 4:119), he deemed such unions to be "unnatural."[10] Numerous early Mâlikî jurists held, reportedly on the authority of Mâlik, that while under normal circumstances a valid marriage contract required that the woman be represented by a male relative (*walî*), this could be relaxed in instances where the woman hailed from lowly origins, was ugly or black.[11] This, they argued, was because Blackness was an affliction that automatically reduced a woman's social standing.[12] Similarly, the twelfth/eighteenth-century Mâlikî jurist, Ahmad al-Dardîr, categorically affirms the unbelief (*kufr*) of any Muslim who claims that the Prophet Muhammad was black.[13]

Nothing would excuse the casual dismissal of such statements from white Americans or Europeans. Nor should their authors' status as Muslim Orientals earn them any such exemption. Holding up such statements for comment, investigation and criticism is not Black Orientalism. On the contrary, it is responsible scholarship whose ultimate aim and effect should be to alert Muslims to the ways in which they have failed to live up to their religious ideals.

Having said this much, we should note that critical references to statements and actions by Muslim Orientals *can* approach Black Orientalism, when they proceed on the uncritical assumption that casual expressions of race or color prejudice in a society that has a history fundamentally different from that of the United States *must* have the same meaning and implications that they would have in America. In other words, Black Orientalism implies not only that Muslim society produced expressions of race- or color-prejudice but that such prejudice fundamentally defined these societies and, in so doing, circumscribed the lives and possibilities of black people within them. In American constitutional law, race is a suspect classification precisely because of the Supreme Court's recognition of the history of *institutionalized* racism in America. Black Orientalism extends this logic by projecting this history onto the Muslim world and from there imputing the same valence to *all* expressions of race or color prejudice.

Among the strongest contentions giving currency to the assumption that black life, *qua* black life, was circumscribed in Muslim society is the misleading insinuation that blacks in Islam were a slave class, as they were in the United States. This not only adds credence to the notion that black life was circumscribed, but it also confers upon all racially biased statements and actions the appearance of being part of the ruling class' effort to justify its domination over its subjugated wards. In point of fact, however, as every historian of Islam knows, most slaves in Muslim society were probably not black but of Turkish origin, and there is no evidence at all that most blacks were slaves.[14] Even, however, if we assume that blacks were a slave class in Muslim society, as Ira Berlin notes in *Many Thousands Gone*, there is a major distinction between "societies with slaves" (e.g., African society) and "slave societies," such as the United States, where color, incidentally, and slavery were coterminous. According to Berlin,

> In societies with slaves, no one presumed the master-slave relationship to be the social exemplar. In slave societies, by contrast, slavery stood at the center of economic production, and the master-slave relationship provided the model for all social relations: husband and wife, parent and child, employer and employee, teacher and student. From the most intimate connections between men and women to the most public ones between ruler and ruled, all relationships mimicked those of slavery . . . "Nothing escaped, nothing and no one." Whereas slaveholders were just one portion of a propertied elite in societies with slaves, they were the ruling class in slave societies; nearly everyone—free and slave—aspired to enter the slaveholding class.[15]

The presumption that blacks under Islam were a slave class in a slave society is a major premise of Black Orientalists and a primary means by which they impose a single interpretation upon every racially tinged statement or action by an Arab or non-Black Muslim. But if views such as Mâlik's regarding Blackness as an affliction are to serve as proof that Arab Muslims were all Jim Crow segregationists, what is to be made of Martin Luther King, Jr.'s statements about dark-skinned women, or Frederick Douglass's reference to the "ape-like appearance of some of the genuine Negroes," or Alexander Crummel's labeling of West Africans as "virile barbarians," or, for that matter, comedian Chris Rock's declaration, "I hate niggers!"?[16] Clearly, Muslims south of the Sahara, who overwhelmingly adopted the Mâlikî school, ignored the view attributed to Mâlik and required a male relative to validate a marriage. Why should the prejudicial view attributed to Mâlik and some Mâlikîs be accepted as the final, definitive word? And are we to impute to the words of King, Douglass, Crummel, and other blacks the same significance as racist statements by southern sheriffs who were clearly committed to the official subjugation of blacks?

We might also ask whether the statement of Ibn Khaldûm quoted previously is necessarily an antecedent to such "scientific" racialist theories as those of Jensen, Schockley, and the authors of *The Bell Curve*.[17] And in making such a determination, how justified would we be in ignoring Ibn Khaldûn's *explicit* statements to the effect that "race" is an imagined social construct, that the notion of black intellectual inferiority is false, that the Old Testament story about Noah cursing his son Ham does not refer to blackness but says only that Ham's sons shall be cursed with enslavement, and that it is climate, not blood, that affects endowments such as intelligence or civilization?[18] According to Ibn Khaldûn's theory, the farther people are removed from the moderate climate of the Mediterranean, the less their intelligence and civilizing potential. Thus, he imputed the same savage status to Africans farthest removed to the south that he did to white "Slavs" (*Saqâlibah*) who were farthest removed to the north.[19] In sum, one must ask why the history of race relations in the United States should be the only prism through which his (and others') statements can be viewed.

It is true that the examples cited, as well as many others, demonstrate that Arab and other non-Black Muslims were afflicted with race- and color-prejudice.[20] The insinuation, however, that such attitudes emerged from the same place psychologically and implied the same all-encompassing social and political reality as that created by white Americans stems more from imagination than from fact. In the year 659/1260, some seven centuries before the U.S. Civil Rights Movement, a black man appeared in Cairo after the sacking of Baghdad by the Mongols and claimed to be a member of the 'Abbâsid House. The Mamlûk Sultan, himself a former slave, ordered the chief justice to make an official inquiry into the claim. After his genealogy was confirmed, this black man took the name "al-Mustansir" and was inaugurated *amîr al-mu'minîn* (Commander of the Faithful), that is, Caliph, temporal successor to the Prophet Muhammad.[21] Clearly, blackness was not an impediment to rising to the highest position not only in Cairo but also in Islam as a global, multiracial faith.

If the real significance of race and color prejudice in Arab/Muslim society is to be understood, facts such as these must be duly recognized and considered. But Black Orientalism deliberately ignores or suppresses such facts, in order to invest race prejudice in the Muslim world with the same significance it has in the United States. The result is that cultural bias and the deliberate, race-based monopoly and abuse of power become so indistinguishable that a cultural idiosyncrasy such

as Rapper Sir Mix-A-Lot's contempt for the gaunt figures and flat buttocks idealized by *Cosmopolitan* magazine takes on the same significance as Jesse Helms's and
the Republican party's traditional opposition to affirmative action.[22]

NATIONALIST BLACK ORIENTALISM: MOLEFI ASANTE

There are at least three types of Black Orientalism: Nationalist, Academic, and
Religious. All three impugn the relationship between Blackamericans and histori-
cal Islam. Because of limitations of space, only one of these, Nationalist Black
Orientalism, will be briefly treated here.[23]

In 1980, Professor Molefi Kete Asante started a fire with the publication of his
provocative work, *Afrocentricity: The Theory of Social Change*.[24] This book became
the manifesto of the "new" Afrocentric movement.[25] It was followed by *The Afrocentric Idea* in 1987 and by *Kemet, Afrocentricity and Knowledge* in 1990.[26] The
purpose of these works was to lay out the aims, ideological underpinnings, and
practical methodology for an approach to historical, cultural, and sociological
studies that viewed the world, especially the African world, from the perspective
of Africa and Africans rather than from the dominant Eurocentric perspective that
claimed objectivity and universality. Asante criticized other approaches, including
those of Africans and African-Americans, that he felt were biased or unduly influenced by uncritically accepted assumptions. Chief among these was the negative
assessment of the achievements of Africa and its contributions to world civilization. Afrocentrism was a clarion call to Africans and African-Americans to free
themselves from these negative stereotypes and return to their true African selves.
It was also an appeal to non-Africans to consider the African, rather than the
reigning European, perspective as an effective tool for rehumanizing the world.

As a professor at Temple University in Philadelphia, a city heavily populated
by Blackamerican Muslims, Asante was well aware that his expression of Afrocentrism would face stiff competition. In this light, he argued preemptively that any
number of lifestyles now popular among Blackamericans were simply inconsistent
with the dictates of Afrocentricity, which alone reflected the *true* African self. Of
Islam in particular, he wrote, "Adoption of Islam is as contradictory to the Dia-
sporan Afrocentricity as Christianity has been. Christianity has been dealt with
admirably by other writers, notably Karenga; but Islam within the African-American community ha [*sic*] yet to come under Afrocentric scrutiny. Understand that
this oversight is due more to a sympathetic audience than it is to the perfection of
Islam for African-Americans. While the Nation of Islam under the leadership of
Elijah Muhammad was a transitional nationalist movement, the present emphasis
of Islam in America is more cultural and religious."[27]

Asante's critique of Islam is neither theological nor philosophical. He does
not attack the foundational beliefs of Islam, such as its monotheism or belief in
an afterlife. His primary focus is rather on what he considers to be the negative,
self-deprecating place of blacks *in* Islam. His appeal in this light is to certain sensibilities presumed to be common among Blackamericans as a result of their New

World experience. His message is essentially that Islam inherently promotes an Arab supremacy that is no less preconscious, pernicious, and injurious to blacks than the white supremacy of the West.

Asante insists that the Arabs have structured Islam in such a way that non-Arabs (read, Blackamericans) are forced to accept the inherent superiority of Arab idiosyncrasies and presuppositions. This leads to "the overpowering submissiveness of Africans and other non-Arabs."[28] The specific means of enforcing this submissiveness are language (i.e., the primacy of Arabic among Muslims); *Hajj*, or pilgrimage to Mecca; the *qiblah*, or the direction in which Muslims must turn when offering ritual prayers; the doctrine that Muhammad was the last prophet; and customs such as dress that were informed by a specifically Arab culture. While space does not permit a full treatment of all these points, what follows should suffice to demonstrate that Asante is a proponent of Black Orientalism. In assessing the validity of this critique, it is important to remember that the determining factor is not whether his list is factually correct but whether the *meaning attributed to it* is grounded in objective analysis or ideological projection.

Asante's first charge is connected with language. In response to the thesis that Arabic spread among Muslim populations because of its "prestige and usefulness," Asante writes, "while this is partially true, it is more correct to say that the language succeeded because of force and punishment."[29] He offers no historical proof from European, African, or Arabic sources. Rather, he relies on his readers' tendency to utilize their Western experience as the analogue for all historical reality. The Arabs, in other words, must have forced Arabic upon their vanquished populations because the loss of African languages among the American slave population proves that white Americans forced their language upon their slaves.

But if the ability to "force and punish" was the primary means by which language spread, Turkish should have wiped out Arabic in all the areas of the Middle East over which the Ottomans ruled for almost half a millennium. And if prestige and usefulness were really marginal as incentives, what accounts for the existence in places as far removed from the Arabs as China, Russia, or Siriname of Muslim populations who continue to learn the language and who pride themselves on their ability to do so? Indeed, even if one concedes that Arabic spread by "force and punishment," would the ultimate effect and meaning of this imposition be the same as the black experience in the New World?

Here we come to a critical failing that virtually compels Asante to projection. He essentially equates whiteness with Arabness and then goes on to assume that the two function identically. This is designed to give the impression that any Arab supremacy in the Arab world would have to have the same effect on blacks as white supremacy had in America, namely that of relegating blacks to a negative, inferior category made inescapable by their skin color. In fact, however, the attempt by the Umayyads (the first Muslim dynasty) in the first/seventh century to perpetuate a system that reduced non-Arabs to second-class citizenship failed. After that, once a people was Arabized, it became equal in its Arabness to its conquerors, as was the case, for example, with the Egyptians, Syrians, and North Africans. This was true whether the adoption occurred through force, choice, or osmosis. In fact, Arabized peoples often eventually superseded the "original" Arabs in intellectual, artistic, and other pursuits, including the acquisition of power, as occurred, for example, with Abû Nawâs in Arab poetry, al-Ghazalî in Muslim theology, Abû Hanifa in Islamic law, and the famous Barmakid family of politicians.[30] By contrast, when the language, religion, and culture of New World Africans were destroyed and replaced by English and Protestantism, these blacks were rendered neither English nor American. The naturalization law passed by the United States Congress in 1790 defined American-ness in terms of whiteness, and whiteness was a boundary that black people could not cross.[31] It is thus misleading to imply, as Asante does, that the experience of subject populations, even under a regime of presumed Arab supremacy, would have to be the same as the experience of New World blacks under a regime of white supremacy. I have often been asked by Arabs who hear me speak Arabic if I am an Arab. I have never been asked by a white person who heard me speak English if I was white. If Arabization, forced or voluntary, expressed a commitment to the principle of *e pluribus unum* (from the many, one), American whiteness emphatically excluded blacks on the principle of *e pluribus duo* (from the many, two).[32]

The remainder of Asante's list implies that black conversion to Islam entails Blackamerican submissiveness to Arabs. We might note, however, that white American Muslims change their names, perform the pilgrimage, offer the daily prayers, modify their customs, and often replace their dress, either based on their understanding of their duty as Muslims or as a preference for traditions deemed to be more identifiably Muslim. But Asante does not speak of white American submissiveness to the culture and religion of the Arabs. And the reason for this is that, in his experience (and that of Blackamericans generally), white people simply do not have culture and religion imposed upon them. Being forced into the role of passive recipient is an exclusively black reality. In the end, it is, again, the force of this projection of the Blackamerican experience that both leads Asante to his submissiveness thesis and sustains its currency among his Blackamerican readership.

Asante's critique reflects a desire to delegitimize Islam in the Blackamerican community. His criticisms, however, are based more on projections from the Blackamerican experience than on an objective assessment of Islam itself. In describing "White" Orientalism, Edward Said noted that it was grounded in the fears, desires, repressions, and prejudices of the West. Asante's Black Orientalism, like that of all Black Orientalists, attempts to cast Islam and the Muslim world in a mold that accommodates Blackamerican imaginings, resentments, prejudices, and difficulties in confronting the intractable problem of American race relations.

THE SIGNIFICANCE OF BLACK ORIENTALISM
FOR ISLAM IN THE UNITED STATES

From the decades following the Civil War, black America has maintained a cultural/political orthodoxy dedicated to policing the boundaries between blacks and "pseudo-blacks." Pseudo-blacks have traditionally been identified as those whose cultural authenticity and/or political loyalty to the Blackamerican community are suspect. This cultural/political orthodoxy has always been part of the mores and sentiments of the folk, and paying homage to it has been the *sine qua non* of success for any serious movement among Blackamericans—even those, such as Elijah Muhammad's, that sought to alter the substance of Blackamerican culture. The early Blackamerican Islamizers' understanding and respect for this tradition facilitated the popularity and growth of their movements. Immigrant Islam, however, arrived in the United States oblivious to this reality and passed on much of this myopia to those Blackamerican Muslims who came under its influence. The result has been a cognitive dissonance in which fossilized doctrines and practices from the Muslim world are imagined to be viable substitutes for effectively engaging American, and particularly urban American, reality. At the same time, the power and status that Islam once enjoyed within the Blackamerican community has been displaced in many quarters by a sense of disappointment and betrayal and a feeling that Islam and Muslims are irrelevant, if not detrimental, to the black cause.

In this context, the rise of Black Orientalism must be viewed not only as a reflection of attempts by Blackamerican Christians and other non-Muslims to regain lost ground. The perspective of immigrant Islam must also be recognized as threatening the status and future of Islam in Black America. Blackamerican Muslims must confront and take concrete steps to overcome their ideological dependency. For they will cease to exist at the mercy of others' definitions only when they acquire the authority to define a properly constituted Islamic life for themselves.

Black Orientalism, however, is not a problem for Blackamerican Muslims alone. Immigrant Muslims are equally affected by the phenomenon, especially in the context of the United States after September 11, 2001. In earlier times, the criticism black leaders and thinkers leveled at Blackamerican Muslims never reached the point of threatening Islam's place in the collective psyche of Blackamericans. In the present atmosphere, however, given the diminished relationship between Islam and Black Religion, on the one hand, and the nationwide rise in anti-Muslim mania, on the other, this danger is far more imminent. Any permanent estrangement between Islam and Blackamericans would be nothing short of disastrous for Muslims of all backgrounds. For it is primarily through Blackamerican conversion that Islam enjoys whatever status it does as a bona fide American religion. Indeed, to date, Blackamericans remain the only indigenous Americans whose conversion to Islam connotes neither cultural nor ethnic apostasy. Without Blackamerican Muslims, Islam would be orphaned in the United States, with virtually nothing to save it from being relegated to the status of an alien, hostile threat. This has obvious implications for anyone associated with Islam.

The threat of Black Orientalism nonetheless lies far more in the refusal of Muslims—Blackamerican and immigrant—to recognize and address the causes that brought it into being than it does in the efforts of Black Orientalists themselves.

Muslims must confront, honestly and energetically, the question of whether the shift in the basis of Islamic religious authority had to result in the kinds of dislocations that contributed to the rise of Black Orientalism. This question is critically important for Blackamerican Sunnis. For they cannot return to classical Black Religion in a manner that privileges it over the historical Sunni tradition. The question for them is ultimately whether they can master and supplement Sunni tradition to the point of being able to speak effectively to their realities as Blacks, as Americans and as Muslims.

As for immigrant Muslims, it may be time for them to recognize that their greatest interest as Muslim Americans lies not in the situation in Palestine or Kashmir but in establishing a sense of belongingness, however problematic, in the collective psyche of Americans as a whole. This may mean devoting more energy to attaching themselves to an already existing tradition of Islamic belongingness in the United States. In such a context, Black Orientalism should reveal itself to be as great a threat to them as it is to Blackamerican Muslims. It is a threat, however, that will only be defeated through practical and attitudinal changes, not the same old rhetorical blue smoke and mirrors.

NOTES

1. Edward Said, *Orientalism* (New York: Pantheon Books, 1978).
2. Said, *Orientalism*, 11.
3. I use the term "Blackamerican" as an alternative to both the "African" and the hyphenation in "African-American." My contention is that blacks in the United States, certainly religiously speaking, are no longer fully or perhaps even primarily African. Politically, the hyphen in "African-American" does not have anything like the efficiency that it does in the case of Jewish American or Italian Americans, the latter's Jewishness and Italianness being essentially protected by their Americanness.
4. Edward E. Curtis, IV, *Islam in Black America: Identity, Liberation, and Difference in African-American Islamic Thought* (Albany: State University of New York Press, 2002).
5. C. Eric Lincoln, *Race, Religion and the Continuing American Dilemma* (New York: Hill and Wang, 1999), 31.
6. Joseph R. Washington, Jr., *Black Religion: The Negro and Christianity in the United States* (Lanham, MD: University Press of America, 1984), 37.
7. Frantz Fanon, *The Wretched of the Earth*, trans. Constance Farrington (New York: Grove Press, 1965).
8. Chancellor Williams, *The Destruction of Black Civilization* (Dubuque, IA: Kendall/Hunt, 1971); Shawna Maglangbayan, *Garvey, Lumumba, and Malcolm: Black National-Separatists* (Chicago: Third World Press, 1972); Yosef Ben-Johnson, *African Origins of Major Western Regions* (New York: Alkebu-Ian Books, 1970); Haki Madhubuti, *Enemies: The Clash of Races* (Chicago: Third World Press, 1978). On these and other works, see the informative article by Y. Hurridin, "African-American Muslims and the Question of Identity Between Traditional Islam, African Heritage, and the American Way," in Yvonne Yazbeck Haddad and John Esposito, *Muslims on the Americanization Path* (Atlanta: Scholars Press, 1993), 282–87.
9. Ibn Khaldûn, 'Abd al-Rahmân b. Khaldûn, *al-Muqaddimah* (Beirut: Dâr wa Maktabat al-Hilâl, 1986), 45. Throughout this chapter, dates are given according to both the Muslim and the Christian calendars.

10. See Muhammad al-Amîn al-Shanqîtî, *Adwâ' al-bayân fî' îdâh al-qur'ân bi al-qur'ân* (Beirut: Dâr al-Kutub al-'Ilmîyah, 1421/2000), 1:330. Al-Shanqîtî refutes the position of Tâ'ûs by referring to several marriages conducted by the Prophet between a black and a white, for example, Zayd b. Hâritha (white) with Barakah, the mother of Usâmah (black); Usâmah b. Zayd (black) with Fâtima bt. Qays (white, from the "royal" tribe of Quraysh); and Bilâl (black) with the sister of 'Abd al-Rahmân b. 'Awf (white).

11. Malîk ibn Anas (c. 713–c. 795), a legal expert in the city of Medina, founded a school of Islamic jurisprudence.

12. *Adwâ'*, 1:33. Al-Shanqîtî, himself a Mâlikî, refutes this view and cites several poems in praise of the beauty of black women.

13. Al-Dardîr, *Al-Sharh al-Kabîr* (Beirut: Dâr al-Fikr, n.d.), 4:309 (on the margin of Muhammad al-Dasûqî, *Hâshîyat al-dasûqî 'alâ al-sharh al-kabîr*).

14. This is obviously not the place for a full treatment of slavery in Muslim history, though the subject certainly deserves a full study, especially given the tendency on the part of Blackamericans to assume that American slavery is the norm that all other systems of slavery followed. They thereby make no distinction between slavery in a capitalist society and slavery in a noncapitalist order, slavery that is race-based and slavery that is race-neutral, slavery that draws slaves under the full orbit of law and slavery that denies slaves any legal rights at all. This makes objective discussions of Muslim or African or Polynesian slavery virtually impossible. It also obscures the fact that it was not slavery but white supremacy that was, and remains, the cause of black subjugation in the United States.

15. Ira Berlin, *Many Thousands Gone: The First Two Centuries of Slavery in North America* (Cambridge, MA: Harvard University Press, 1998), 8.

16. See M. E. Dyson, *I May Not Get There With You: The True Martin Luther King, Jr.* (New York: Free Press, 2000), 193–94; Frederick Douglass, quoted in W. J. Moses, *Afrotopia: The Roots of African American Popular History* (Cambridge, UK: Cambridge University Press, 1993), 80; Alexander Crummel, quoted in Moses, *Afrotopia*, 69. Chris Rock, comedian, used the line, "I love black people . . . but I hate niggers," in one of his comic routines.

17. Richard Hernstein and Charles Murray, *The Bell Curve: Intelligence and Class Structure in American Life* (New York: Free Press, 1994).

18. Ibn Khaldûn', *al-Muqaddimah*, 89, 63, 61. Ibn Khaldûn states that al-Mas'ûdî took the fallacious notion of black intellectual inferiority from the Arab philosopher al-Kindî, as cited by Galen.

19. Ibid., 60. But see the entire discussion for a full exposé of the theory of climate, 44ff. This is confirmed by St. Clair Drake in his *Black Folks Here and There: An Essay in History and Anthropology* (Berkeley: University of California, 1987), 2:157–59. Drake relies on the French translation of Ibn Khaldûn. In my view, Drake was not a Black Orientalist. Indeed, the fact that he relies exclusively on Orientalist writings but is still able to avoid Black Orientalism shows the extent to which this phenomenon is far more conscious than unconscious. Black Orientalists, in other words, tend to find only what they are looking for.

20. See, for example, St. Clair Drake, *Black Folks Here and There*, 2:77–184.

21. See Shâfî' b. 'Alî, *Husn al-manâqib al-sirrîyah al-muntaza'ah min al-sîrah al-zâhirîyah*, 2nd ed., ed. 'A. Khowaytar (Riyadh, 1410/1989), 79. There are numerous other instances of black rulers in the central Arab Islamic lands.

22. See Sir Mix-A-Lot's hit single "Baby Got Back" on the album *Mack Daddy* (Universal, 1992).

23. All three forms of Black Orientalism are discussed in Sherman A. Jackson, *Islam and the Blackamerican: The Third Resurrection* (Oxford: Oxford University Press, 2005).

24. Molefi Kete Asante, *Afrocentricity: The Theory of Social Change* (Trenton, NJ: Africa World Press, 1988).

25. As Wilson Jeremiah Moses points out, Afrocentric thought dates back at least to the nineteenth century and was even championed in the twentieth century by a number of white scholars, most notably Melville Herskovitz in *Myth of the Negro Past* (Boston: Beacon, 1958) and Martin Bernal in *Black Athena: The Afroasiatic Roots of Classical Civilization* (Camden, NJ: Rutgers University Press, 1987). The term "Afrocentrism" was used by W. E. B. Du Bois as early as 1962. See Moses, *Afrotopia: The Roots of African American Popular History* (Cambridge, UK: Cambridge University Press, 1998), 1–2, 11–12.

26. Molefi Kete Asante, *The Afrocentric Idea* (Philadelphia: Temple University Press, 1987); Molefi Kete Asante, *Kemet, Afrocentricity, and Knowledge* (Trenton, NJ: Africa World Press, 1990).

27. Molefi Kete Asante, *Afrocentricity* (Trenton, NJ: Africa World Press, 1996), 2. This was the eighth printing of the work that originally appeared in 1988.

28. Asante, *Afrocentricity*, 3.

29. Molefi Kete Asante, *Kemet, Afrocentricity and Knowledge* (Trenton, NJ: Africa World Press, 1998), 131. This is a reprint of the work first published in 1990.

30. Though born of a Persian mother, Abû Nawâs (130/747–c. 195/8100) was, and is considered to be, among the greatest of all Arab poets. See *The Encyclopedia of Islam*, ed. E. J. Brill (Leiden: 1913), 1:102. Al-Ghazalî (450–505/1058–1111), also of Persian lineage, is thought by many to be the most famous Muslim after the Prophet Muhammad himself. His most influential works were written in Arabic. *The Encyclopedia of Islam*, 2:146–49. Abû Hanifa, again of Persian ancestry, was the eponym of the Hanafi school of law, numerically the largest in all of classical Islam. He died in 150/767, and even today, many, if not most, Muslims believe he was a pure Arab. *The Encyclopedia of Islam*, 1:90. The Barmakid family, originally a Buddhist priestly family from Balkh, rose to power as government ministers under the Abbasid Caliph. *The Encyclopedia of Islam*, 1:663–66.

31. In the Act of March 26, 1790, Congress authorized naturalization for "free white persons" who had resided in the United States for at least two years and swore loyalty to the U.S. Constitution. The racial requirement remained on the federal books until 1952, though naturalization was opened to members of some Asian nationalities in the 1940s.

32. For more on this theme, see Matthew F. Jacobson's important *Whiteness of a Different Color: European Immigrants and the Alchemy of Race* (Cambridge, MA: Harvard University Press, 1998), 109–35.

Chapter 3

Islamism and Its African American Muslim Critics

Black Muslims in the Era of the Arab Cold War

Edward E. Curtis IV

Rather than treating African American Muslims as marginal Muslims, a species of Muslim largely separate from immigrant Muslims, this chapter adopts black Muslim perspectives on the history of Islamism, the twentieth-century transnational ideology that sees Islam as both a political system and a religion. I argue that the contours of Islamic identity and practice among African Americans after the World War II developed partly in response to nascent Islamist missionary efforts led by ideological participants in the so-called Arab Cold War. During this era, Islamic missionary activity became a well-funded and well-organized component of Saudi Arabia's foreign policy. Several international organizations, local Islamic centers, and tract societies targeted U.S. blacks as potential allies in the struggle to construct Islamic religion as a response to Arab socialism and nationalism. As foreign and immigrant Muslim missionaries reached out to African American Muslims in the 1960s, they claimed the authority to interpret what constituted legitimate Islamic practice, encouraged African American Muslims to join their missionary organizations, and, in some cases, challenged the Islamic authenticity of indigenous African American Muslim groups and leaders.

This contact and competition with the missionaries had far-reaching implications and important repercussions for African American Muslim religious practice and political identity. In one sense, African American Muslim reactions to the Islamist call reflected the ideological and cultural diversity of the thousands of

African Americans who called themselves "Muslim." It is no surprise—given that there were more than a dozen different African American Muslim networks and groups by 1960[1]—that African American Muslim responses would differ. Rather than attempting to describe all of these groups' reactions to the new missionary activity, however, I will limit my discussion to three strains of African American Muslim hermeneutics, exemplified respectively by Malcolm X, the Nation of Islam, and Shaikh Daoud Ahmed Faisal, the founder of the State Street Mosque. I show how Shaikh Daoud Ahmed Faisal aligned his community of believers with Islamist ideology; how Malcolm X became the student and ally of these new foreign and immigrant missionaries, though he resisted their politicized interpretation of Islam; and finally, how members of Elijah Muhammad's Nation of Islam rejected the missionaries' claims to ultimate religious authority and instead defended Elijah Muhammad's prophetic voice.

While pointing out the different reactions of African American Muslims to Islamism, this account also argues for some shared repercussions. As a result of the increased immigrant and foreign Muslim presence in the United States, many more African American Muslims began to use canonical Islamic texts, including the Qur'an and, in some cases, the *hadith* (the sayings and deeds of the Prophet Muhammad and his companions) to articulate ethical, theological, political, and socioeconomic visions for themselves and other U.S. blacks. This adoption of sacred texts altered not only the aesthetics of African American Muslim religious practices but also the communal identity of persons who now thrust themselves into an age-old, transnational conversation about the meaning of these texts. Increasing African American Muslim identification with the rest of the Muslim world also became manifest in African American Muslim visual art and poetry. Many African American Muslims literally drew and rhymed themselves closer to the imagined worldwide community of Muslim believers. Finally, I discuss how this shift in African American Muslim consciousness had significant but diverse political implications, as African American Muslims came to hold differing interpretations of their obligations to the worldwide community of Muslims and the heritage of Islam.

TWENTIETH-CENTURY AFRICAN AMERICAN ISLAM IN TRANSNATIONAL CONTEXT

During the late nineteenth century, Islam became linked in black international English-language discourse with the politics of anticolonialism. In the first decades of the twentieth century, African American Islam in the United States developed partly as an international discourse shaped by actual contact and exchange with persons from Muslim-majority lands, including Muslims from Asia. After the World War II, such religious and cultural exchange between indigenous African American Muslims and foreign Muslims, especially from the Middle East, expanded dramatically. During the era of decolonization and the "rising tide of color," African American interests in the link between black nationalism and

Islam became even more prominent. More and more persons in the African American diaspora identified Islam and Muslims as potential allies in the struggle against European neocolonialism and white supremacy, often framing the domestic struggle for civil rights as part of a global struggle for the self-determination of all persons of color.[2] Malcolm X, for example, famously spoke of the 1955 Afro-Asian Conference of nonaligned countries in Bandung, Indonesia, as a turning point in the affairs of the world, as people of color everywhere vowed not only to reject the yoke of neocolonial political control but also to eschew a colonized consciousness.

After the 1956 Suez Crisis, Egyptian president Gamal Abdel Nasser emerged as a powerful symbol of victory in this third-world struggle against imperialism and inspired admiration among many African Americans, especially those associated with Elijah Muhammad and the Nation of Islam. When Nasser hosted the Afro-Asian conference in 1958, Elijah Muhammad telegrammed the Egyptian president to assure him that "freedom, justice and equality for all Africans and Asians is of far-reaching importance, not only to you of the East, but also to over 17,000,000 of your long-lost brothers of African-Asian descent here in the West."[3] Some members of the Nation of Islam hung Nasser's picture in their homes. Nasser was received enthusiastically by Muslims and non-Muslims alike when he visited Harlem in 1960.[4]

In turn, Nasser cultivated at least symbolic ties with African American Muslims, responding to Elijah Muhammad's 1958 telegram with his "best wishes to our brothers of Africa and Asia living in the West." Elijah Muhammad visited Egypt in 1959, and one of his sons, Akbar Muhammad, studied there during the 1960s.[5] These contacts between Egypt and the NOI begin to indicate the extent to which African American Muslims became potential foreign policy allies and symbols of political struggle, not only in the grand struggle for the freedom of all formerly colonized peoples but also in the more local and regional struggles waged by differing interest groups in the Middle East.

To understand foreign and immigrant Muslim engagement, especially Arab Muslim engagement with African American Muslims, it is necessary to explore more deeply how the financial, diplomatic, and cultural outreach of Arab Muslims was often colored by these local and regional interests. In the postwar era, the neocolonial elites and newly empowered military juntas who had seized authority within political boundaries initially drawn to serve European and U.S. interests were forced to negotiate the interference of superpowers, the pull of regional desires for Pan-Arab unity, and the challenge of the Palestinian-Israeli conflict.[6] Part of their struggle was ideological, and although various national elites may have celebrated the solidarity of all formerly colonized peoples and touted both Pan-Arabism and Pan-Islamic unity, such rhetoric was often deployed to buoy their own national legitimacy and manage popular opinion.[7]

One seldom mentioned, but pivotal, crucible of African American Muslim and foreign Muslim interaction was what Malcolm Kerr famously referred to as the "Arab Cold War." The Arab Cold War was a conflict waged primarily between the Kingdom of Saudi Arabia and Nasser's Egypt, which were locked in both ideological and military struggles from 1958 through the 1960s. The battle of ideas commenced shortly after Nasser successfully entered into a political union with Syria in the winter of 1958. The United Arab Republic (UAR), as the two states became known, signaled the growth of both revolutionary socialism and Pan-Arabism, the movement to unite all Arab peoples into one political entity expressing their shared historical and linguistic roots. A few months later, when revolution overturned the Iraqi monarchy and an uprising occurred against President Sham'un in Lebanon, monarchs throughout the Middle East feared that Nasserism might actually succeed. The Arab Cold War continued into the 1960s, perhaps reaching its apex in 1962, when Egypt sent troops to support the leftist revolution in Yemen. Saudi Arabia threw its financial and political clout behind the Yemeni monarchy.[8]

But Saudi Arabia and its allies also forged a secondary front in this war, an ideological effort designed to bolster its legitimacy in the West and among Muslim states and persons. This was a battle for hearts and minds, and it was joined through Saudi Arabia's generous support and careful organization of global missionary activities. Up to this point, missionary societies such as the Muslim Brothers in Egypt and the Society of the Call and Guidance in South Asia aspired mostly to change Islamic practices within historically Islamic lands.[9] By the 1960s, however, Saudi Arabia's aid allowed these groups' ideas to be broadcast, printed, and distributed around the world, including in the United States. In 1961, Saudi Arabia established a new university in Medina committed to the training of Muslim missionaries. The following year, as tension over Yemen escalated, the government also supported the founding of the Muslim World League, whose statement of purpose included a commitment to global missionary work. Not surprisingly, the conference was strongly anti-Nasser, promoting a vision of Pan-Islam that hoped to counter the powerful Arab populist.[10] An impressive array of Muslim personages attended the organization's inaugural meeting, including Mawlana Mawdudi of Pakistan and Said Ramadan of Egypt. One of the most influential intellectuals of Islamic reform and revival in the twentieth century, Pakistani ideologue and Jama'at-i Islami (Islamic party) founder Mawdudi argued that the *shari'a*, or Islamic law and ethics, provides God's blueprint for all human societies, which should be organized into an Islamic state.[11] Egyptian representative Said Ramadan was the son-in-law of Hasan al-Banna, the founder of the Muslim Brothers, an Islamist organization that Nasser came to oppose as he consolidated power in postrevolutionary Egypt. Seeking refuge from Nasser's repression in 1958, Said Ramadan immigrated to Switzerland, where he established, with Saudi assistance, the Centre Islamique des Eaux-Vives, an institution that became one of the nodes in a transnational Islamist intellectual network.[12] Like other Muslim Brothers, Ramadan insisted that Islam was a total way of life, applicable as much to public affairs as to private morality.[13]

Students, visitors, and refugees from the Middle East brought these Islamist ideas with them to the United States during the late 1950s and early 1960s. Confronting what they considered to be overly assimilated American Muslims, some Arab students immediately challenged the "liberal" and "Westernized" practices of various Muslim persons and organizations.[14] At the Islamic Center of New England, which had been built partly with a 1962 donation from King Saud, students made available various pieces of Islamic literature not previously translated into English.[15] In 1963, students from a variety of Muslim-majority countries gathered at the University of Illinois, Urbana-Champaign, to establish the Muslim Students Association (MSA). Among the founding members were three Muslim Brothers from Egypt. Using their positions on college campuses, these activists helped to make the MSA one of the most successful immigrant-led organizations in propagating Islamist ideas throughout North America.[16] In some cases, student advocates met African American Muslims who were already proponents of Islamist ideologies; in other cases, they confronted African American Muslims who had never heard the Islamist message.

AFRICAN AMERICAN ISLAMISM
IN THE ERA OF THE ARAB COLD WAR

One African American pioneer who had already articulated Islamist ideas in print by 1950 was Caribbean immigrant Shaikh Daoud Ahmed Faisal (d. 1980), reportedly the son of a Moroccan father and a Jamaican mother. Echoing the Islamist call that all societies must be governed by the Islamic *shari'a*, his book, *Islam the True Faith: The Religion of Humanity*, proclaimed that human beings should submit their societies to the authority of God, the Prophet, and the *shari'a*.[17] Shaikh Daoud was a pioneer who successfully converted hundreds, if not thousands, of African Americans to a Sunni interpretation of Islam at his Brooklyn-based State Street Mosque. His adventuresome spirit led him to establish a short-lived Muslim village in rural New York State, but he achieved his greatest success as leader of the mosque in Brooklyn. In 1939, he leased a brownstone at 143 State Street in Brooklyn Heights, just a block away from the heart of the Arab American community on Atlantic Avenue. He called his congregation the Islamic Mission of America in New York.[18] Shaikh Daoud welcomed both indigenous and immigrant Muslims to his mosque, where he warned them not to let the allure of the material world take them away from their Islamic practice.[19]

Before most Saudi-funded Muslim missionaries arrived in America, Shaikh Daoud's intellectual life bore the influence of Islamic reform and renewal movements. His publications, which often borrowed from other Muslim missionary tracts, sought to inform the American public on the basics of Islamic religion. They described the holy cities of Mecca and Medina, reproduced large excerpts from the Qur'an, taught believers how to make the *salat*, or daily prayers, and detailed and praised contemporary Muslim heads of state. As a New Yorker, Shaikh Daoud came to know Muslims who traveled to the city from various countries, especially diplomats who worked at the United Nations.[20] In the 1950 edition of *Islam the True Path*, he acknowledged the assistance not only of his wife,

Khadijah, but also M. A. Faridi of Iran, Bashir Ahmed Khan of Pakistan, and others from Afro-Eurasia.[21] The sheikh was also pictured in this volume wearing light-colored Arab robes, sitting cross-legged on a prayer rug or oriental carpet. In his hands he held the Qur'an, deeply contemplating its contents in the manner of an Old World Islamic scholar.

His mission was devoted to converting everyone in the United States to what he considered to be the only true religion of humankind. The profession of Islamic faith, he said, was a prerequisite to peace and security, as was Islam's implementation as a form of government and law based on the Qur'an. The holy book, he argued, "contains the complete Revelations of 'Allah,' the 'Almighty God,' the Lord of the worlds with the complete Laws for the government and guidance of humanity and as a protection for us from evil. The Criterion of all Laws is enclosed in the Holy Quran."[22] Shaikh Daoud's old-time missionary techniques, evangelical in tone, were harshly critical of Jews and Christians who ignored the truth that would set them free. He practically begged them to convert to Islam, warning that no one would be saved, no one would have "true religion" unless, and until, they became Muslims. Attacking the growing ecumenism among some American monotheists after World War II, Shaikh Daoud explicitly rejected the idea that all Abrahamic faiths were equally valid paths to salvation. [23] His criticism of Jews and Christians was grounded both in an Islamic critique of Jewish and Christian religious claims and in his experience as an African American New Yorker who had faced discrimination at the hands of some Jews and Christians. "The Jews of America are the proudest of all the people," he claimed. "If by chance a man of colour would move into their neighborhood they would raise such a rumpus which would give one cause to believe that that one person had committed murder."[24] He also criticized Christianity as a form of white supremacy. "Christianity," he claimed, is but a "social order, a philosophy, based on certain principles of White Supremacy, that White people are superior to their human brethren who are not White." Like other twentieth-century African American converts to Islam, he viewed Christianity as "an instrument of conquest" [25]

On the one hand, Shaikh Daoud's comments about Jews and Christians sound similar to other Islamic traditions of anti-Jewish and anti-Christian polemic grounded in a sense of religious superiority and particularity—a rhetorical mode hardly unique to Islam.[26] On the other hand, Faisal's critique also reflects a reading of history shaped by the racist contexts in which he was living—in 1950 and after, white Jews and Christians, from this black man's point of view, did indeed seem like godless creatures when they discriminated against people of color in their own neighborhoods and in foreign lands.[27] In asserting that Islam was the solution to such problems, Shaikh Daoud was participating in multiple discourses, international and local in scope. He did not resort to violence to achieve his end; instead, he relied on quintessentially American missionary techniques—he wrote missionary tracts, preached of divine justice and the chance of salvation, and established a successful congregation devoted to his teachings.[28] According to one scholar, he assiduously avoided any politically subversive activities and instructed his followers to follow all U.S. laws.[29]

In fact, some of his followers left the Islamic Mission of America precisely because they believed Shaikh Daoud to be overly supportive of the political status quo.[30] Many of them found their fellow congregants at the State Street Mosque to be morally lax and insufficiently pious. In the 1960s, some of Faisal's followers broke away to form the Ya-Sin (pronounced "yah-seen") mosque, which sought to separate from mainstream society so that believers could adhere as strictly as possible to *shari'a*. They were influenced in part by Hafis Mahbub, a Pakistani member of the Tablighi Jama'at, a Muslim reform group known mainly for its emphasis on spiritual purification and world renunciation over political entanglements. In 1960, Faisal reportedly hired Mahbub as a religious teacher. African American followers Rijab Mahmud and Yahya Abdul-Karim adopted Mahbub's call for "personal transformation" by living a life in strict adherence with the ethical example of the Prophet Muhammad of Arabia.[31] Women at Ya-Sin often covered themselves with both a head scarf and a face veil. Some men practiced polygamy.[32] Their movement would become known as Darul Islam, or the abode of Islam. By the 1970s, leader Yahya Abdul-Karim would declare that Muslims should avoid participation in U.S. politics and eschew friendships with all Americans, non-Muslim and Muslim alike, if they did not practice the "correct" form of Islam. Eventually, the movement spawned other African American Muslim groups, including the network of twenty mosques led by Jamil al-Amin, the former H. Rap Brown, who became known for his urban revitalization work in Atlanta.[33] All of these African American Islamist groups dreamed of a morally revived Islamic society, but they sought to realize that goal largely by personal example and the organization of utopian communities, not through violent *jihad*. Though many of these African American Muslims shared the same basic Islamist ideas, they were fractured into different groups with different leaders and they often translated Islamism into an American idiom.

MALCOLM X IN THE MISSIONARY MAELSTROM

Despite the growing power of Islamist ideas, many African American Muslims in the age of the Arab Cold War rejected the arguments and authority of the Muslim missionaries and the ideology of Islamism. This was especially true in Elijah Muhammad's Nation of Islam (NOI), which was harshly criticized by Sunni Muslims of all stripes for its black separatist version of Islam. The rise of the NOI's profile as the most popular African American Muslim movement in the United States coincided with the increased presence and impact of foreign and immigrant Muslim missionaries. Most Americans, including immigrant Muslims, knew little about the NOI until the late 1950s. Then, in 1959, New York's WNTA-TV aired a five-part series about the movement, hosted by Mike Wallace, titled "The Hate That Hate Produced."[34] Following that program, stories about the NOI appeared in national magazines such as *Time* and *U.S. News and World Report*. This coverage was generally negative, criticizing the movement as an anti-American or black supremacist organization. African American civil rights leaders, including Roy Wilkins of the National Association for the Advancement of Colored People (NAACP), denounced the NOI as a hate group.[35]

As negative portrayals in the mainstream press and criticism from black leaders increased, more and more Muslims in the United States joined to condemn, dispute, and reject the teachings of Elijah Muhammad and the NOI. Their criticism of the NOI was a public performance of Muslim identity that expressed the growing cultural power of foreign and immigrant Muslims. By making such public pronouncements, whether they had been formally trained in the Islamic religious sciences or not, these self-appointed spokesmen for Islam attempted to define the doctrinal boundaries of Islamic religion. As Malcolm X, the chief spokesperson for Elijah Muhammad, made his way around the college lecture circuit, he was constantly hounded by Muslim students and others who considered themselves the guardians of "true" Islam. He mustered Qur'anic verses and his best exegetical rhetoric to defend Elijah Muhammad's unique Islamic mythos but to little avail among his critics. Malcolm's inability to bring them over to his side seemed to bother him or at least to intrigue the famous debater. In 1962, for example, one Muslim student at Dartmouth College, Ahmed Osman, traveled to NOI Mosque No. 7 to question Malcolm about Islam. After grilling Malcolm in the question-and-answer section of his talk, Osman came away "unsatisfied." When Osman began to send Malcolm literature from the Centre Islamique des Eaux-Vives in Geneva, Malcolm read it and asked for more. In another incident, Arab students from the University of California, Los Angeles, surrounded Malcolm after a March 1963 appearance on the *Ben Hunter Show* in Los Angeles. After hearing the students argue that his belief in white devils was un-Islamic, Malcolm became quite disturbed, according to journalist Louis Lomax, who was accompanying Malcolm at the time.[36]

These students, and the larger trend of which they were a part, had a profound influence on Malcolm's religious life.[37] Perhaps their criticism of Elijah Muhammad's Islamic legitimacy was one contributing factor in Malcolm's defection from the NOI in 1964. But even if Malcolm left for other reasons—such as Elijah Muhammad's moral failings and the NOI's lack of direct political action[38]—his subsequent understanding of Islam certainly adopted the ideas and symbols of the new Muslim missionaries. After Malcolm X broke away, he turned to Dr. Mahmoud Youssef Shawarbi, a University of Cairo professor and Fulbright Fellow in the United States who was teaching at Fordham University.[39] Malcolm knew of Shawarbi through the numerous immigrant Muslims, especially students, who had been confronting him after his various lectures and appearances. "Those orthodox Muslims whom I had met, one after another, had urged me to meet and talk with a Dr. Mahmoud Youssef Shawarbi," he said in his *Autobiography*.[40] Shawarbi encouraged Malcolm to make the *hajj*, the annual pilgrimage to Mecca, and instructed him in the fundamental elements of Sunni Islam.[41] After training Malcolm in the rudiments of Sunni Islamic thought and practice, Shawarbi gave

Malcolm a letter of recommendation, a copy of *The Eternal Message of Muhammad* by the renowned Pan-Islamist Abd al-Rahman Azzam, and the phone number of Azzam's son, who happened to be married to the daughter of Saudi Prince Faysal.[42] The elder Azzam was one of Pan-Islam's most important figures. A father of Arab nationalism and a distinguished Egyptian diplomat, Azzam was a chief architect of the Arab League and served as its first secretary general from 1945 to 1952. But like so many others, he lost favor after Nasser came to power, finding refuge in Saudi Arabia, where he became a leading polemicist and author.[43] His *Eternal Message of Muhammad*, available in a 1993 edition, was a prime example of a popular modern Islamic polemic that both defended Islam against Western critics and advocated a vision of the ideal Islamic nation-state. Islam, Azzam said, was a "faith, a law, a way of life, a nation, and state." Contrary to Western assumptions, Azzam implied, Islam was a highly modern religious tradition that promoted tolerance, removed superstition, and encouraged mercy, charity, industriousness, fairness, and brotherhood in the hearts and minds of its adherents.[44]

Malcolm read the book while flying over the Atlantic Ocean on his way to Mecca, and then met Azzam himself during the pilgrimage. On April 13, 1964, Malcolm departed JFK International Airport with a one-way plane ticket to Jidda, Saudi Arabia.[45] When Malcolm arrived on the Arabian Peninsula, Saudi authorities detained him for special interrogation. After fretting for some time, Malcolm telephoned Azzam's son, Dr. Omar Azzam. The Azzams immediately interceded with the proper authorities, vouching for Malcolm when he faced an examination by the *hajj* court, the legal entity that decides whether one is a legitimate Muslim able to participate in the pilgrimage. In addition, the elder Azzam insisted that Malcolm stay in his suite at the Jedda Palace Hotel. Later, the Saudi government officially extended its welcome when the deputy chief of protocol, Muhammad Abdul Azziz Maged, gave Malcolm a private car for his travels around the kingdom.[46]

Because of the Saudis' hospitality, and because of what he witnessed during the pilgrimage rites, Malcolm issued a strong endorsement of Sunni Islam. He argued, famously, that Islam was a religion of racial equality and brotherhood—which is what his Saudi hosts hoped to hear. After Malcolm completed the pilgrimage, Prince Faysal invited him for an audience. The prince quizzed Malcolm about the Nation of Islam, carefully suggesting that if what he had read in Egyptian papers were true, they did not practice the real Islam. Further, the prince reminded Malcolm that due to the abundance of English literature on Islam, "there was no excuse for ignorance, and no reason for sincere people to allow themselves to be misled."[47] Prince Faysal, constructing himself as an authority on proper Islamic practice, apparently wanted to make sure that Malcolm understood the meaning of his royal hospitality.

Malcolm sustained these relationships with various Saudi-financed missionary groups until his untimely death. During September 1964, Malcolm left for another pilgrimage to Mecca.[48] During this *'umra*, or "lesser" pilgrimage, Malcolm underwent training as an evangelist by the Muslim World League, the organization that had been established in 1962 to propagate an Islamist interpretation of Islam around the world. Shaykh Muhammad Sarur al-Sabban, secretary general of the organization and a descendant of black slaves, supervised his education. The University of Medina granted Malcolm several scholarships for U.S. students who wanted to study there. According to Richard W. Murphy, then second secretary at the U.S. embassy in Jidda, Malcolm granted an interview to a Jiddan newspaper, *al-Bilad*, in which he "took pains . . . to deprecate his reputation as a political activist and dwelt mainly on his interest in bringing sounder appreciation of Islam to American Negroes."[49]

Though these missionaries had a profound effect on Malcolm, he had a serious disagreement with some of them about the question of black political liberation in the United States. One of Malcolm X's last press interviews was given to *Al-Muslimoon*, a journal published by Said Ramadan's Centre Islamique des Eaux-Vives in Geneva. Malcolm had visited the Islamic center's director, Said Ramadan, in 1964. Ramadan was one of several persons who helped to establish the league with the support of Saudi Arabia. Like Dr. Omar Azzam, Ramadan strongly asserted the view that Islam was both a religion and a state, the solution to all of humanity's economic, cultural, and political problems, including the oppression of black persons in the United States. In his written questionnaire, Ramadan challenged Malcolm about his continued focus on racial identity and the need for black liberation, asserting that the conversion of Americans to Islam would solve such problems. Malcolm X completed his answers to Ramadan's questions on February 20, 1965, one day before his assassination. Malcolm disagreed with Ramadan's view and insisted that while he would always be a devout Muslim, his first duty in life was to work for the political liberation of all black persons around the globe.[50] While Malcolm was no less committed to Islam as a religious and spiritual path, he rejected the view that Islam could offer a specific solution to every political problem. "My fight is two-fold, my burden is double, my responsibilities multiple . . . material as well as spiritual, political as well as religious, racial as well as non-racial," he told a crowd in Cairo. "I will never hesitate to let the entire world know the hell my people suffer from America's deceit and her hypocrisy, as well as her oppression."[51] Malcolm remained as committed as ever to a program of political liberation that remained outside the purview of his commitments as a Sunni Muslim. Though he had received Arab Muslims' financial support, their religious imprimatur, and their friendship, Malcolm resisted the Islamist view of his allies and sponsors.

Instead, Malcolm's break with the Nation of Islam freed him to articulate a powerful Pan-Africanist politics.[52] By the time he had declared his independence from Elijah Muhammad in 1964, Malcolm had already cultivated ties to other black nationalists in New York and had met several leaders of the nonaligned movement, including President Sukarno of Indonesia, President Castro of Cuba, and President Nasser of Egypt.[53] He furthered such connections throughout his travels in 1964, much of which he spent in Africa. During his May visit to Nigeria, he proudly acquired a new title, "Omowale," or, the son who had come home.

That year, Malcolm also met with Kwame Nkrumah, president of Ghana; Milton Obote, president of Uganda; and Jomo Kenyatta, president of Kenya. He also attended two Africa Summit Conferences, where he represented his own Organization of Afro-American Unity, a group he had modeled on the Organization of African Unity.[54] Until his death in February 1965, Malcolm also sharply criticized U.S. foreign policy toward Africa, especially U.S. support for Moise Tshombe's regime in the Congo. He stated—correctly, it turns out—that the United States had supported the overthrow of Patrice Lumumba, who had helped to expel Belgian forces, and he wondered aloud whether he should recruit African American freedom fighters to fight Tshombe's regime.[55] In a similar way, his domestic politics focused on finding black solutions to black problems, and his rearticulations of black nationalist themes included calls for racial solidarity in the face of white supremacy. His politics may have been radical, but they were not Islamist. Malcolm X did not believe Islamism was the solution to the problems of black people. Only black people, reaching across continents and across confessional lines, could solve black problems.

TRANSNATIONAL REVERBERATIONS IN THE NATION OF ISLAM

The rejection of the Islamists' authority was even more profound in the NOI, which would not permit other Muslims to define what it meant to be an authentic Muslim, as Elijah Muhammad himself made clear: "Neither Jeddah or Mecca have sent me! I am sent from Allah and not from the Secretary General of the Muslim League," he said, referring to the Muslim World League, created in 1962.[56] The NOI had its own system of rituals and code of ethics, which relied more on the prophecies of Elijah Muhammad than on the *shari'a*. But intellectuals in the NOI were not indifferent to the criticisms of other Muslims. Elijah Muhammad and his lieutenants were extremely sensitive to public opinion and vulnerable to attacks from other Muslims, as outlined in the examination of Malcolm X's efforts to defend Elijah Muhammad. Like Malcolm X, they were also influenced by the new ideas, texts, pamphlets, translations, stories, and symbols circulated by the Muslim missionaries in the United States. The NOI's reactions show just how important the missionaries and their ideas were to the development of African American Islam and American Islam as a whole.

For example, *Muhammad Speaks*, the NOI's newspaper of record from 1961 until the middle 1970s, frequently published endorsements of Elijah Muhammad and the NOI by mainstream Asian Muslim leaders. Like other so-called new religious movements, the NOI used appeals to traditional religious authorities in an effort to legitimate their movement.[57] In addition to citing foreign authorities to counter charges of illegitimacy, a whole cadre of intellectuals inside the organization, including NOI ministers and newspaper columnists, responded to Muslim criticisms of Elijah Muhammad's Islamic bona fides by constructing him as a qur'anically sanctioned prophet. For example, the prominent Nation of Islam cartoonist Eugene XXX, or Eugene Majied, frequently incorporated images and passages from the

Qur'an into his drawings of Elijah Muhammad, who was variously depicted as a doctor healing the "deaf and dumb Negro," as a Daniel fighting off critics, or as a Moses plaguing Pharaoh Lyndon Johnson.[58] Others, including Minister Abdul Salaam and columnist Tynnetta Deanar, fiercely defended their prophet by citing the Qur'an.[59] A few NOI intellectuals sought ideological rapprochement with the critics by reinterpreting the mission of Elijah Muhammad in terms more suitable to Sunni orthodoxy or by asking the critics to adopt a more sympathetic, theologically pluralistic view of Elijah Muhammad's claims.[60]

This engagement with Elijah Muhammad's critics represented a remarkable moment of contact and confrontation. As a result, NOI intellectuals increasingly read Islamic texts, especially the Qur'an. In some cases, this deeper engagement with Islamic texts and traditions led to dissension and outright rebellion among NOI intellectuals, as was the case with Elijah Muhammad's son, Wallace D. Muhammad.[61] However they answered the question of Elijah Muhammad's legitimacy, all of these intellectuals, card-carrying members of the NOI and defectors alike, had something in common. They had become part of an old Islamic tradition—a transnational conversation about the meaning of the Holy Qur'an.[62]

This moment of contact and confrontation also led to a shift in the mental geography of many African American Muslims in the NOI, as artists and poets incorporated new Islamic names, places, dates, figures, and ideas in their historical imaginings of black identity. This deployment of Islamic signs was part of a larger trend in black American culture that represented a reorientation of African American politics and religion toward the Middle East more generally, as Melani McAlister has argued.[63] But for African American Muslim members of the NOI, this was a collective reorientation not only toward the Middle East but also toward other places and times in which blacks and Muslims had lived. African American Muslims in the NOI located the story of black/Muslim people in many epochs and locales, including ancient Egypt, Muslim West Africa, Asia, a mythical Arabia, and the classical period of Islam during and immediately after the time of Prophet Muhammad of Arabia. These black Muslims "moved across" time and space, constructing their contemporary identities by imagining who they had been in the past.[64]

In 1967 William E. X published a poem titled "Black Man" in *Muhammad Speaks*, which references Egypt, Persia, Central Africa, and the Himalayas, showing how black/Muslim geography came to include Asia, which had been constructed as a racial home by African American Muslims at least since the 1927 publication of Noble Drew Ali's *Holy Koran of the Moorish Science Temple*. Brother William's poem recognized the diverse linguistic, phenotypical, and geographic roots of blacks/Muslims but insisted that, on the whole, blacks are still one and the same. In his poetic romp around Asia and Africa, Brother William sees the Muslim "black man" as possessing a common god, language, and character—all of which are viewed as Islamic. Whether the black man lives in the Middle East, China, or even Tibet, he is, at heart, a Muslim.

This reorienting of African American mental geography toward the Muslim world had important implications for NOI members' sense of religious community. Though some of their symbols, texts, and narratives had been adopted from foreign and immigrant missionaries, members of the NOI reappropriated such raw materials in their own understandings of what it meant to be a Muslim. Black Muslims in the NOI looked beyond the black Atlantic world to form their communal identities and created narratives that linked the history of black people to this history of Islam. They felt allegiance not only to the black nation but also to a community of Muslims who might be members of several different nations. Elijah Muhammad and many of his followers did not define the collective identity of blacks *exclusively* in terms of a desire for a separate nation or polity. Many in the NOI constructed black identity in terms of a shared history that was defined by its Islamic character. Of course, members offered differing understandings of their shared black/Islamic heritage, and they sited Islam in multiple times and places. Refusing to locate the history of blacks in one country or even on one continent, these stories adopted a transnational perspective toward black identity that rested upon its common Islamic roots. The radical implications of such identity making would become clear in the 1970s and beyond, when some black intellectuals, especially Chancellor Williams, Molefi Asante, and later, Henry Louis Gates, Jr., came to depict Islam as an enemy to, or at least a foreign element within, African cultures and civilizations.[65] Such reactions, dubbed "black orientalism" by Sherman Jackson, are difficult to imagine in the absence of a culturally influential and institutionally successful African American Islam. At the least, it is clear that Islam, however constituted, had become a potent signifier of black identity for some African Americans and that other African Americans resisted this remaking of black identity in Islamic terms.

CONCLUSION: POLITICAL REFRACTIONS OF TRANSNATIONAL ENCOUNTERS

The political implications of African American reorientations toward Islam were more ambiguous. Several scholars of the NOI have insisted, for decades now, that the insular, "cultic" qualities of Elijah Muhammad's NOI, in addition to the organization's millennialism and Victorian gender relations as well as its Puritanism and embrace of the Protestant work ethic, rendered the group an unwitting, decidedly conservative agent of the political status quo.[66] Such approaches seem to reflect the view, stated famously by Sacvan Berkovitch, that groups like the Nation of Islam, though appearing, at first, to look like manifestations of dissent, have actually functioned in U.S. history as vehicles of social control, since their teachings have not attacked the root causes of oppression.[67]

This argument, that the absence of a direct and organized assault on the political economy and patriarchy of the United States effectively sustains the status quo, reflects only a partial view of political action and resistance. Rebellion, as Robin D. G. Kelley argues, also includes cultural acts of resistance that reject the values and expectations of the powerful. [68] In this sense, the NOI's activities, and the growth of Islam more generally among black Americans, were extremely rebellious in the 1960s. During the 1950s, as Penny Von Eschen has pointed out, "the Nation of Islam permitted a space—for the most part unthinkable in the Cold War era—for an anti-American critique of the Cold War."[69] During the Vietnam era, in the middle of a civil rights movement that was an important component of U.S. foreign policy, many in the Nation of Islam and other African American Muslims rejected American nationalism, refused to serve in Vietnam, criticized the Civil Rights Movement as hollow, and challenged the legitimacy of the nonofficial state religion, Christianity. The way these Muslims dressed and talked—in addition to the pictures they drew and the poems they wrote—questioned the cultural foundations of the state and its legitimacy to rule. The fact that U.S. government officials associated members of the NOI with violent revolution, despite the lack of any organized effort in the movement to confront authorities with violence, indicates the extent of the ideological challenge.[70] According to one observer, the NOI was among the most watched organizations in the government's Counter-Intelligence Program (COINTELPRO).[71] Surely, this evidence indicates that the message of the Nation of Islam and its members was politically dangerous in some way.

The political implications of this Islamist flowering were also multivalent. The radical call for God's sovereignty over all the earth and the establishment of the United States as an Islamic nation was reflected and refracted in an array of community programs and political platforms. Shaikh Daoud, as we have seen, preached the necessity of establishing God's rule over all the earth but relied on nonviolent missionary work as his means to accomplish this end. The younger critics of Daoud's supposed passivity cried even more loudly about the moral bankruptcy of the West and its evil ways, but in most cases, they sought separation from mainstream society, not violent revolution. For example, Sheik Tawfiq (d. 1988), an African American from Florida, founded the Mosque of the Islamic Brotherhood in Harlem, New York. Stressing the call of Islamic universalism—the idea that Islam crossed all racial barriers—African Americans, Hispanics, and others prayed together, established housing and education programs, and ran small businesses in the heart of the largest U.S. city. In 1971, Yusuf Muzaffaruddin Hamid, who was a student of Pakistan's Jama'at-i Islami, established the Islamic Party of North America, which advocated the creation of an Islamic state in North America through a mass religious revival.[72] These African American Sunni Muslim leaders regularly mingled with foreign and immigrant *imams*, who were now entering the United States in larger numbers due to 1965 reforms in immigration policy and financial support from Saudi Arabia and other Gulf countries. After the 1973 and 1974 OPEC oil embargo, the price of oil skyrocketed, and at least some of those petrodollars were used to support the missionary efforts created during the Arab Cold War. By the early 1980s, according to one scholar, twenty-six communities "were receiving the services of leaders provided by the Muslim World League."[73] While infused with a powerful jeremiad and the call for an Islamic revolution, however, their rhetoric led mainly to a moral revival, not violent *jihad*. Even when African American Muslims engaged in violent *jihad* during this era, it was directed toward other African American Muslim groups, not the government or larger society.[74] There is simply no hard evidence indicating that their radical rhetoric led to organized terrorism or violent revolution on a mass scale.

As in the case of the Nation of Islam, it might be tempting to conclude that the political impact of these utopian groups was conservative, but such arguments would once again ignore the ideological and cultural resistance that the groups offered. Some African American Islamists became effective spokespersons in the United States for foreign Muslim causes, including those in Palestine, Afghanistan, Bosnia, and Chechnya, and while some funds were raised for these causes, American Muslim support was mainly moral in nature, as supporters gave fiery speeches from the pulpit on Fridays and discussed the issues in conferences and study groups.[75] Author Daniel Pipes is right that some American Muslims, including African American Muslims, would eventually become entangled in violent jihadist networks. A few African American Muslims have been convicted of aiding al-Qaʿida or other terrorist groups, but these are rare, if dramatic, cases.[76]

In some ways, the culture of African American Islam from the 1970s until today has borne the imprint of the contact and confrontation with foreign and immigrant Muslims during the 1960s. In the 1970s, W. D. Muhammad, the son of Elijah Muhammad, would change the name of the Nation of Islam to the World Community of al-Islam in the West and ask his followers to observe the ethical, theological, and ritual directives of Sunni Islam.[77] He also reached out to foreign Muslims, accepted Egyptian president Anwar Sadat's offer of scholarships to attend Egyptian universities, and placed a Sudanese shaykh educated at the University of Medina as the prayer leader of the Chicago temple.[78] Despite these ties, he would tailor Sunni Islamic teachings to the African American experience, advocating a platform of political, economic, and social reforms that were transnational in style but local and national in their content.[79] In a sense, he had truly internalized the reformist attitudes of the Muslim missionaries by focusing so intently on the scriptures of Islam; but his close readings did not always agree with theirs. The same was true for other African Americans who sought to apply the texts of Islam to their own circumstances. By the 1990s, some of these efforts produced progressive political positions on a variety of issues, especially on gender. Islamic studies professor Amina Wadud issued an academic manifesto called *Qur'an and Woman*.[80] Eventually, she would challenge the taboo of a woman leading a mixed-gender prayer.[81] Similarly, African American women at the grass roots interpreted the Qur'an and hadith as documents of womanist liberation.[82] More and more African American Muslims would travel abroad to study the classical Islamic sciences in the aftermath of the Arab Cold War, but when they came back from foreign *madrasas* and Muslim universities, their interpretations of Islam were still infused with an African American sensibility focused on the problems facing Muslims in the United States.[83]

The interaction of African American Muslims with ideological players in the Arab Cold War may have changed the contours of African American Islamic culture, but it did not undermine African American Muslim religious or political agency. On the contrary, African American Muslims often appropriated the cultural and intellectual resources of the missionaries into an Islam that reflected their own interests. Greater transnational ties between African American Muslims and Muslims abroad have led to an ever-larger variety of Islamic religious expression in black America. Some African American Muslims have joined both new and traditional Sufi orders, the mystical groups of Islam, including the West African-based Tijaniyya, and the Philadelphia-based fellowship of Shaykh M. R. Bawa Muhaiyaddeen.[84] There is an important, if small, number of black Shi'a Muslims as well—the result of Shi'i outreach from the 1970s until today.[85] Minister Louis Farrakhan, who reconstituted the Nation of Islam in 1978, has incorporated more and more Sunni Islamic texts and traditions into his religious practice and sought strong ties to foreign Muslim leaders, but he has also continued to claim a special place for black nationalism, Pan-Africanism, and his own interpretations of Elijah Muhammad's teachings within the Nation of Islam.[86] All of these groups, and their differing religious and political outlooks, reflect the vitality of an African American Islam both transnational and local.

NOTES

1. See Aminah Beverly McCloud, *African American Islam* (New York: Routledge, 1995), 9–40.
2. See Melani McAlister, *Epic Encounters: Culture, Media, and U.S. Interests in the Middle East since 1945*, updated ed. (Berkeley: University of California Press, 2005).
3. C. Eric Lincoln, *The Black Muslims in America*, 3rd ed. (Grand Rapids, MI: William B. Eerdmans, 1994), 225.
4. See Brenda Gayle Plummer, *Rising Wind: Black Americans and U.S. Foreign Policy, 1935–1960* (Chapel Hill: University of North Carolina Press, 1996), 257–66, 285.
5. Lincoln, *Black Muslims*, 226, 227; Claude Andrew Clegg III, *An Original Man: The Life and Times of Elijah Muhammad* (New York: St. Martin's, 1997), 135–136, 189.
6. See Fawaz A. Gerges, *The Superpowers and the Middle East: Regional and International Politics, 1955–1967* (Boulder, CO: Westview, 1984).
7. See Malik Mufti, *Sovereign Creations: Pan-Arabism and Political Order in Syria and Iraq* (Ithaca, NY: Cornell University Press, 1996).
8. Malcolm H. Kerr, *The Arab Cold War, 1958–1964: A Study of Ideology in Politics*, 2nd ed. (London: Oxford University Press, 1965), 21–22, 53. See also Michael C. Hudson, *Arab Politics: The Search for Legitimacy* (New Haven, CT: Yale University Press, 1977).
9. Reinhard Schulze, "Institutionalization [of *da'wa*]," in *Oxford Encyclopedia of the Modern Islamic World*, ed. John L. Esposito (New York: Oxford University Press), 1:346–50; James P. Piscatori, *Islam in a World of Nation-States* (Cambridge: Cambridge University Press, 1986).
10. Reinhard Schulze, "Muslim World League," in *Oxford Encyclopedia of the Modern Islamic World*, 3:208–10.
11. See Charles J. Adams, "The Ideology of Mawlana Mawdudi," in *South Asian Politics and Religion*, ed. Donald E. Smith (Princeton, NJ: Princeton University Press, 1966), 371–97; Hamid Enayat, *Modern Islamic Political Thought* (Austin: University of Texas Press, 1988),

101–10; and John L. Esposito, *The Islamic Threat: Myth or Reality*, 3rd ed. (New York: Oxford University Press), 129–35.

12. Hans Mahnig, "Islam in Switzerland: Fragmented Accommodation in a Federal Country," in *Muslims in the West: From Sojourner to Citizens*, ed. Yvonne Y. Haddad (New York: Oxford University Press, 2001), 75–76. And, for a contrast, see Said Ramadan's son, Tariq Ramadan, *Western Muslims and the Future of Islam* (New York: Oxford University Press, 2005).

13. See Geneive Abdo, *No God but God: Egypt and the Triumph of Islam* (New York: Oxford University Press, 2000); Nazih N. Ayubi, *Political Islam: Religion and Politics in the Arab World* (London: Routledge, 1991); and Gilles Kepel, *Muslim Extremism in Egypt: The Prophet and the Pharaoh* (Berkeley: University of California Press, 1985).

14. Yvonne Y. Haddad and Jane I. Smith, eds., *Muslim Communities in North America* (Albany: State University of New York Press, 1994), xxi.

15. Mary Lahaj, "The Islamic Center of New England," in Haddad and Smith, *Muslim Communities*, 299–300.

16. Larry Poston, *Islamic Da'wah in the West* (New York: Oxford University Press, 1992), 79.

17. See Shaikh Daoud Ahmed Faisal, *Islam the True Faith: The Religion of Humanity* (Brooklyn, NY: Islamic Mission of America, 1965), n.p.

18. Marc Ferris, "To 'Achieve the Pleasure of Allah': Immigrant Muslims in New York City, 1893–1991," in Haddad and Smith, *Muslim Communities*, 212.

19. See McCloud, *African American Islam*, 21–24; and Dannin, *Black Pilgrimage to Islam* (New York: Oxford University Press, 2002), 63–67.

20. Ferris, "To 'Achieve the Pleasure of Allah,'" 214.

21. Shaikh Daoud Ahmed Faisal, "Author's Note," in *Al-Islam: The Religion of Humanity* (Brooklyn, NY: Islamic Mission of America, 1950), 7–8.

22. Ibid., 51.

23. Ibid., 15–16. Compare Will Herberg, *Protestant, Catholic, and Jew: An Essay in American Religious Sociology* (Garden City, NY: Doubleday, 1955).

24. Ibid., 49–50.

25. Ibid., 60.

26. See, for example, Regina M. Schwartz, *The Curse of Cain: The Violent Legacy of Monotheism* (Chicago: University of Chicago Press, 1997).

27. For accounts of the relationships between blacks and Jews in the United States, see V. P. Franklin et al., eds., *African Americans and Jews in the Twentieth Century: Studies in Convergence and Conflict* (Columbia: University of Missouri Press, 1998); Cheryl Lynn Greenberg, *Troubling the Waters: Black-Jewish Relations in the American Century* (Princeton, NJ: Princeton University Press, 2006); and Eric J. Sundquist, *Strangers in the Land: Blacks, Jews, Post-Holocaust America* (Cambridge, MA: Harvard University Press, 2005).

28. For classic accounts of these "American" missionary strategies, see, for example, William R. Hutchison, *Errand to the World: American Protestant Thought and Foreign Missions* (Chicago: University of Chicago Press, 1987); William G. McLoughlin, Jr., *Revivals, Awakenings, and Reform* (Chicago: University of Chicago Press, 1978); and Edith Blumhofer and Randall Balmer, eds., *Modern Christian Revivals* (Urbana: University of Illinois Press, 1993).

29. Dannin, *Black Pilgrimage to Islam*, 64.

30. R. M. Mukhtar Curtis, "The Formation of the Dar ul-Islam Movement," in Haddad and Smith, *Muslim Communities in North America*, 54.

31. Dannin, *Black Pilgrimage to Islam*, 66–68.

32. McCloud, *African American Islam*, 71.

33. See Sherman A. Jackson, *Islam and the Blackamerican: Looking Toward the Third Resurrection* (New York: Oxford University Press, 2005), 48–49.

34. Louis E. Lomax et al., "The Hate That Hate Produced," on *Newsbeat*, WNTA-TV, July 23, 1959, a transcript of which is available in a declassified FBI report. See Federal Bureau of Investigation, "SAC, New York, office memorandum to director, FBI, July 16, 1959," in *Malcolm X: The FBI Files*, http://wonderwheel.net/work/foia/ (accessed May 1, 2007).

35. E. U. Essien-Udom, *Black Nationalism: A Search for an Identity in America* (Chicago: University of Chicago Press, 1962), 73–74.

36. Louis A. DeCaro, Jr., *On the Side of My People: A Religious Life of Malcolm X* (New York: New York University Press, 1996), 159–60, 201–2.

37. See DeCaro, *On the Side of My People*, esp. 159–293. And compare Turner, *Islam in the African-American Experience*, 174–237, and Curtis, *Islam in Black America*, 85–105.

38. See Malcolm X and Alex Haley, *The Autobiography of Malcolm X* (New York: Ballantine, 1965), 266–317.

39. Ferris, "To 'Achieve the Pleasure of Allah,'" 215.

40. Malcolm X and Haley, *Autobiography of Malcolm X*, 318.

41. Bruce Perry, *Malcolm: The Life of a Man Who Changed Black America* (Barrytown, NY: Station Hill, 1991), 261–64; DeCaro, *On the Side of My People*, 202–3.

42. Malcolm X and Haley, *Autobiography of Malcolm X*, 320.

43. Yaacov Shimoni, *Political Dictionary of the Arab World* (New York: Macmillan, 1987), 105–6.

44. Abd al-Rahman 'Azzam, *The Eternal Message of Muhammad*, trans. Caeser E. Farah (Cambridge: Islamic Texts Society, 1993).

45. DeCaro, *On the Side of My People*, 206.

46. Malcolm X and Haley, *Autobiography of Malcolm X*, 331–33.

47. Ibid., 348.

48. Perry, *Malcolm*, 322.

49. DeCaro, *On the Side of My People*, 336.

50. The interview is reproduced in Steve Clark, ed., *February 1965: The Final Speeches* (New York: Pathfinder, 1992), 252–55, and on the Web at Malcolm-X.Org: http://www.malcolm-x.org/docs/int_almus.htm (accessed August 28, 2006).

51. DeCaro, *On the Side of My People*, 233, 238–39.

52. See "Notes on the Invention of Malcolm X" and "Malcolm X and the Failure of Afrocentrism" in Gerald Early, *The Culture of Bruising* (Hopewell, NJ: Ecco, 1994), 233–58.

53. See George M. Fredrickson, *Black Liberation: A Comparative History of Black Ideologies in the United States and South Africa* (New York: Oxford University Press, 1995), 277–97; and Gomez, *Black Crescent* (New York: Cambridge University Press, 2005), 348–55.

54. Malcolm X and Haley, *Autobiography of Malcolm X*, 323–72; Gomez, *Black Crescent*, 372; Curtis, *Islam in Black America*, 96–99.

55. See Clark, ed. *February 1965: The Final Speeches*, 20–21, and Clayborne Carson, *Malcolm X: The FBI File* (New York: Carroll and Graf, 1991), 79–80.

56. "Mr. Muhammad Answers Critics: Authority from Allah, None Other," *Muhammad Speaks*, August 2, 1962, 3.

57. James R. Lewis, *Legitimating New Religions* (New Brunswick, NJ: Rutgers University Press, 2003), 13–14. For an NOI example, see Sylvester Leaks, "The Messenger of Allah as Seen by an Islamic Leader from Pakistan," *Muhammad Speaks*, May 8, 1964, 3.

58. See "Our Great Physician," *Muhammad Speaks*, June 1962, 14; "As It Was in the Days of Daniel, So It Is Today," *Muhammad Speaks*, June 19, 1964, 9; and "As It Was with Pharaoh So It Is Today," *Muhammad Speaks*, July 17, 1964, 9.

59. See "First Printing of Holy Qur'an in U.S.," *Muhammad Speaks*, August 17, 1973, 23, and "Women in Islam: Is the Honorable Elijah Muhammad the Last Messenger of Allah?" *Muhammad Speaks*, September 16, 1965, 19.

60. See "Allah and His Messenger," *Muhammad Speaks*, January 1, 1965, 1; and "Where Others Fail, Our Messenger Succeeds," *Muhammad Speaks*, May 14, 1965, 3.

61. See Zafar Ishaq Ansari, "W. D. Muhammad: The Making of a 'Black Muslim' Leader (1933–1961)," *American Journal of Islamic Social Sciences* 2, no. 2 (1985): 248–62.

62. See further Curtis, *Black Muslim Religion*, 65.

63. McAlister, *Epic Encounters*, 86.

64. See Thomas A. Tweed, *Crossing and Dwelling: A Theory of Religion* (Cambridge, MA: Harvard University Press, 2006); Jonathan Z. Smith, *Map Is Not Territory: Studies in the History of Religions* (Chicago: University of Chicago Press, 1978).

65. Jackson, *Islam and the Blackamerican*, 99–129.

66. One argument for the movement's gender conservatism can be found in E. Frances White, *Dark Continent of Our Bodies: Black Feminism and the Politics of Respectability* (Philadelphia, PA: Temple University Press, 2001), 43. For counterarguments and qualifications, see Cynthia S'thembile West, "Nation Builders: Female Activism in the Nation of Islam, 1960–1970" (PhD diss., Temple University, 1994); and Curtis, *Black Muslim Religion*, 95–130. For critiques of the NOI on other scores, see Essien-Udom, *Black Nationalism*, 286–87, 339; and Hans A. Baer and Merrill Singer, *African-American Religion in the Twentieth Century: Varieties of Protest and Accommodation* (Knoxville: University of Tennessee Press, 1992), 143.

67. See further Sacvan Bercovitch, *The American Jeremiad* (Madison: University of Wisconsin Press, 1978).

68. Bercovitch, *The American Jeremiad*.

69. Penny M. Von Eschen, *Race against Empire: Black Americans and Anticolonialism, 1937–1957* (Ithaca, NY: Cornell University Press, 1997), 174.

70. See "Nation of Islam: Cult of the Black Muslims," May 1965, available through the FBI's Web site at http://foia.fbi.gov/nation_of_islam/nation_of_islam_part02.pdf (last accessed May 1, 2007).

71. In 1967, the FBI included "black nationalist hate groups" in COINTELPRO. See Frank T. Donner, *The Age of Surveillance: The Aims and Methods of America's Political Intelligence System* (New York: Knopf, 1980), 178, 212–13.

72. See McCloud, *African-American Islam*, 64–72; and Dannin, *Black Pilgrimage to Islam*, 66–71.

73. Poston, *Islamic Da'wah in the West*, 39.

74. See Dannin, *Black Pilgrimage to Islam*.

75. See Jackson, *Islam and the Blackamerican*, 73.

76. See, for example, United States District Court, Western District of Washington at Seattle, *United States of America v. Earnest James Ujaama*, at http://fl1.findlaw.com/news.findlaw.com/hdocs/docs/terrorism/usujaama82802ind.pdf (accessed February 18, 2009).

77. For background, see Clifton E. Marsh, *The Lost-Found Nation of Islam in America* (Lanham, MD: Scarecrow, 2000), 67–78, 101–28.

78. Curtis, *Islam in Black America*, 115, 120–21.

79. See, for example, W. Deen Mohammed, *Focus on Al-Islam* (Chicago: Zakat, 1988).

80. Amina Wadud, *Qur'an and Woman* (New York: Oxford University Press, 1999).

81. See Amina Wadud, *Inside the Gender Jihad: Women's Reform in Islam* (New York: Oxford University Press, 2006).

82. See Carolyn Rouse, *Engaged Surrender: African American Women and Islam* (Berkeley: University of California Press, 2004).

83. See Jackson, *Islam and the Blackamerican*, and Imam Zaid Shakir, *Scattered Pictures: Reflections of an American Muslim* (Haywood, CA: Zaytuna Institute, 2005).

84. See McCloud, *African American Islam*, 88–94, 248; Dannin, *Black Pilgrimage to Islam*, 248, 255–56; and Jackson, *Islam and the Blackamerican*, 50–51, 191–98. For one example of the influence of the Bawa Muhaiyaddeen Fellowship on an African American Muslim, see Gwendolyn Zoharah Simmons, "Are We Up to the Challenge? The Need for a Radical Re-ordering of the Islamic Discourse on Women," in *Progressive Muslims: On Justice, Gender, and Pluralism*, ed. Omid Safi (Oxford: Oneworld, 2003), esp. 235–39.

85. Poston, *Islamic Da'wah in the West*, 108–9.

86. See further Mattias Gardell, *In the Name of Elijah Muhammad: Louis Farrakhan and the Nation of Islam* (Durham, NC: Duke University Press, 1996).

EAST OF THE SUN (WEST OF THE MOON)

ISLAM, THE AHMADIS, AND AFRICAN AMERICA

MOUSTAFA BAYOUMI

SEPIA TONES

TRAVELING SOMEWHERE BETWEEN LIVING IN A RACIALIZED STATE AND STATING the life of a race lies the story of African American Islam. Found in narratives of struggle and spirit, of edification and propagation, of incarceration, incarnation, and ideology, and of blacks, Asians, and Middle Easterners, this is a tale seldom told and even less often heard. When it does get some play, the way is in a single key. Separation is sounded brassily as the dominant chord, modulating being minor into a major ideology. The dissonances of dissidence. From Moorish Science to Garveyism, from Elijah's honor to Malcolm's rage, Islam is understood as a tool of politics, pliant to complaint and made to speak a language of plain truth against the tricknology of white folk. The soul almost disappears, replaced with an iconography of militarized Islam, boots and bowties battling white supremacy, dividing One Nation Under God with the Nation of Islam.

The fate of Malcolm concludes this narrative by necessity. Epiphanies of a universal spirit clash with narrow-minded parochialisms in a death match of blood and assassination. Malcolm is lionized and history, tragically, marches on. But did this battle between the particular and the universal, between Islam as a unique expression of African American political aspirations for separation, and Islam as a universal religion of belonging first find its articulation with Malcolm's rupture with Elijah Muhammad, or has the customary story we have, up until now, been unable to comprehend the complexity of Islam in the African American experience? Is the divide between the universal and the particular so easily drawn as a picture in black and white, or are there sepia tones of black, brown, and beige that call out to be seen? This chapter is an examination of the browns and beiges, a look at the notes and tones of the Muslim experience.

I would like to start with three tableaus, one involving an Asian immigrant, another looking at Brother Malcolm, and the third, a study in sound. All three are signifying the idea of Islam in the United States, finding a context in which to belong along with a place to disagree, and providing me a text with which to continue.

THE MUFTI

Islam in African America has a history as long as memory, when Muslim slaves from Africa wrapped their faith tightly around them as invisible armor against daily degradation. But the practice does not seem to continue. Religious revivalists in the early part of the twentieth century, mostly in the North, where large numbers of new migrants sought the strength of a community, found populations willing to listen and eager to believe. In 1913, Timothy Drew donned a fez and claimed Moroccan heritage for his people in the Moorish Science Temple. For all its imaginative reconstruction, the Moorish Science Temple has little under the surface to connect it to worldwide Islam. But its spirit of displacing the term "Negro" from blacks, of thinking of darker-skinned peoples as Asiatics and Moroccans, of allying Drew Ali with "Jesus, Mohamed, Buddha, and Confucius,"[1] is part of the productive tension between separatism and universalism that will follow all African American Islam throughout the rest of the century. But it would be in the next decade, with the growth of the Ahmadiyya community, that the Asian connection forges ahead.

One night in January 1920, a gentle and bespectacled Muslim by the name of Mufti Muhammad Sadiq left London for New York to become one of the first "Pioneers in the spiritual Colonization of the Western world."[2] This phrase, conveyed by the then leader of the Ahmadiyya movement in India, Mirza Mahmud Ahmad, to the Mufti's work, interestingly linked Ahmadiyya missionary activity with British rule and with its own missionary activity, along with the pioneer mythology of the New World. The Ahmadis had objected to the manner in which British missionaries were defaming Islam by reviling the Prophet Muhammad and set out not just to correct this error but also to illustrate how Jesus was a prophet of Islam. They had observed how missionaries in the East had succeeded in misrepresenting Islam and felt that a proactive agenda of missionizing was needed to counteract this damage. Recent Hindu-only movements in India also fueled the drive to survive in a world of plural faiths. "Reason itself revolts against this exclusiveness," wrote Ahmadi founder Ghulam Ahmad.[3]

The Ahmadiyya community began in late nineteenth-century India with the figure of Mirza Ghulam Ahmad, a charismatic reformer who believed he had received divine revelations, starting in 1876, requiring him to promote the unity of all religions as manifest through Islam, whose chief object is "to establish the unity and majesty of God on earth, to extirpate idolatry and to weld all nations into one by collecting all of them around one faith."[4] It is a particular universalism. In seeking this unity, Ahmad would call himself "the Mahdi of Islam . . . the Promised Messiah of Christianity and Islam, and an avatar of Krishna for the Hindus,"[5] a claim that would ultimately oust him and his movement from the mainstream Muslim establishment. We should note how Ahmad's ideas are an attempt to confront communal feelings in India of his day and how this relationship between faith and nation would resonate in the American Ahmadiyya movement.

We can note, then, the links between the putative universalism of colonialism, which saw the spread of Western values as a mission manifest in direct and indirect colonial rule (*la mission civilisatrice*), to the missionary activities of the Ahmadis. Ahmadi missionizing, particularly in its pioneering New World aspects, thus borrows heavily from the script of European expansion and accepts modernity's commonplace division between the spiritual and secular worlds ("the spiritual colonization") where the East is spiritual and the West, material. A significant difference, however, divides the methodologies of Western expansionism and Ahmadi missionary activity, for the Ahmadis were addressing the rest of the world as a colonized people and the religious foundation of their work is thus by definition a minority religion, unencumbered by state apparatuses or ideology. Its universalism percolates from below rather than being dusted from above, thus achieving a kind of dissident political flavor separate from the tastes of dominant rule.

In 1920, the movement, fresh from its missionary successes around the world (including England and West Africa) and full of the optimism that the new world is supposed to hold, sent its first missionary to the United States. Mufti Muhammad Sadiq boarded his ship in London and, each day, entertained his fellow passengers with his erudition. "Say, if you love Allah, follow me; then will Allah love you," he is reported to have intoned. Before the end of the trip, Sadiq is said to have "converted four Chinese men, one American, one Syrian, and one Yugoslavian to Islam."[6]

The American authorities were hardly as sanguine with Sadiq's sagacity. They seized him before he could leave the ship, accusing him of coming to the United States to practice polygamy, and placed him in a Philadelphia detention house. So began a dark hour for the gentle Sadiq. Seven weeks later, he was eventually released but not before making nineteen other converts in jail, from Jamaica, British Guyana, Azores, Poland, Russia, Germany, Belgium, Portugal, Italy, and France.

What Sadiq found when he reached the welcoming shores of the United States was a history of institutional racism and Asian exclusion laws for which he was unprepared. White nationalism would already be working against the Mufti's message. Later he would write that "if Jesus Christ comes to America and applies for admission to the United States under the immigration laws, [he] would not be allowed to enter this country because:

1. He comes from a land which is out of the permitted zone.
2. He has no money with him.
3. He is not decently dressed.
4. His hands have holes in the palms.
5. He remains bare-footed, which is a disorderly act.
6. He is against fighting for the country.
7. He believes in making wine when he thinks necessary.
8. He has no credential to show that he is an authorized preacher.
9. He believes in practicing the Law of Moses [polygamy].[7]

Originally conceiving of his work as broad-based, ecumenical, multiracial missionary activity, Sadiq soon realized that whites were bitter and fearful of his message and African Americans interested and open. Early reports indicate that several Garveyites attended his lectures and were among his first converts, and the white press seemed generally baffled and lost in its own prejudices when considering the movement. One account tells us that "all the audience has adopted Arabic names . . . There is the very dark Mr. Augustus, who used to belong to St. Marks church in this city [Chicago], but who now sings a pretty Arabic prayer and acts rather sphinx-like. Half a dozen Garvey cohorts are counted, one in his resplendent uniform. There is one pretty yellow girl and another not so pretty."[8]

The fact is that the Ahmadiyya movement attracted women and men. It formed a community made up of black, brown, and white people in a scattering of cities across the eastern half of the country (and St. Louis). But it mostly attracted African Americans, who were also given early leadership roles.[9] Participating in Islam vitally meant discovering the history of black contributions to Islam, a topic generating some interest broadly in the black press at the time. In these years, articles appeared in *The Crisis* (1913), *The Messenger* (1927), and *Opportunity* (1930) about Islam, notably about Bilal, the Abyssinian slave freed by Prophet Muhammad and Islam's first muezzin, illustrating Islam's historic connection with Africa.[10] It is important to underline that Islam within the Ahmadiyya community was not considered a religion just for blacks but a religion in which blacks had an alternative universal history to which to pledge allegiance. Christianity and narrow nationalisms allowed no such thing, as *The Moslem Sunrise*, the Ahmadi journal, argued. In 1923, it printed a half-page exhortation on "the real solution of the Negro Question," calling on African Americans to see that "Christian profiteers brought you out of your native lands of Africa and in Christianizing you made you forget the religion and language of your forefathers–which were Islam and Arabic. You have experienced Christianity for so many years and it has proved to be no good. It is a failure. Christianity cannot bring real brotherhood to the nations. So, now leave it alone. And join Islam, the real faith of Universal Brotherhood."[11]

Universal brotherhood, of course, sounds similar to Universal Negro, as in the Improvement Association, and links should be made between the philosophy of Garveyism and the Ahmadis, but, again, not simply through the lens of separatism but a reconfigured universalism. Considering the racial and religious divisions in the world, the Ahmadis reinterpreted the Islamic concept of *tawheed*, the one-ness of God, as unifying the world, people, and faith around Islam (as Ghulam Ahmad wanted for India). In the American context, then, Ahmadi thought opened a critical space for race in the realm of the sacred. In this way, African Americans could metaphorically travel beyond the confines of national identities. They could become "Asiatics" and remain black, could be proud of their African heritage and feel a sense of belonging to, and participation with, Asia. Being plural in this scheme meant not having to feel the psychic tear of double consciousness but a way of living wholly in the holy. This ecumenicalism could be very powerful, both spiritually and politically. By being opened-palmed about life when the secular world is clenching fists at you meant that your pluralist unity viewed the divisions of the world as contemptibly parochial.

By 1940, the movement could claim around ten thousand converts. Its impact would be wider still, and, in his early years, it would reach the ears of Malcolm X.

BROTHER MALCOLM

Malcolm X, the eloquent minister of information for Elijah Muhammad, is commonly seen as speaking the fire of separatism and black pride until his fateful *hajj* in 1964 tamed his message, as he discovered the true universal spirit of Islam. Conventional as this story is, with its Augustinian turns of the will, it fails when confronted with history. The rise and development of Malcolm's message is a story of the conflict between the particular universalism of Ahmadi-type Islam against the more narrow confines of Nation of Islam creed.[12] When we understand this, we can view the intellectual development of Malcolm as a way of thinking through the role of faith in determining consciousness and that that activity itself for Malcolm was hardly a settled issue.

Consider, for example, the fact that, early in his life and while considering the value of Islam while in prison, Malcolm was visited by an Ahmadi, Adbul Hameed, who was on his outreach to local populations. Abdul Hameed even sent Malcolm a book of Arabic Muslim prayers, which Malcolm memorized phonetically.[13] This contact may help to explain why, after being released from Charlestown prison on parole, Malcolm, too, identifies himself at least once as an "Asiatic," which I have been arguing is not false consciousness of African American history or self-hatred but a strategic belief in the particular universal of Islam. The incident was as follows.

In 1953, Malcolm, who was now a fully fledged Muslim and member of Elijah Muhammad's flock, was pulled aside one day at his work at the Gar Wood factory in Wayne, Michigan, by the FBI. He had failed to register for the Korean War draft, the agent needled him, and was thereby jeopardizing his parole. Malcolm heeded the warning and registered but how he registered is noteworthy. Under the section on citizenship, which read, "I am a citizen of . . . ," Malcolm inscribed

"Asia." In his form on being a conscientious objector, he stated his belief that "Allah is God, not of one particular people or race, but of All the Worlds, thus forming All Peoples into One Universal Brotherhood." Asked to identify his religious guide, Malcolm wrote "Allah the Divine Supreme Being, who resides at the Holy City of Mecca, in Arabia."[14]

Unlike orthodox Nation of Islam creed, which would connect Allah with W. D. Fard and the religious guide as Elijah Muhammad, Malcolm identifies Allah with the God of Islam and, like the Ahmadis, stresses the universal character of God. We could perhaps cynically see this move as a means to defeat the draft by identifying with a more orthodox religion than the Nation of Islam, but to do so is to miss the manner in which Malcolm would later repeatedly seek to integrate the Nation of Islam into the fold of worldwide Islam. In 1960, after the scholar C. Eric Lincoln coined the term "Black Muslims" for Nation followers, Malcolm objected vehemently. "I tried for at least two years to kill off that 'Black Muslims,'" he said. "Every newspaper and magazine writer and microphone I got close to [I would say] 'No! We are black people here in America. Our religion is Islam. We are properly called "Muslims"!' But that 'Black Muslims' name never got dislodged."[15]

This tension, between the Ahmadi vision of a particular universal vision of Islam and the Nation's notion of an Islam for black people, underscores the conflict between two very different roles for religion in the political sphere. Admittedly, the Ahmadi spirit is less confrontational, less public, less typical of the struggle we have come to recognize as identity politics, and yet it is still revolutionary in its own way by providing a radical ontology of self. To reorient one's body toward the Orient means a refusal to engage with the first principles of white America's definitions of blackness and instead to cut to the heart of an old American principle, the freedom of worship. Yet unlike the primary demand placed upon American religion—that religion be relegated solely to the private sphere—Islamic faith is seen as enveloping and thereby surpassing national belonging.

Reverberating through the African American community, this notion that a reconfigured universal faith can free your mind and body gained ground. While the Nation used the media (and the media used the Nation) to promote its belief, this other vision of Islam was quietly seeping into the pores of African American communities around the country, giving them a spiritual place to repudiate the nation of America not with the Nation of Islam but with a new universalism. Genealogically, this idea should be seen as descending from the Ahmadiyya movement, and, musically, it had a soundtrack that large segments of the American public were listening to. Many of the major figures of mid-century jazz were themselves directly influenced by the Ahmadiyya movement, and the yearning for a universal and spiritual sound was in large part a result of Ahmadiyya labor.

A LOVE SUPREME

In 1953, *Ebony* magazine felt that the rise of Islam among the jazz musicians of the era was sufficiently important to publish its article on "Moslem Musicians." "Ancient Religion Attracts Moderns," spoke its headline, and it centered on the importance of jazz among musicians. Drummer Art Blakey, we are told, "started looking for a new philosophy after having been beaten almost to death in a police station in Albany Ga., because he had not addressed a white policeman as 'sir.'"[16] Talib Dawood, a former jazz player and Ahmadi, introduced Blakey to Islam. Blakey's house was a known center for Islamic learning, and in an important engagement at Small's Paradise in Harlem, he organized a seventeen-member band, all Muslim, as the Messengers. Later, the band's personnel would change, as would the name (to the Jazz Messengers), but the Islamic influence in jazz would continue.[17]

Other important figures of the period also converted to Islam. Yusef Lateef, Sahib Shihab, Ahmed Jamal, and McCoy Tyner would all convert, and Dizzy Gillespie, Miles Davis, and John Coltrane would all be significantly influenced by its spirit. It is with John Coltrane that I want to conclude this chapter, since his influence has been so remarkable in the jazz sound and because his debt to other Eastern philosophies is relatively well known. But his relationship to Islam has not, to my knowledge, been sufficiently acknowledged despite the fact that it can be heard in his most famous work.

To have a soundtrack to a movement does not mean to play an anthem. Rather than indicating a representational scheme of signifying a specific community, I am interested in listening for the ways in which the yearning for a new kind of community, one based on a new universalism that has a (but not, by necessity, the only) base in Islam, can be heard in the ways in which the music is pushing itself. Coltrane's search for a tone that could extend the saxophone is well known, as is the critics' initial bewilderment to his pitch. He himself talked about his desire to incorporate the fullness of expression in his music. "I want to cover as many forms of music that I can put into a jazz context and play on my instruments," he wrote in his notebooks. "I like Eastern music; Yusef Lateef has been using this in his playing for some time. And Ornette Coleman sometimes plays music with a Spanish content."[18] In an unreleased session from his Village Vanguard recordings, Coltrane is also playing with Ahmed Abdul Malik, a Sudanese bass and oud player who was part of Monk's band, a regular partner to Randy Weston, and an innovator in incorporating Middle Eastern modal organization in jazz improvisation. Coltrane's sidemen regularly included Muslim musicians from Philadelphia, and he himself, married to Naima (a Muslim) and, after 1957, increasingly interested in all things spiritual, regularly engaged his friend, piano player Hassan Abdullah, in discussions about Islam.

Space prevents me from etching in detail the milieu in which Coltrane repeatedly encountered and considered Islam. Instead, I want to move toward a conclusion in a musical note by considering the ecumenical sound of Islam found in Coltrane's most commercially successful recording, *A Love Supreme*. Significantly,

Coltrane was often portrayed by the media of his day as blowing the sounds of black rage. The Angry Young Tenor was the musical equivalent of the angry Malcolm X. But Coltrane never saw his music this way. Responding to his critics, he said, "If [my music] is interpreted as angry, it is taken wrong. The only one I'm angry at is myself when I don't make what I'm trying to play."[19] Later he would be quoted as saying this about the philosophy of his music: "I think the main thing a musician would like to do is to give a picture to the listener of the many wonderful things he knows of and senses in the universe. That's what music is to me—it's just another way of saying this is a big, beautiful universe we live in, that's been given to us, and here's an example of just how magnificent and encompassing it is."[20]

If there is a tendency to view this wisdom as apolitical, liberal claptrap, it is, I think, misplaced. Searching for the universal in a minor key is less about escape or about colonizing the spiritual experiences of the dark world to rejuvenate an exhausted Western sensibility, in the mode of Richard Burton through George Harrison. Coltrane's universal is a search for a big philosophy of sound, which repudiates the thin, reedy existence of American racial politics, and it does so, often, by an invocation of Islam.

"During the year of 1957, I experienced, by the grace of God, a spiritual awakening which was to lead me to a richer, fuller, more productive life." So wrote Coltrane in the famous liner notes for *A Love Supreme*. The notes continue in this tenor, and anyone with an ear attuned to Islamic language will hear its echoes: "NO MATTER WHAT . . . IT IS WITH GOD. HE IS GRACIOUS AND MERCIFUL. HIS WAY IS IN LOVE, THROUGH WHICH WE ALL ARE. IT IS TRULY—A LOVE SUPREME. Al-rahman, al-raheem. The Gracious, the Merciful." The two qualities which follow God everywhere in the Muslim tradition are invoked by Coltrane, who ends his text with "ALL PRAISE TO GOD." Alhamdulillah. Consider the first track, "Acknowledgement." Built around a simple, four-note structure, this piece is an attempt to unify and capture the rapture of the divine. Listen how, two-thirds of the way through, Coltrane meanders around the simple theme in every key, as if to suggest the manner in which God's greatness truly is found everywhere, and then the ways in which the band begins to sing the phrase "A Love Supreme," like a roving band of Sufi mendicants singing their dhikr. The words could change. As the love is extolled, the phrase begins to include the sounds of "Allah Supreme," another Arabic expression, Allahu Akbar. Coltrane makes the connection from "A Love Supreme" to Allah Supreme for his entire listening audience, forever delivering a sound of Islam to the world of American music.

To appreciate the depth of mutual involvement between blacks and Asians means acknowledging not just how histories of faith exist to be excavated, which illustrates a level of shared struggle toward an acceptable ontology for living in the racialized United States, but it also means investing the sacred with the possibilities for radical thought, even if its effects are less visible to us than the legacy of political activism through ideologies of separatism. Ahmadi Islam was the space where this place was opened up for many African Americans. It defines certain aesthetics of living, where the text to life is in a language white America cannot read and the sounds of existence flutter beyond white America's ears. This is not about being "Omni-American," to use a phrase associated with Albert Murray, but it is about assimilating into the omnipresence of a just universal order. It is where blacks become Asians and Asians, black, under color of divine law.

NOTES

1. C. Eric Lincoln, *The Black Muslims of America* (Grand Rapids, MI: Erdmanns, 1994), 49.
2. *The Moslem Sunrise* (July 1921), 3.
3. Quoted in Yvonne Haddad and Jane Smith, *Mission to America: Five Islamic Sectarian Communities in North America* (Gainesville: University Press of Florida, 1993), 55.
4. Hazrat Mirza Ghulam Ahmad, *Message of Peace* (1908; Columbus, OH: Ahmadiyya Anjuman Isha'at Islam Lahore, 1993), 23.
5. Quoted in Richard Brent Turner, *Islam in the African American Experience* (Bloomington: Indiana University Press, 1997), 112.
6. Richard Brent Turner, *Islam in the African American Experience* (Bloomington: Indiana University Press, 1997), 115.
7. "If Jesus Comes to America," *The Moslem Sunrise* (April 1922), 55–56.
8. Roger Didier, "Those Who're Missionaries to Christians: Prophet Sadiq Brings Allah's Message Into Chicago and Makes Proselytes," reprinted in *The Moslem Sunrise* (October 1922), 139.
9. Aminah McCloud reports that eventually, dissension arose among Ahmadis over the fact that more African Americans were not appointed to leadership positions and that the Indian customs of the missionaries and the immigrant Muslims eventually clashed with the African American desires to apply the faith to domestic situations. See Aminah McCloud, *African American Islam* (New York: Routledge, 1995), 21. However, in the early years, the community was certainly highly multiracial in many ways, including in its leadership roles. *The Moslem Sunrise* contains many such photographs and examples, including highlighting the role of one early "zealous worker for Islam, appointed a Sheikh to work among his people in the district of St. Louis and vicinity," named Sheikh Ahmad Din (formerly P. Nathaniel Jonson). See, for example, *The Moslem Sunrise* (July 1922), 119.
10. J. A. Rogers, "Bilal Ibn Rahab—Warrior Priest," *The Messenger* 9 (July 1927): 213–14. Rogers states, "When the Christian Negro points with pride to St. Augustine, the Numidian Negro, and tells what he did to advance Christianity, the Mohammedan one can point to Bilal, and tell what he did for Christianity's greatest rival. The Mohammedan Negro is, however, hardly likely to do as Islam not only in theory, but in actuality, knows no color line. This probably accounts for its success in Africa." See also A. T. Hoffert, "Moslem Propaganda: The Hand of Islam stretches out to Aframerica," *The Messenger* (May 1927), 141, 160. Hoffert describes "a woman convert who had belonged to various churches spoke of her previous life like that of a dog or cat before its eyes are opened; they are going to have their share of good things and stand on their own feet. She spoke of the universality of Islam, its way of life, one God, one aim, one destiny." Blanche Watson, "The First Muezzin," *Opportunity* (September 1930), 275.
11. "True Salvation of the American Negroes: The Real Solution to the Negro Question," *The Moslem Sunrise* (April–July 1923), 184.

12. It should be stressed that the dichotomy I am establishing here, between the particular-
    ism of the Nation and the ecumenicalism of the Ahmadis, is obviously more complicated
    in many circumstances, and that the Nation has, at its heart, the ability to see itself as a
    universal theology in certain respects, just as Ahmadi creed can be (and is often, by the
    mainstream Muslim community) understood as a narrower and more particular vision,
    especially since the Ahmadis themselves are marginalized by the mainstream Muslim
    establishment. The Nation also often employed Sunni Muslims as advisors and teachers,
    such as Abdul Basit Naeem, editor of a couple of small publications (*Moslem World & the
    USA* and *The African-Asian World*) and author of the introduction to Elijah Muhammad's
    The Supreme Wisdom (Atlanta: Messenger Elijah Muhammad Propagation Society, n.d.),
    2:3. These advisors and, later, Elijah Muhammad himself recognized the radical differ-
    ences between the Nation of Islam creed and mainstream Sunni beliefs yet justified the
    Nation's theology as being the best way to bring African Americans to Islam. At the very
    end of his life, it appears that even Elijah Muhammad believed in mainstream Islam.
    Similarly, Louis Farrakhan, now facing his mortality as he battles cancer, has made sig-
    nificant gestures towards reforming Nation of Islam creeds towards an acceptable form of
    mainstream Islam.

13. Louis DeCaro, *On the Side of My People: A Religious Life of Malcolm X* (New York: New
    York University Press, 1996), 136.

14. DeCaro, *On the Side of My People*, 97–98.

15. Malcolm X and Alex Haley, *The Autobiography of Malcolm X* (New York: Grove Press,
    1964), 247.

16. "Moslem Musicians Take Firm Stand Against Racism," *Ebony* (April 1953): 111.

17. Charley Gerard, *Jazz in Black and White: Race, Culture, and Identity in the Jazz Commu-
    nity* (Westport, CT: Praeger, 1998), 75

18. C. O. Simpkins, *Coltrane: A Biography* (Baltimore: Black Classic Press, 1975), 118.

19. Simpkins, *Coltrane*, 84.

20. Ibid., 151.

CHAPTER 5

REPRESENTING PERMANENT WAR

BLACK POWER'S PALESTINE AND THE END(S) OF CIVIL RIGHTS

KEITH P. FELDMAN

> In the past few weeks, the Arab-Israeli conflict exploded once again into all-out war as it did in 1956 and as it had done in 1948, when the State of Israel was created. What are the reasons for this prolonged conflict and permanent state of war which has existed between Arab nations and Israel? . . . Since we know that the white American press seldom, if ever, gives the true story about world events in which America is involved, then we are taking this opportunity to present the following documented facts on this problem. These facts not only affect the lives of our brothers in the Middle-East, Africa and Asia, but also pertain to our struggle here. We hope they will shed some light on the problem.
>
> —Student Nonviolent Coordinating Committee

THUS OPENS WHAT QUICKLY BECAME AN INFAMOUS ARTICLE PUBLISHED IN THE summer of 1967 by the Student Nonviolent Coordinating Committee (SNCC).[1] SNCC's "Third World Round Up: The Palestine Problem: Test Your Knowledge" does groundbreaking rhetorical and imaginative work to remap the relationship between domestic movements for racial justice in the United States and transnational struggles for liberation in Israel-Palestine. This chapter sketches a genealogy of this remapping project. The social, imaginative, and rhetorical spaces of what we might call "Black Power's Palestine" clarify and contest the shifting imperial formation emerging in the confluence of the end of the U.S. Civil Rights Movement and the beginning of Israel's occupation of the West Bank and Gaza.

While SNCC's article has been routinely condemned for contributing to the tragic fracture of the so-called black-Jewish civil rights alliance, in the spaces it opened up in the months and years that followed its publication, a critique of the occupation could be lodged in discussions about the future of New Left racial politics; members of the Black Panther Party could cultivate durable allegiances with the Palestine Liberation Organization from Oakland to Algiers; Andrew Young, the first African American ambassador to the United Nations, could sacrifice his vaunted position by entering discussions with the PLO; organizations like *Freedomways* could pair with the Association of Arab American University Graduates to expose the 1982 massacres in the Sabra and Shatila refugee camps; and poets like June Jordan could bear witness to these massacres, imagining being "born a Black woman" and in the presence of Sabra and Shatila having "become Palestinian." Indeed, some of the most durable critiques of the colonial occupation of Palestine have emerged through the embattled post–civil rights spaces imagined by SNCC in 1967. That is the article's oft-forgotten future. But what of its past?

Let us first consider the analytical purchase SNCC captures in the phrase "permanent state of war." How does this phrase operate? One finds this figure in a range of literary and polemical works in the post-World War II black freedom movement. In the preface to the 1953 edition of *The Souls of Black Folk*, for instance, W. E. B. Du Bois revises his famous thesis about the "world problem of the color line." "Back of the problem of race and color," writes Du Bois, "lies a greater problem that both obscures and implements it." This is a problem articulated through the functional register of permanent war as, he claims, it maintains through violence the material privileges of "so many civilized persons" so that "war tends to become universal and continuous, and the excuse for this war continues largely to be color and race."[2] More than a decade later, picking up on the Maoist conception of a people's war, Huey P. Newton turns to the figure in his famous June 1967 essay "In Defense of Self-Defense." "The laws and rules which officials inflict upon poor people prevent them from functioning harmoniously in society . . . We are advocates of the abolition of war," Newton argues, echoing Mao. "We do not want war, but war can only be abolished through war. In order to get rid of the gun it is necessary to pick up the gun."[3]

From a seemingly distant remove, almost ten years later, Michel Foucault outlines this figure's analytical function, tracing its relation to race struggle, class struggle, and the horizon of law. In his quest to rethink history as a project delinked from the historiographical and juridical constraints of the state, Foucault inverts German military theorist Claude Von Clausewitz's famous aphorism that "war is the continuation of politics conducted by other means." Instead, Foucault asks, "If we look beneath peace, order, wealth, and authority, beneath the calm order of subordinations, beneath the State and State apparatuses, beneath the laws, and so on, will we hear and discover a sort of primitive and permanent war?"[4] The answer, Foucault suggests, lies in understanding the emergence of the modern state. It was Clausewitz, he argues, who inverted a historical principle whose *longue durée* shows how modern state sovereignty was conceived and maintained through the violence of battle, the drawing of sides, and through vicious and protracted struggle. While the forces of order, the state, the law—and the discursive practices of history that such forces produce—claim supremacy, cohesiveness, even abstract and universal form, in actuality, they are always in the process of formation. These struggles, Foucault continues, obfuscate their necessary production of racism, the "biological-type caesura within a population" that demarcates "what must live and what must die."[5]

The task Foucault sets out is an inquiry into strategies of "counterhistory" that make legible this far messier narrative. He does so not simply to refine an abstract theory of the origins of modern politics or historiography; nor is it a theory that emerges in a vacuum. Rather, Foucault was attuned to the social and political transformations of the late 1960s in France as in many other parts of the globe, from anticolonial movements to the "rights revolution." He found modern political theory—with its analytical grounding in the state—unable to capture adequately the radically imaginative performances, portrayals, and projections of power produced by so-called nonstate actors in these day-to-day battles. During these years, not only had Foucault begun reading Clausewitz, alongside Leo Trotsky, Che Guevara, and Rosa Luxemburg; he was also paying close attention to the Black Panthers in the United States, remarking that "they are developing a strategic analysis that has emancipated itself from Marxist theory."[6] And in the months surrounding the June War, it is worth recalling that Foucault was in Tunis and witnessed, first hand, the extraordinary anti-Jewish show of Pan-Arab solidarity by an anticolonial student movement confronting its own government's support of U.S. war policy in Vietnam and, more quietly, its détente with Israel.[7] In reflecting on the route of his own scholarship, Foucault locates 1968 as a crucial turning point:

> The way power was exercised—concretely and in detail—with its specificity, its techniques and tactics, was something that no one attempted to ascertain . . . This task could only begin after 1968, that is to say on the basis of daily struggles at [a] grass roots level, among those whose fight was located in the fine meshes of the web of power. This is where the concrete nature of power became visible, along with the prospect that these analyses of power would prove fruitful in accounting for all that had hitherto remained outside the field of political analysis.[8]

We should note Foucault is referring at least as much to March 1968 in the Tunisian postcolony, in the midst of a resurgence of student activism around the June War, as he is that far more famous May in the metropole.

Given this theorization, SNCC's act of framing an article on the origins of the Arab-Israeli conflict as a genealogy of the "documented facts" of permanent war asks us to take seriously the counterhistorical strategies used by artists, scholars, and activists to articulate substantive forms of freedom, equality, and self-determination "located in the fine meshes of the web of power." This should come as no surprise; culture workers within the black freedom movement have long attempted to ascertain power's specificities and develop strategies that pushed against the forms of "rights"-based discourse that had gained hegemony in the "short" civil rights era.[9] How, then, in the immediate wake of the June 1967 War, did SNCC expose the permanent war being waged through and against a Zionist ideology steeped in the settler colonial logic it shared with the United States? How did this work reveal both the formal limitations of the Civil Rights Movement and the deep entrenchment of U.S. popular, ideological, and material support for Palestine's colonial occupation?

Accurately depicting this terrain requires moving beyond the borders of the nation-state and into the colonial world in ways SNCC, Du Bois, Newton, and many others in the black freedom movement routinely elucidated. As historians, political theorists, and literary critics alike are increasingly excavating, the long trajectory of the black freedom movement advanced a wide-ranging global lens, committed to representing the linkages between antiracist domestic decolonizing struggles and those taking place across the globe. From Du Bois, C. L. R. James, and Aimé Césaire to Claude McKay, Richard Wright, Langston Hughes, and James Baldwin to Stokely Carmichael, Jack O'Dell, Eldridge Cleaver, and David Graham Du Bois, this global lens enabled a critical purchase on forms of history writing predicated on the disembodied abstractions of rights, order, and the law could. These abstract concepts could from this vantage point be seen for what they were: if not flat-out ruse, then part of a complex set of contextually specific relations of power relying upon and producing colonized and racialized subjects.

The concept "imperial formation" captures the mobile terrain on which these battles for historical legibility have been waged. Imperial formations, writes Ann Laura Stoler, are "macropolities whose technologies of rule thrive on the production of exceptions and their uneven and changing proliferation." They "produce shadow populations and ever-improved coercive measures to protect the common good against those deemed threats to it . . . giv[ing] rise both to new zones of exclusion and new sites of—and social groups with—privileged exemption."[10] This theory of the shifting cartography of empire as one built on differential forms of exclusion operating through racist social structures begins to help us see how SNCC and, increasingly, many others involved in the black freedom movement began to see in Palestine "facts . . . that pertain to our struggle here." A critique of the widespread discourse of U.S. support for Palestine's occupation could challenge the staid exceptionalist arguments that the United States and Israel were somehow unique in achieving their philosophical commitments and political practices of freedom and democracy.

The recently hard-fought struggles of activists, artists, and scholars in many locations around the world have revealed where and how these sites of exception in Palestine have been produced: in the "permanent state of war" that operates through the military and administrative occupation begun in 1967; the limited forms of sovereignty for Palestine's local and "state" governments; the Israeli state's fluid and flexibly produced borders, which Palestinians have no guaranteed substantive rights to cross; Palestinians' widespread encampment in military prisons, refugee camps, and recently behind the so-called apartheid wall; and the epidemic underdevelopment of the Palestinian economy.[11]

But such sites of exception have routinely been exempt from scrutiny; indeed, their form and meaning are still hotly contested in the U.S. public sphere. Even within the black freedom movement's intellectual tradition, with its origins in an internationalist Pan-Africanism that had, for over a century, confronted U.S. exceptionalism, seeing this kind of imperial formation in Palestine had hardly been predetermined. For the most influential thinkers in this tradition, from Blyden to Garvey to Makonnen, Du Bois to Robeson to Nkrumah, Palestine was legible only through the lens of Jewish Zionism. Jewish Zionism provided an analogy to think diasporic black political consciousness rooted in an imaginative articulation of ancient scriptural reference and modern nationalist ideology. For Du Bois and Robeson, as for a young Stokely Carmichael, Zionism also offered a set of leftist economic and political commitments that could be deployed in a shared black-Jewish struggle against U.S. capitalist hegemony and white supremacy.

With the revelation of genocide conducted in Nazi concentration camps, the enmeshing of an internationalist black imaginary and Jewish nationalism became even tighter. Jewish settlement in Palestine became the touchstone for Afro-Zionist responses to Auschwitz, even when it ran against the impulses of an anti-imperialist black imaginary. The National Association for the Advancement of Colored People, for instance, passed a resolution stating that "the valiant struggle of the people of Israel for independence serves as an inspiration for all persecuted people throughout the world."[12] Du Bois would pen the impassioned "Case for the Jews," deploying conventional rhetoric in support of a Jewish national home that fused a sense of Western civilizational progress with a pointed redress for Nazi genocide. Robeson regarded the violent struggles of the Haganah against British mandate rule as the culmination of a centuries-old Jewish national liberation movement.

If seeing an imperial formation in Palestine was blocked by this linkage of Zionism with Pan-Africanism and the impassioned humanitarian response to Nazi genocide, it was further complicated by an emergent "racial liberalism" in the United States.[10] For many, World War II effectively laid bare the racial violence that had structured Western imperial projects in the Americas, Africa, and Asia. Auschwitz provided concrete evidence for wide-ranging critiques of modernity, including those offered by Du Bois, Césaire, and Hannah Arendt, who saw in the death camps the crude doppelganger of Jim Crow lurking at the intersection of racial science, fascism, and industrial capitalism. The publication, in 1944, of Gunnar Myrdal's *American Dilemma: The Negro Problem and Modern Democracy*, a mammoth study of U.S. race relations, significantly subsumed these antiracist analyses in a U.S. exceptionalist framework. Myrdal's definitive theory of what became known as racial liberalism shaped how race was understood in the United States from the end of World War II through the mid-1960s. "At racial liberalism's core was a geopolitical race narrative," writes Jodi Melamed. "African American integration within U.S. society and advancement toward equality defined through a liberal framework of legal rights and inclusive nationalism would establish the moral legitimacy of U.S. global leadership."[13] In other words, the state could internalize, legislate, and manage antiracism as a means of codifying a singular nationalism in which race purportedly no longer mattered. This process was meant to elevate the moral status of the United States above the European colonial powers and, at the same time, formally erase the residues of racial violence—made manifest in Auschwitz—that these powers shared.

Under the racial liberal regime, U.S. discourse shaped interpretations of settlement in Palestine as Jewish immigrants "integrated into democracy" under the rubric of rights at last guaranteed by an internationally recognized sovereign state. Just as "America" could erase the ethnic and national difference of its immigrant populations, so, too, could the state of Israel with the possibility of an "open" immigration policy—albeit one applied to only a very particular population. The revelation of Israel's "ethnocracy," that is, its stark refusal of democratic integration for all Palestine's residents, including its long-time Arab residents and its significant non-European Jewish population, would only gain legibility in the U.S. public sphere as the racial liberal regime came under widespread scrutiny.[14]

The 1950s offered glimpses of a reconceptualized relationship between the black freedom movement and Israel-Palestine, though these possibilities were largely underelaborated. The 1955 Asian-African Conference in Bandung, Indonesia, and the 1956 Suez Crisis began to reveal the shared concerns of an Afro-Arab movement for culture- and class-based forms of international solidarity, with Egyptian president Gamal Abdel Nasser instrumental in shaping a broad understanding of an anticolonial Pan-Arabism. In a poetic response to the European support of Israel against the Egyptian nationalization of the Suez Canal in 1956, Du Bois saw Israel's actions as a hallmark example of the "West," which "betrays its murdered, mocked, and damned." Nasser, by contrast, waves "that great black hand," which "grasps hard the concentrated hate / of myriad million slaves."[15] Earlier in the decade, Du Bois had attempted to add an entire paragraph meant to add complexity to a chapter in *Souls of Black Folk* on the forms of exploitation by Jewish landowners of black labor that emerged in the post-Reconstruction rural South. The paragraph was meant to clarify his "inner sympathy with the Jewish people" and "illustrates how easily one slips into unconscious condemnation of a whole group." Du Bois' editors did not include the paragraph because they saw it as a belated "response to changes in the political climate."[16] Such a paragraph, though, might well have opened up an inquiry into the distinction between anti-Semitism and a critique of Zionism, not to mention the continuities across the historical landscape of practices of racialized exploitation operative in Palestine and their manifestation in forms of "universal and continuous" war.

Given these multiple blockages, then, we might think of the imperial coordinates in Palestine prior to the late 1960s as an image never substantively captured by the black freedom movement. If, as Michael Williams has argued, "the harmony of interests between Zionism and world imperialism did not become apparent until the era of decolonization," what was the imaginative and rhetorical frame that enabled the slow process of bringing Palestine's sites of exception into view in the United States?[17] To what ends?

GHETTOS, PRISONS, COLONIES, AND ANALOGICAL FORCE

In April 1967, after having spent nearly two decades writing on the forms of psychic, sexual, and economic violence brought to bear on African Americans, James Baldwin published "Negroes Are Anti-Semitic Because They're Anti-White" in the *New York Times Magazine*. The *Times* essay, like his first major essay, "The Harlem Ghetto," published nineteen years earlier, maps the relationship between racism and ghettoization at a time of heightened tension in Israel-Palestine. "It is bitter to watch the Jewish storekeeper locking up his store for the night, and going home," Baldwin writes. "Going, with *your* money in his pocket, to a clean neighborhood, miles from you, which you will not be allowed to enter." If, in 1948, Baldwin saw Jews living in the midst of Harlem's ghetto, by 1967, Baldwin suggests that anti-Semitism emerged because not only had American Jews become assimilated into a national ideology of exclusion predicated on race—the "American pattern"—but, in doing so, they had embraced a spatially stratified whiteness.

"The Jew is a white man," writes Baldwin. Against the backdrop of the urban unrest of the 1960s, Baldwin clarifies the differential racialized practice of imagining social struggle, one that heroifies, the other that criminalizes: "When white men rise up against oppression they are heroes: when black men rise they have reverted to their native savagery. The uprising in the Warsaw ghetto was not described as a riot, nor were the participants maligned as hoodlums: the boys and girls in Watts and Harlem are thoroughly aware of this, and it certainly contributes to their attitudes toward the Jews." The Holocaust-era analogy of Jewish resistance, replete with its implications of creeping fascism and genocide in the United States, is incommensurable with the differential forms of exclusion that distinguish Jews and blacks. "[I]f one is a Negro in Watts or Harlem," Baldwin continues, "and knows why one is there, and knows that one has been sentenced to remain there for life, one can't but look on the American state and the American people as one's oppressors. For that, after all, is exactly what they are. They have corralled you where you are for their ease and their profit, and are doing all in their power to prevent you from finding out enough about yourself to be able to rejoice in the only life you have."[18] Just as the ability to identify with the Jewish diaspora through scriptural reference was severely curtailed by the material realities of black existence (as Baldwin described in 1948), the spatial logic of the ghetto as a corral for a criminalized underclass was made illegible in the context of racial liberalism.

A widely mimeographed 1968 statement authored by New York Black Panther Zayd Shakur, titled "America is the Prison," echoes Baldwin's sentiments: "prisons are really an extension of our communities."[19] Scholarship on criminality, space, and political economy has, in recent years, documented the broader trends of the process Baldwin and Shakur are capturing. Sociologist Loïc Wacquant has argued persuasively that by the end of the 1960s, there emerged a "peculiar institution," which "operated to define, confine, and control African Americans." Shaped by the residues of institutionalized racial slavery, Jim Crow segregation, and ghettoized urban space, the late 1960s saw the emergence of a "novel institutional

complex formed by the *remnants of the dark ghetto and the carceral apparatus*"—
what Wacquant terms the "deadly symbiosis" of a "single *carceral continuum*."
Wacquant describes the ghetto as a "relation of ethnoracial control and closure
[which] . . . operates as an *ethnoracial prison*: it encages a dishonoured category
and severely curtails the life chances of its members." Shifts from an urban indus-
try-based economy to one that was decentered and service-based, buttressed by
a post-1965 boom in laboring-class immigration, made black workers living in
urban settings functionally obsolete. With the federal passage of civil rights legislation—and the corresponding revelation that such juridical reforms substantively
did little to alter a landscape built on centuries of institutionalized racism—many
African Americans saw fit to take to the streets, often sparking violent contestations with law enforcement and property owners. But "as the walls of the ghetto
shook and threatened to crumble," Wacquant writes, "the walls of the prison were
correspondingly extended, enlarged and fortified, and 'confinement of differen-
tiation,' aimed at keeping a group apart . . . gained primacy over 'confinement of
safety' and 'confinement of authority.'" *De jure* segregation was outlawed, but *de
facto* segregation became entrenched: the ghetto became more like a prison and
the prison became more like a ghetto.[20]

At the time, this race-making transformation of space became understood in
terms of the "colonial analogy," a rhetoric meant to contest the racial liberal discourse that had previously rendered these processes invisible or exceptional. This
rhetoric drew in part on the Comintern theorization of the "Negro problem" in
the 1920s, which itself had been attuned to the Pan-Africanist organizing of Marcus Garvey's Universal Negro Improvement Association. As early as 1962, Harold
Cruse argued that "the revolutionary initiative passed to the colonial world and
in the United States is passing to the Negro." In much of his work in the 1960s,
Cruse saw in African American cultural politics the potential to translate into
the U.S. context the organizational, philosophical, and rhetorical effectivity of
decolonization and anticolonial nationalism gleaned from Third World liberation
struggles. In 1972's *Racial Oppression in America*, sociologist Robert Blauner confronted the facile liberal multiculturalist arguments for pluralism and assimilation
offered by the likes of Nathan Glazer by framing his scholarly inquiry into U.S.
racial formation with a theory of internal colonialism. Jack O'Dell, the longtime
coeditor of the magazine *Freedomways*, who would lead delegations of black leaders in solidarity with Palestine to the West Bank, Egypt, and Lebanon in the late
1970s and early 1980s, stressed that black proletarian life was shaped by a "special
variety of colonialism." In mid-1968, Eldridge Cleaver began his discussion of
the "land question" by asserting, "The first thing that has to be realized is that
it is a reality when people say that there's a 'black colony' and a 'white mother
country.'" In an evocative analogy, Cleaver then draws on the "parallel situation of
the Jews at the time of the coming of Theodore Herzl": "The Jewish people were
prepared psychologically to take desperate and unprecedented action. They saw
themselves faced with an immediate disastrous situation. Genocide was staring
them in the face and this common threat galvanized them into common action.

Psychologically, black people in America have precisely the same outlook as the Jews had then."[21]

Of those works elaborating the "colonial analogy," perhaps the most widely read and regarded was Stokely Carmichael and Charles Hamilton's book *Black Power: The Politics of Liberation in America. Black Power* was published in September 1967 in the midst of Carmichael's wide-ranging tour of London, Cuba, Moscow, Beijing, Vietnam, Algeria, and Guinea, where Carmichael would meet the likes of Shirley Graham Du Bois, Sekou Toure, Kwame Nkrumah, and his future partner, the exiled South African singer, Miriam Makeba. The book took up Fanon's argument in *Wretched of the Earth* to assert that institutional racism in the United States "has another name": colonialism. Black people formed an internal colony in the United States, and the first step toward black liberation was to emulate the decolonizing struggles under way across the Third World. Countering Myrdal's thesis, Carmichael and Hamilton "put it another way": "There is no 'American dilemma' because black people in this country form a colony, and it is not in the interest of the colonial power to liberate them."

Carmichael and Hamilton continue: "Obviously, the analogy is not perfect." After all, there is no geographically distant "Mother Country" from which colonial sovereignty emanates; nor are raw materials produced in the colony and exported to the "Mother Country." What concerns Carmichael and Hamilton, though, and what preoccupies Cruse, Blaune, O'Dell, and Cleaver is "not rhetoric . . . or geography" but the "objective relationship" of blacks to racist rule.[22] Such analytical limitations of the colonial analogy appear throughout discussions of the relationship between U.S. race-making and the structures of global capital, limitations that rightly focus attention on the contextually specific forms of colonizing dominance and subjection, anticolonial resistance and struggle in "actually existing colonialisms." To elide these material specificities is itself to perform a certain epistemological colonizing violence.

And yet analogies have extraordinary rhetorical force precisely because their form keeps these limitations in full view. An analogy can never be "perfect" in any simple sense. The "likeness" or "parallel" of Zionism to Pan-Africanism, Warsaw to Watts, the Jewish Holocaust to racial slavery, the wandering Jew to the black diaspora: these analogies juxtapose unique historical formations, ideological concepts, or geographies, which are then linked together via the radically unstable "like" or "as." With an analogy, one cannot escape difference. That is both its danger and its force. As a rhetorical figure it is at its core a difference *always on the verge* of collapse into an identity, one always socially produced under contextually specific conditions *always on the verge* of erasure. These indelible conditions hold an analogy together and produce its effectivity as an articulation.

In this way, the colonial analogy builds on the many analogical constructions already discussed, operating as a rhetorical and geographic figure to reveal the contradictions of post–civil rights U.S. imperial formation. Edward Said persuasively argued that dominant forms of colonial discourse perform an imaginative geography that maps a coherent, rational, liberal, modern West "by dramatizing the distance and difference between what is close to it and what is far away." The colonial analogy, by contrast, has provided scholars, artists, and activists of the black freedom movement an imaginative terrain to perform a radically contestatory remapping of geography. The rhetoric of colonialism gave a name to the actually existing inequalities the civil rights legislative process was seemingly meant to ameliorate. For many in the black freedom movement, the uneven development of deindustrialized urban space had its representational correlates in other colonized sites in the Third World including, significantly, Israel-Palestine after the June War.

"TEST YOUR KNOWLEDGE"

Published in mid-August, 1967, "Third World Round Up: The Palestine Problem: Test Your Knowledge" initiated a long process of reframing the question of Palestine as one of black liberation. Its publication—and widespread condemnation—emerged during the tumultuous months of an extraordinary historical transformation in movements for liberation. In a very short period of time, a confluence of events and their discursive residues renewed and revised an imaginative geography first broached in the interwar years—by the likes of Robeson, Bunche, Du Bois, and others on the Left—that connected struggles for black freedom in the United States with decolonizing movements around the world. In April, Martin Luther King, Jr., delivered his "Beyond Vietnam: A Time to Break the Silence" speech at New York City's Riverside Church, for the first time depicting the "very obvious and almost facile connection" between struggles for racial equality at home and struggles against the unjust war being conducted by the United States in Vietnam. Several days later, the *New York Times Magazine* published Baldwin's "Negroes are Anti-Semitic Because They're Anti-White." On May 2, Bobby Seale and thirty members of the Black Panther Party sported guns and uniforms and staged a major protest at the California State Capitol in Sacramento. Later that month, SNCC elected a new chairman, H. Rap Brown, declared itself a human rights organization, and redirected its purpose to "encourage and support the liberation struggles against colonialism, racism, and economic exploitation" around the world. The months of June, July, and August saw the widespread mimeographed circulation of Huey P. Newton's theory of permanent war "In Defense of Self Defense." At the end of August, the National Conference for New Politics in Chicago inadvertently continued the process of disarticulating interracial coalitions for social change. Harold Cruse released *Crisis of the Negro Intellectual*. Some 164 "civil disorders" in twenty-eigth U.S. cities transformed U.S. urban and suburban landscapes. And FBI Director J. Edgar Hoover received approval to redirect the bureau's Counter Intelligence Program to operate against "black nationalist, hate-type organizations," launching a "secret war against Black Power activists . . . that featured the systematic, illegal harassment, imprisonment, and, at times, death, of black militants."[23]

SNCC's "Third World Round Up: The Palestine Problem: Test Your Knowledge" is clearly informed by all this. However, most responses to "The Palestine Problem" have focused narrowly on whether the text was deploying anti-Semitic tropes as well as its political ill advisability. The *Times* devoted an entire front-page article to "The Palestine Problem," titled "S.N.C.C. Charges Israel Atrocities: Black Power Group Attacks Zionism as Conquering Arabs by 'Massacre.'" Tracing the legacy of the shift in SNCC activism from domestic broad coalition-building to Black Power internationalism as routed through Fanon and Malcolm X, the article does not take up whether such charges were indeed warranted or accurate. Instead, it eulogizes a prior time of solidarity as it chastises SNCC for its "hate-filled" rhetoric. "It is a tragedy that the civil rights movement is being degraded by the injection of hatred and racism in reverse," noted the general counsel of the Anti-Defamation League. The next day, the *Times* printed a story dedicated to recounting a series of "angry statements" by "civil rights leaders" against SNCC's "Israel Stand," running the gamut of representatives of the fracturing black-Jewish coalition. Martin Luther King, Jr., declined to comment specifically on SNCC's article, saying only that he was "strongly opposed to anti-Semitism and 'anything that does not signify my concern for humanity for the Jewish people.'"

The secondary literature discussing "The Palestine Problem" is likewise preoccupied by charges of anti-Semitism and the ostensibly fraught political efficacy of publishing such a polemic. Rarely is the article treated with a substantive reading. The fullest has been in Kwame Turé, né Stokely Carmichael's recently published autobiography, *Ready for Revolution*. According to Turé, who had just been replaced by H. Rap Brown as SNCC chairman when the newsletter appeared, the document originated in a reading group organized by "one courageous activist sister." (The origin story of the article is considerably more complex than Turé describes here, as will be discussed below.) Turé refuses to refer to this organizer by name, though other accounts suggest her name was Ethel Minor, an activist involved in Latin American liberationist organizing and the Nation of Islam. The reading group convened first in the wake of Malcolm X's assassination in 1965 and proceeded to read and discuss one book a month over the course of two years. The reading list, according to Turé, included "not just pro-Palestinian or anti-Zionist materials" but "Jewish writers who, from the perspective of the moral traditions of Jewish thought, opposed the militaristic expansionism of Zionist policies." They also read writings from "Herzl, Ben-Gurion, Begin, documents from the Stern Gang, etc., etc." The turning point for the reading group was realizing "the close military, economic, and political alliance between the Israeli government and the racist apartheid regime in South Africa." "I have to say," Turé avers, "discovering that the government of Israel was maintaining such a long, cozy, and warm relationship with the worst enemies of black people came as a real shock. A kind of betrayal. And, hey, we weren't supposed to even talk about this? C'mon."[27] Turé claims that drafting "The Palestine Problem" with Minor was his last act as chairman, meant primarily to take the pulse of SNCC's leadership through "the form of sharp questions against a background of incontestable historical facts." The systematic study was "short-circuited," though, when the newsletter was handed over to mainstream journalists. Turé concludes that "had the process not been short-circuited, I'm sure the overwhelming sentiment would have been to make a statement, a moral statement, on justice for the Palestinian people while trying hard not to offend or alienate our Jewish friends on a personal level. Such a statement, one intended for public distribution, would almost certainly have been more nuanced. In properly diplomatic language, which the talking paper definitely was not. But you crazy if you think the language would have made any difference politically. This was an orchestrated declaration of war, Jack."[24]

The "talking paper" was comprised of thirty-two "documented facts," two archival photographs, and two cartoons. It is clear from the headnote that the piece serves as a knowledge project meant to "shed some light" on the conditions of the decolonizing world in ways that "pertain to our struggle here." Its "documented facts" suggest that such knowledge is based in objective historical reality, and, when phrased in terms of a "test," complete with interspersed headers repeating the phrase "Do you know," these facts do complex rhetorical work. The article's readership, directly addressed through the second person "you," is presumed unaware of these facts as it condemns the "white American press" for obfuscating the "true story about world events in which America is involved." Each fact is phrased in terms of a question. With the present-tense "do you know"; however, each fact demands that not only are these "documented facts" crucial for framing an understanding of the post-June War conjuncture but one either "knows" or one does not.

While the questions are generally organized chronologically, by breaking each "fact" into its own distinct number and juxtaposing these questions with photographs and cartoons, this formulation captures the synchrony of historical knowledge. The post-June War occupation is represented as the culmination of a trajectory begun in 1897, when "Zionism, a world-wide nationalistic Jewish movement," formulated a program to "create for the Jewish People a home in Palestine according to Public law." This program, according to the article, received "maximum help, support, and encouragement from Great Britain, the United States, and other white Western colonial governments." With 1917's Balfour Declaration, Britain subsequently "took control of Palestine," creating a "world problem." But, according to the article, there were very few "native" Palestine Jews, and only 56,000 Jews in total, most of who had recently immigrated to the British colony. By 1947, when "Britain passed the Palestine problem on to the United Nations," "Zionists owned no more than 6 per-cent of the total land area in Palestine" and had a population roughly half the size of the Arab Palestinians. The "formal beginning of the Arab-Israeli War" commenced after the "formal end of British rule" on May 15, 1948, when "Arab States had to send in their poorly trained and ill equipped armies against the superior western trained and supported Israeli forces, in a vain effort to protect Arab lives, property and Arab rights to the land of Palestine."

Question 16 of "The Palestine Problem" accuses "Zionist terror gangs . . . [of having] deliberately slaughtered and mutilated women, children and men, thereby causing the unarmed Arabs to panic, flee and leave their homes." Question 20 illustrates the polarizing vote in the UN for the 1947 Partition Plan, asserting, in all capital letters, that "ISRAEL WAS PLANTED AT THE CROSS-ROADS OF ASIA AND AFRICA WITHOUT THE FREE APPROVAL OF ANY MIDDLE-EASTERN, ASIAN OR AFRICAN COUNTRY!" Questions 25 and 26 provide evidence of racist practices within the post-1948 state of Israel, where Arabs are "segregate[d] . . . , live in 'Security Zones,' under Martial Law, are not allowed to travel freely within Israel, and are the victims of discrimination in education, jobs etc." Further, "dark skinned Jews from the Middle-East and North Africa are also second-class citizens in Israel, that the color line puts them in inferior position to the white, European Jews." The last two questions bring to the fore the perceived relationship between Israel and African neocolonialism. Question 31 asserts that not only were "the famous European Jews, the Rothschilds" involved "in the original conspiracy with the British" to found Israel but they "ALSO CONTROL MUCH OF AFRICA'S MINERAL WEALTH." Question 32 contends that Israel has "gone into African countries, tried to exploit and control their economies, and sabotaged African liberation movements, along with any other African movements or projects opposed by the United States and other white western powers."

Many of these "documented facts" have been corroborated by subsequent reputable scholarly research, often located within the Israeli academy; others stretch the historical archive, while others are built on unsubstantiated myth. That the article is silent on the documents from which these facts were drawn—do they emerge from the systematic course of comparative reading led by Ethel Minor, as suggested by Turé, were they hastily cobbled together at the last minute, or do they come from another source?—leaves open the question of what constitutes "proper" knowledge of Israel-Palestine in the United States.

In fact, the article is strikingly similar to a pamphlet prepared in the immediate aftermath of the June War at Kuwaiti professor Fayez Sayegh's newly founded PLO Research Center in Beirut. Sayegh was particularly instrumental in passing the 1975 UN resolution condemning Zionism as a "form of racism and racial discrimination." His post–June War pamphlet, titled "Do You Know? Twenty Basic Facts About the Palestine Problem" takes identical form as "The Palestine Problem," complete with the repeated question "Do You Know?" Fully sixteen of the pamphlet's facts appear verbatim (or nearly so) in the SNCC article. And near-identical repetitions of SNCC's "Palestine Problem" (or Sayegh's "Do You Know") were produced in the immediate aftermath of the October 1973 War in the *Black Panther Intercommunal News Service* (under the editorial eyes of David Graham Du Bois, W. E. B. Du Bois' stepson) and are said to be drawn from the Middle East Coordinating Committee's "Did You Know? . . . Facts about the Middle East."

The textual affinities of "The Palestine Problem" to Sayegh's pamphlet (and the Panthers' repetitions with a difference) are obscured in tales of its origin. But since circulation of something like Sayegh's pamphlet had little access to a broad U.S. public sphere, we can see how SNCC's article transformed the rhetorical landscape in its ability to transpose and supplement a Palestinian counterhistory through its connection to the U.S. Third World Left. It likewise relentlessly asserts that what is being depicted is knowable at all, should be known, is required knowledge for apprehending the present, and, even more, that it is in fact knowledge at all. The effect was to present knowledge of occupation and confinement across which the black freedom movement could begin to draw analogies, alliances, and allegiances that were deepened in the years to come.

The visual elements of the textual affinities of "The Palestine Problem" make these imaginative linkages clear. One such linkage is the structural one between the genocidal violence of the Nazi Holocaust and the foundation of the State of Israel. The typical portrayal of this relationship has considered the latter the ethical and just solution to the former. An archival photograph of a dozen men kneeling with their hands on their heads and guns blazing behind them cautions otherwise. The photograph's caption reads, "Gaza Massacres, 1956," a reference most likely to the massacre at Khan Yunis during the opening moments of Israel's 1956 incursion into the Sinai peninsula. "Zionists lined up Arab victims and shot them in the back in cold blood," the caption continues. "This is the Gaza Strip, Palestine, not Dachau, Germany." Juxtaposing these two frames makes legible a counterhistorical structural linkage, one that sees the genocidal violence of World War II echoed by the Israeli state. Just as Baldwin's citation of the Warsaw Ghetto was a lever to distinguish how one understands rebellion, Dachau's metonymic status condenses the racial violence practiced across the Nazi concentration camps and intimates that the conditions under Israeli rule are, at times, some of its most pernicious effects.

Another structural linkage is captured in SNCC artist Kofi Bailey's cartoon portrait of Israeli minister of defense Moshe Dayan. Dayan had achieved heroic status in the U.S. for orchestrating much of the June War and bringing the Western Wall under Israeli sovereignty by "uniting" Jerusalem; one comedian joked, for instance, that "Dayan should be hired to put a quick end to the fighting in Vietnam." For example, an image circulated that depicted Dayan with U.S. dollar signs on his epaulets. Some have accurately interpreted this caricature as anti-Semitic, suggesting it plays on the stereotype of Jewish financial dominance; but it also condenses the close material linkages between the U.S. and Israeli militaries described elsewhere in the article.

"The Palestine Problem's" most complex visual image is another image by Kofi Bailey that juxtaposes three forms of articulation, linking a history of U.S. racial violence, imperial military engagements in Vietnam and the Arab world, and a broad Afro-Arab struggle for freedom. At the top of the image is a disembodied hand with a six-pointed Jewish Star of David overlaying a dollar sign. Charges of anti-Semitism have been brought against this image as well, though given the

Dayan cartoon, one might understand this juxtaposition to also signify U.S. financial support for the Israeli military. The hand grasps the middle of a rope dangling downward on both sides. At one end is the likeness of Egyptian president Gamal Abdel Nasser drawn from the chest up, donning a dark suit jacket, white shirt, and tie. During the high point of Third World nonalignment, manifesting itself in the 1955 Bandung Conference, Nasser consistently advocated for Palestinian freedom from imperial rule as part of a larger Pan-Arab nationalism. His likeness thus condensed "an emotionally explosive convergence of anti-colonial defiance and global racial consciousness." At the rope's other end, also depicted from the chest up in similar garb, is Muhammad Ali, the U.S. heavyweight boxing champion, whose embrace of the Nation of Islam and concomitant antiwar stance had brought his boxing career under fire. In 1966, Ali had refused induction into the U.S. Army. On June 20, 1967, Ali was convicted of draft evasion, sentenced to prison pending numerous appeals (including to the U.S. Supreme Court, which unanimously overruled the conviction), and was barred from boxing. This illustration of a double lynching imaginatively links the fates of Nasser and Ali, cast as they are as twinned victims of a common racially coded form of Jim Crow–style extralegal violence. In the background of the lynching is a disembodied dark-skinned arm bent at the elbow, labeled "THIRD WORLD," wielding a scimitar—itself commonly perceived as rooted in Persian history—labeled "LIBERATION MOVEMENT." The force of Third World struggle emanates from the Middle East, the image suggests, with transnational repercussions; its horizon sees the liberation of the Arab world, African Americans, and practicing Muslims from the intertwined violence of the U.S. and Israeli states.

Understood in this way, "The Palestine Problem" deploys a variety of strategies, transpositions, and reproductions in its attempt to represent the material, ideological, and epistemological linkages between struggles for black liberation in the United States and the historically embedded colonial conditions in Palestine. As a Black Power intervention, it reveals such linkages on the terrain of a much larger decolonizing knowledge project. Its reception reveals the fault lines of a dawning post–civil rights public sphere that continues to haunt contemporary U.S. engagements with Israel-Palestine, one still structured by "prolonged conflict and permanent state of war."

Given our contemporary moment's resurgence of discourses of permanent war, SNCC's article leaves us with a series of pressing questions about historiography, representation, and racial justice movements. How might we begin to articulate and disrupt contemporary modes of race-making that link, for instance, mass incarceration in the United States and Gaza's open-air prison, the militarizing of the U.S.-Mexico border and Israel's "separation barrier," or the production of a constellation of shared racialized figures organized under the instrumental category "Islamofascism"? What would such performances of imaginative geography look like and how and to whom would they be legible? How would these reconfigure the political? SNCC's article does extraordinary work to open up these questions for us but it hardly answers them. The coordinates have clearly changed, and we presume otherwise at our peril. Perhaps, then, the notion of "tragedy" used in 1967 to describe SNCC's article is apt but in a much different sense. Perhaps the narrative logic of tragedy, as explored by anthropologist David Scott in conceiving of the global terrain and broad historical sweep of movements for black freedom, helps us "reorient our understanding of the politics and ethics of the postcolonial present." The tragic narrative of colonial enlightenment is in this sense not "a flaw to be erased or overcome . . . [but] a permanent legacy that has set the conditions in which we make of ourselves what we make and which therefore demands constant renegotiation and readjustment."[25] The emergence of "Black Power's Palestine" at the end of the civil rights era clearly makes such demands on us in the present.

NOTES

1. Student Nonviolent Coordinating Committee, "Third World Round-up: The Palestine Problem: Test Your Knowledge," *SNCC Newsletter* 1, no. 2 (July–August 1967): 5–6.
2. Quoted in Brent Hayes Edwards, "Late Romance," in *Next to the Color Line: Gender, Sexuality, and W. E. B. Du Bois*, ed. Susan Gillman and Alys Eve Weinbaum (Minneapolis: University of Minnesota Press, 2007), 124–49, 130.
3. "In Defense of Self-Defense." *The Black Panther*, June 20, 1967.
4. Michel Foucault, *Society Must Be Defended": Lectures at the Collège de France, 1975–1976*, trans. David Macey (New York: Picador, 2003), 46–47.
5. Ibid.
6. Ibid., 282.
7. Randall E. Auxier, "Foucault, Dewey, and the History of the Present," *The Journal of Speculative Philosophy* 16, no 2 (200): 75–102.
8. Michel Foucault, "Truth and Power," in *Power/Knowledge: Selected Interviews and Other Writings, 1972–1977* (New York: Pantheon, 1980), 109–33.
9. Nikhil Pal Singh distinguishes the "long civil rights era" from what he calls the dominant national narrative of the "short civil rights era," 1955–1965. The short civil rights narrative narrowly focuses on a liberal capitalist, domestic, and integrationist struggle, contained in the U.S. South, and uncritically celebrates the federal passage of Civil Rights and Voting Rights legislation. The narrative of the long civil rights era, by contrast, emerges as early as the 1930s and continues into the 1970s, has an internationalist lens shaped by anticolonial, liberationist, and anticapitalist movements, and views struggles for black freedom in the United States as part of a broader global struggle. Nikhil Pal Singh, *Black is a Country: Race and the Unfinished Struggle for Democracy* (Cambridge: Harvard University Press, 2004).
10. Ann Laura Stoler, "On Degrees of Imperial Sovereignty" *Public Culture* 18, no. 1 (2006): 125–47.
11. The literature on Palestine's sites of exception is vast. Achille Mbembe's reworking of Foucault performs the extraordinary intellectual labor of theorizing these sites of exception as "the most accomplished form" of late modern colonial occupation, with antecedents in racial slavery and Nazi genocide. Mbembe draws directly from ethnographic accounts of conditions in the Gaza Strip. See Achille Mbembe, "Necropolitics," *Public Culture* 15, no. 1 (2003): 11–40.
12. Quoted in Melani McAlister, *Epic Encounters: Culture, Media, and U.S. Interests in the Middle East, 1945–2000* (Berkeley: University of California Press, 2001), 89.

13. Jodi Melamed, "The Spirit of Neoliberalism: From Racial Liberalism to Neoliberal Multiculturalism," *Social Text* 24, no. 4 (2006): 1–24.

14. See Oren Yiftachel, *Ethnocracy: Land and Identity Politics in Israel/Palestine* (Philadelphia: University of Pennsylvania Press, 2006).

15. Melani McAlister, *Epic Encounters: Culture, Media, and U.S. Interests in the Middle East, 1945–2000* (Berkeley: University of California Press, 2001), 85.

16. Qtd. in "Note on the Text," W. E. B. Du Bois, *The Soulds of Black Folk* (New York: Vintage, 1990), 220.

17. Michael W. Williams, "Pan-Africanism and Zionism: The Delusion of Comparability," *Journal of Black Studies* 21, no. 3 (1991), 348–71.

18. James Baldwin, "Negroes are Anti-Semitic Because They're Anti-White," *New York Times Magazine* (April 9, 1967) 27, 93–96.

19. Zayd Shakur, "America is the Prison," in *Off the Pigs!: The History and Literature of the Black Panther Party*, ed. G. Louis Heath (Metuchen, NJ: Scarecrow, 1976).

20. Loïc Wacquant, "From Slavery to Mass Incarceration: Rethinking the 'Race' Question in the U.S.," *New Left Review* 13 (2001): 41–60.

21. Eldridge Cleaver, "The Land Question and Black Liberation," in *Eldridge Cleaver: Post-Prison Writings and Speeches*, ed. Robert Scheer (New York: Random House, 1969).

22. Stokely Carmichael and Charles V. Hamilton, *Black Power: The Politics of Liberation in America* (New York: Vintage, 1967).

23. See Peniel E. Joseph, *Waiting 'Til the Midnight Hour: A Narrative History of Black Power* (New York: Henry Holt, 2006).

24. Stokely Carmichael, with Ekwueme Michael Thelwell *Ready for Revolution: The Life and Struggles of Stokely Carmichael (Kwame Ture)* (New York: Scribner, 2003).

25. David Scott, *Conscripts of Modernity: The Tragedy of Colonial Enlightenment* (Durham, NC: Duke University Press, 2004).

PART II

SOLIDARITY AND RESISTANCE

CHAPTER 6

FROM HARLEM TO ALGIERS

TRANSNATIONAL SOLIDARITIES BETWEEN THE AFRICAN AMERICAN FREEDOM MOVEMENT AND ALGERIA, 1962–1978

SAMIR MEGHELLI

Consciousness of Africa mounted again as more and more African nations regained their independence. The inhuman atrocities of the French colonialists against the Algerian people, who were struggling valiantly for their independence, aroused widespread sympathy and fraternal support among the people of Harlem.[1]
— Richard B. Moore, "Africa Conscious Harlem"

We saw Algeria in terms of our pasts and what our futures might be. I saw, see, and am feeling it that way now. And what I write is an attempt to make that experience more available to me, to you, to us.[2]
— Michele Russell, one of the many young African Americans who attended the 1969 Pan-African Cultural Festival in Algeria

IN 1959, AFRICAN AMERICAN INTELLECTUAL AND ACTIVIST HOYT FULLER MADE a brief stopover in Algeria on his way to visit the newly independent African republic of Guinea. In an excerpt from a journal of his experiences, he recounts, "Algeria was an armed camp, with the French colonial masters firmly in control. Soldiers and gendarmes were everywhere, arms at the ready, and many of the public buildings were 'protected' from guerilla assault by layers of barbed wire." He continues, "I had entered the city from the liner, *Foch*, with two young Africans, one from Abidjan, the other from Brazzaville, and we had moved about with relative freedom until we reached the famed Casbah, the incredible labyrinthine quarter made famous over the world by Hollywood's film, *Algiers*, starring Charles Boyer and Hedy Lamarr. The armed guard at the entrance to the Casbah politely but firmly turned the three black visitors away, offering no explanation beyond the simple statement that entrance was forbidden." After his two African companions decided to return to the passenger ship, Fuller again attempted to enter the Casbah by himself, this time successfully so. As he explains, "In Paris, an Algerian friend had given me the name of a young freedom fighter in the Casbah and I set out to locate him." Fuller eventually found this young freedom fighter and his comrades, with whom he "drank coffee and talked of African liberation. Afterwards," he relays, "they walked with me down a twisting 'street' to the entrance above the great plaza. We said goodby. As I strolled out, a guard stopped me. He asked me for my papers and I showed him my American passport. What was I doing in the Casbah? he asked. Didn't I know that it was closed to tourists and that it was dangerous? 'But, M'sieur,' I said to him, mustering my best French, 'I am a black man. The Algerians have no need to harm me. We are fighting the same war.'"[3]

Exactly ten years later, Hoyt Fuller, who, by that time, was editor of the important *Negro Digest* (which he renamed *Black World* shortly thereafter), returned to Algeria for the historic First Pan-African Cultural Festival, alongside more than 10,000 official delegates and visitors from more than thirty African nations, North and South America, the Caribbean, Asia, and Europe, including politicians, musicians, writers, scholars, filmmakers, actors, visual artists, and liberation movement leaders, as well as many others who held personal, professional, or political interests in the realization of a liberated and united Africa.[4] And, as many attendees noted, there were few others places that would have seemed as appropriate for such an event as Algeria. Making this point rather explicitly, African American poet Ted Joans wrote, "Algeria, the largest country in North Africa. Algeria, the country that fought the enslaver and won. Algeria, the revolutionary stronghold of African nationalists. With these and many other black references, Algeria was 'the place' to stage the First Pan African Cultural Festival."[5]

Indeed, in the decade leading up to the 1969 festival, Algeria became a powerful symbol of revolutionary struggle and was looked to as a model of revolutionary success for radicals around the world. Through widespread favorable coverage of its revolution and independence in the African American press, the many local screenings of the popular film *The Battle of Algiers*, and Frantz Fanon's writings, Algeria came to hold a critical place in the iconography, rhetoric, and ideology of key branches of the African American freedom movement. By 1959, when Hoyt Fuller noted that the African American freedom movement and the Algerian independence movement were fighting "the same war," African Americans had, for a long time, been identifying closely with Africa and African anticolonial movements. As historian James Meriwether notes, this trend "can [at least] be traced back to black America's responses to the Italo-Ethiopian War, which had energized widespread African American interest in the continent and had broadened many black Americans' notions of ethnicity to include contemporary Africans."[6] Beginning during this period, greater numbers of African Americans sought to frame transnational identities for themselves, coming to an understanding of the connectedness of their struggle for civil rights and the struggles of African nations for independence. Meriwether explains, "to advance their objectives, African Americans protested, lobbied, and worked with national governments and international organizations, thereby internationally politicizing their expanded constructions of identity."[7]

The recent body of works on the international and transnational dimensions of African American radicalism during the Long Civil Rights Movement and Black Power era has had such an impact that one historian has suggested that it "forms by itself a new canon and direction in the history of African Americans."[8] The First Pan-African Cultural Festival of 1969, held in Algeria, represents an important moment in the history of African and African American political linkages but also begs larger historical questions about Algeria's role as a site and symbol of revolutionary significance. And yet, Algeria is curiously absent from the secondary historical literature—when discussed, it has only been cursorily so.

Thus, the aim of this chapter is to trace the emergence of transnational solidarities between the African American freedom movement and Algeria during the Civil Rights and Black Power eras and to examine the place of Algeria in the African American political imaginary, using the 1969 Pan-African Festival as a moment through which to understand how these linkages were forged, tested, and contested. The story opens in the late 1950s and early 1960s when African Americans began drawing parallels between Algeria's revolution for independence from France and their own freedom movement in the United States. African Americans strengthened their critique of a racist American society in their comparisons between French colonialism in Algeria and American segregation, and between the necessity for Algerian and African American armed resistance. But these linkages took on new meaning as African American radicals began traveling to and, in some cases, living in, Algeria, collaborating officially and unofficially with the Algerian government and the many revolutionary movements housed there. By the mid-1970s, however, Algeria had fallen from its position as a site and symbol of Third World solidarity. As Algeria found itself facing an increasingly dire economic situation, and as the African American freedom movement struggled to maintain momentum in the face of ever more severe repression by American authorities, transnational solidarity became less and less plausible. Ultimately, symbolic and practical ties between African America and Algeria were severed on account of Cold War geopolitics and economic interests, coupled with a mutual lack of understanding between African Americans and Algerians of the concrete challenges each other faced.

"EVERY BROTHER ON A ROOFTOP CAN QUOTE FANON": ALGERIA IN THE AFRICAN AMERICAN IMAGINARY

In October of 1962, shortly after Algeria had emerged victorious from a long and brutal revolution for independence, Ahmed Ben Bella, one of the leaders of the revolutionary National Liberation Front (FLN) and the newly appointed president of Algeria, made his way to New York City where he attended the United Nations' induction ceremony for his young nation. During his stay at the Barclay

Hotel in midtown Manhattan, Ben Bella granted an exclusive, one-hour interview to Charles P. Howard, an accredited United Nations correspondent who played an important role in raising awareness about the connections between the situation of African Americans and that of oppressed people throughout the world through his informative articles in *Muhammad Speaks*, the nationally distributed weekly paper of the Nation of Islam.

Ben Bella, the subject of Howard's interview, received front-page coverage in *Muhammad Speaks*—his picture was accompanied by the headline: "Drive On To Free All Africa!" This is significant and symbolic in that, in part as a result of the relatively widespread coverage of the Algerian revolution and its leaders (like Ben Bella) in the Black press, Algeria became firmly linked to the discourse around the African American freedom movement. This kind of coverage placed Algeria alongside the Mau Mau rebellion and Ghanaian independence as helping keep issues of armed struggle and transnational political linkages in the minds and hearts of the African American masses, intellectuals, and activists.[9] As *Muhammad Speaks* was the most widely circulating African American paper for much of the 1960s,[10] it became an important appendage to the African American freedom movement, and especially the effort to internationalize that struggle and forge solidarities with African independence movements. Nearly every issue carried news and information on the state of African affairs and the Muslim world. And, as one historian has noted, the Black press often had greater reach than the strict circulation numbers reveal, for papers "passed from family to family and could be found in barbershops, churches, lodges, and pool parlors," just as "their contents passed by word of mouth among those who could not read."[11]

Ben Bella's visit to the United States received coverage in the mainstream American press as well, as the *New York Times* ran, among other articles on Ben Bella's visit, one on his historic meeting with Dr. Martin Luther King, Jr. Meeting at the Barclay Hotel, Dr. King and Ben Bella spoke with one another, with the aid of a translator, for nearly two hours. The *New York Times* headline read, "Ben Bella Links Two 'Injustices,'" along with the subheading, "Tells Dr. King Segregation Is Related to Colonialism." Thus, chief among the issues they discussed was the nature of the relationship between the segregation that African Americans were facing, the colonialism that the Algerian people faced under French rule, and Europe's continuing colonial and neocolonial domination of much of Africa and of the so-called Third World.[12] Arriving at a press conference following their meeting, Dr. King was described as having "emerged sounding more like Malcolm X than the civil rights leader reporters knew."[13] King explained that "Ben Bella had made it 'very clear' that . . . he believed there was a direct relationship between the injustices of colonialism and the injustices of segregation here [in the U.S]."[14] King went on to say that he agreed with Ben Bella and that "the struggle for integration here was 'a part of a larger worldwide struggle to gain human freedom and dignity.'"[15] Ben Bella followed King by noting that the African American struggle was "widely publicized in Algeria, and in Africa more generally,"[16] and concluded by declaring that "the United States could lose its 'moral and political voice' in the world if it did not grapple with segregation problems here in a forthright manner."[17]

After his meeting with Ben Bella, King wrote an article himself for the widely read Black newspaper *New York Amsterdam News* titled "My Talk With Ben Bella" in which he detailed the nature of their conversation. He described Ben Bella and Algeria in these terms: "A few days ago I had the good fortune of talking with Premier Ben Bella of the New Algerian Republic. Algeria is one of the most recent African nations to remove the last sanction of colonialism. For almost two hours Mr. Ben Bella and I discussed issues ranging from the efficacy of non violence to the Cuban crisis. However, it was on the question of racial injustice that we spent most of our time."[18] King continued, apparently surprised and encouraged, "the significance of our conversation was Ben Bella's complete familiarity with the progression of events in the Negro struggle for full citizenship. Our nation needs to note this well. All through our talks he repeated or inferred, 'We are brothers.' For Ben Bella, it was unmistakably clear that there is a close relationship between colonialism and segregation. He perceived that both are immoral systems aimed at the degradation of human personality. The battle of the Algerians against colonialism and the battle of the Negro against segregation is a common struggle."[19]

Before returning to Algeria to assume his role as president, Ben Bella went on to meet with Adam Clayton Powell, Jr., as well as Malcolm X—both significant figures in the emerging Black Power movement—at the well-known Absynnian Baptist Church in Harlem.[20] As Malcolm X embarked on a trip to the Middle East and Africa some two years later, he would again meet with Ben Bella during his stop in Algeria. The impact that his experience there had on him became evident when, just after returning from his trip, he spoke at the Militant Labor Forum in May 1964. In responding to the allegations that there existed some sort of "hate-gang" called the "Blood Brothers" that was based in Harlem and calculatedly committed crimes against whites, Malcolm declared,

> I visited the Casbah . . . in Algiers, with some of the brothers—blood brothers. They took me all down into it and showed me the suffering, showed me the conditions they had to live under while they were being occupied by the French . . . They showed me the conditions that they lived under while they were colonized by these people from Europe. And they also showed me what they had to do to get these people off their back. The first thing they had to realize was that all of them were brothers; oppression made them brothers; exploitation made them brothers; degradation made them brothers; discrimination made them brothers; segregation made them brothers; humiliation made them brothers . . . The same conditions that prevailed in Algeria that forced the people, the noble people of Algeria, to resort eventually to the terrorist-type tactics that were necessary to get the monkey off their backs, those same conditions prevail today in America in every Negro community.[21]

In this speech, and in others that he made after this time, Malcolm X drew important parallels between the Algerian revolution and the African American freedom movement. As a result, he helped spread awareness of the Algerian struggle but simultaneously advocated for a global perspective on the situation and conditions of African Americans in America. In particular, his comparison of the Casbah in Algiers to Harlem in New York City was to become a familiar one, especially with the release of the film *The Battle of Algiers* in 1966.

Francee Covington, a student in political science at Harlem University in the late 1960s, penned an essay titled "Are the Revolutionary Techniques Employed in *The Battle of Algiers* Applicable to Harlem?" that appeared in the anthology *The Black Woman*, an important product of the Black Arts Movement edited by Toni Cade Bambara. Covington noted, "The Chinese Revolution, the Russian Revolution, and even the Kenyan Revolution labeled 'Mau Mau' have not been given the attention that the Algerian Revolution has. This is primarily because of the great extent to which the public has been made aware of this specific revolutionary instance through the writings of Fanon and the more graphic motion-picture illustration, *The Battle of Algiers*."[22] She also made the following point: "In the past few years the works of Frantz Fanon have become widely read and quoted by those involved in the 'Revolution' that has begun to take place in the communities of Black America. If *The Wretched of the Earth* is the 'handbook for the Black Revolution,' then *The Battle of Algiers* is its movie counterpart." [23] As the title of the piece makes clear, Covington's essay evaluated the relevancy of the strategies and techniques used in the film by the Algerian revolutionaries against the French colonizers to the situation of African Americans in urban communities throughout the United States.

After drawing parallels and then pointing out important differences, Covington concluded that "the idea of importing the techniques of revolution that were successful in one place may prove disastrous in another place,"[24] implying that it would be misleading to assume that because the Algerians were successful, the same approach could succeed in Harlem and urban Black America more broadly. Despite her rejection of the possibility of the relevancy of *The Battle of Algiers*, the fact of her meditation upon that possibility is in itself a testament to the degree to which the Algerian revolution became an important point of reference and to which there was an understood relationship between the African independence movements and the African American freedom movement.

Frantz Fanon, the Martiniquan psychiatrist and intellectual who joined the Algerian struggle for independence, also became a revolutionary point of reference and his writings helped cultivate a generation of Black liberation theorists and activists in America: "Fanon's ideas were unleashed at a moment truly coincidental with the phenomenal impact of the Black Power Movement in the United States which transformed the Civil Rights Movement into the Black Liberation Movement, subdued more moderate black organizations and leaders by transforming some into Black Power organizations and spawning new ones."[25] The publication of Fanon's *The Wretched of the Earth* in America in 1965 was hailed by Eldridge Cleaver, Minister of Information for the Black Panther Party, as "itself a historical event,"[26] and he referred to the book from then on as "the Black Bible."[27] The Black Panther Party was known for conducting teach-ins for

The Wretched of the Earth at chapter meetings throughout the country.[28] And, at moments, Stokely Carmichael, famous for popularizing the term Black Power and for acting as the honorary prime minister of the Black Panther Party, saluted Frantz Fanon as his own personal "patron saint."[29]

Although Fanon's writings may have first been popular only among a select few, by 1970, *The Wretched of the Earth* alone had sold 750,000 copies.[30] As anecdotal evidence of the significance of Fanon to the everyday African American Black Power proponent, when in conversation with Jimmy Breslin of the *Chicago Sun-Times*, and in what is now an oft-quoted testimonial by Dan Watts, the editor of *Liberator* magazine, Watts told Breslin that, "You're going along thinking all the brothers in these riots are old winos . . . Nothing could be further from the truth. These cats are ready to die for something. And they know why. They all read. Read a lot. Not one of them hasn't read the Bible." Breslin questioned, "The Bible?" And Watts responded, "Fanon . . . you'd better get this book. Every brother on a rooftop can quote Fanon."[31]

Fanon's place in the African American freedom struggle cannot be overstated, for although he did not write much himself about the African American situation, his words and ideas were adopted and adapted and generally were thought to be of great relevance to revolutionary struggles around the world. Through his writings, as well as through the popular film *The Battle of Algiers* and the ever-influential African American press, Algeria and Algerian leaders like Ben Bella were crystallized as symbols of revolutionary significance. They came to hold great meaning for many who were themselves engaged in a protracted struggle for freedom. However, Algeria's significance was not simply to be found in its being a symbol; Algeria also became a very real supporter of many African American radicals who had the opportunity to travel to Africa and visit the nation.

"THERE WAS A BATTLE IN ALGIERS . . . ": THE FIRST PAN-AFRICAN CULTURAL FESTIVAL OF 1969

There was a battle in Algiers in late July . . . The troops came together, African generals and footsoldiers in the war of words and politics that splashed against the calm waters of the Mediterranean Sea—in the First Pan-African Cultural Festival—from everywhere in greater numbers than ever before; from San Francisco to Senegal, from Dakar to the District of Columbia.[32]
—Nathan Hare, "Algiers 1969: A Report on the Pan-African Cultural Festival"

Upon hearing of the assassination of Martin Luther King, Jr., Eldridge Cleaver, the Minister of Information for the Black Panther Party, proclaimed,

That there is a holocaust coming I have no doubt at all. I have been talking to people around the country by telephone—people intimately involved in the black liberation struggle—and their reaction to Dr. King's murder has been unanimous: the war has begun. The violent phase of the black liberation struggle is here, and it will spread. From that shot, from that blood. America will be painted red. Dead bodies will litter the streets and the scenes will be reminiscent of the disgusting, terrifying, and nightmarish news reports coming out of Algeria during the height of the general violence right before the final breakdown of the French colonial regime.[33]

Cleaver's words capture the profound anger and disillusionment as well as new-found determination for liberation that was birthed by King's death. The days following the tragic event were met head-on by violent uprisings in cities throughout United States, including a series of incidents which resulted in Cleaver's apprehension by the police.

Although the First Pan-African Cultural Festival could not be said to have any direct relationship to King's death, the festival certainly occurred while the stench of King's death and all that it brought in its aftermath still lingered in the air. And Cleaver's words were prophetic both in his call to arms for African Americans (evident in the riots that occurred around the country in the ensuing days and weeks) and in his evocation of Algeria, for he would find himself in the very city of *The Battle of Algiers* only one year later. After clandestinely leaving the United States to evade being sent back to prison for a charge of parole violation, Cleaver headed to Cuba where he hoped to receive "backing to establish a base for Black Panther political and military action against the United States."[34] He soon found out that the Cuban authorities were not going to provide such support, and he was eventually transported to Algeria, where he initially received a more sympathetic welcome. Amidst the preparations for the festival, Algerian authorities offered Cleaver and his fellow Panther members official invitations to the forthcoming festivities. Cleaver did not formally announce the fact of his residence in Algeria until some time later, at the opening of the festival, when he held a press conference specifically for that purpose.

The idea for the Pan-African Cultural Festival originally came about in September 1967 at an Organization of African Unity (OAU) Council of Ministers meeting held in Kinshasa, Congo, where the decision was made that "there is an urgent need to undertake common measures that would assist in the popularizing, development and refinement of the various cultures obtaining in Africa."[35] As a result, the OAU passed a resolution to "sponsor an All-African Festival of African Drama, Folk Song, and Instrumental Music."[36] Not long afterward, the Algerian government, with aid from the OAU, went about publicizing the July-August 1969 gathering. Anticipation was quick to build, as the African press—as varied as *Fraternité Matin* of the Ivory Coast, *L'Effort Camerounais* of Cameroon, *Jeune Afrique*, a French language Pan-African magazine published out of Paris, *La Semaine* of Congo-Brazzaville, *El-Ayam* of Sudan—and many more periodicals throughout the continent, Europe, America, and the Caribbean announced the coming of and significance of the First Pan-African Cultural Festival.[37]

In America, the African American press played an important role in publicizing the festival. The Nation of Islam's *Muhammad Speaks* ran several articles in anticipation of the festival. One article in particular, titled "Algerian Festival to Spotlight Africa's Vast Cultural Heritage," reported, "Massive preparations are under way for the First Pan-African Cultural Festival . . . Black artists from

America and 15 African nations will attend and museums throughout the continent are sending works of art—some hundreds of years old—to the cultural festival, which promises to be the greatest event in the history of Africa, if not the entire world."[38] Alongside print publicity, travel agencies began organizing and advertising group tours of Africa, which centered around the Pan-African Festival in Algiers. The *Africa Tourist and Travel Agency*, based in New York City, printed brochures for a tour they organized specifically for the festival, which they called the "Organization for African Unity Cultural Festival Tour."[39] The tour, to be hosted by the "tour personality," renowned jazz pianist Ahmad Jamal, was to last a total of twenty-one days, from July 21 through August 11, 1969, and would include travel through Morocco and Algeria. It is likely that travel agencies similar to the *Africa Tourist and Travel Agency*—which most likely catered to a largely middle-class, African American clientele as well as social and political organizations whose focus was the African American and African freedom movement(s)—would have organized comparable group trips to the Pan-African Festival.

The eagerness of the participants and attendees of the festival mirrored the triumphant language of the newspaper and magazine announcements of the approaching event. Dave Burrell, a young jazz pianist who was invited to play at the festival, described he and his bandmates' preparations for their departure:

> Before we went, we got as African as we could get in New York . . . the dashikis had just come into vogue, and we were sort of very much in the vanguard of the movement in New York . . . [we had] to run around New York and get African material and make the dashikis and the different skull caps and to have the Black Power sign . . . I remember [fellow musician] Sunny Murray saying to me, "Hey, I joined the Panthers." I said, "What?" He said, "Yeah, don't tell anybody." He showed me a little membership card. I said "Oh, where did you do this?" He said, "I did it in Philadelphia."[40]

Sunny Murray most likely joined the Panthers knowing full well that Cleaver had surfaced in Algiers. The anticipation was felt among all who were planning to attend. Henri Lopes, at the time a budding writer and the Minister of Education for Congo-Brazzaville who was helping lead the country's delegation to the festival, was also quite eager, as he recounted:

> My arrival in Algeria was met with great emotion, in large part because our political consciousness . . . was cultivated as a result of our knowledge of the Algerian war for independence. And so, for us, it was the country that had obtained genuine independence, more so than most African countries, and to arrive there, it was like going to a Rome or a Mecca . . . Secondly, the artists, the creators in Africa, each one of us was isolated in our countries . . . And, the Festival gave us all the opportunity to meet one another, to exchange ideas with one another, and try to get to know one another . . . Thus, the Festival represented a tangible image of what could one day be a united Africa.[41]

The Festival opened on July 21, 1969, with a speech from Algerian president Houari Boumedienne, which began,

> Algeria is happy to welcome the First Pan-African Cultural Festival on behalf of our entire continent. The importance of this event and the joy and enthusiasm which it has aroused and is still arousing, the diversity and quality of the manifestations to which it will give rise, should not make us forget to what an extent this first Pan-African Cultural Festival is concerned, not only with our values and sensitivities, but also with our very existence as Africans and our common future. This Festival, far from being an occasion for general festivities which might momentarily distract us from our daily tasks and problems, should rather be related to them and make a direct connection to our vast effort of construction. It constitutes an intrinsic part of the struggle we are all pursuing in Africa—whether that of development, of the struggle against racialism, or of national liberation.[42]

El Moudjahid was quick to find peoples' first impressions of the Pan African Festival as they questioned people on the street and published the responses in an article titled "Le Coeur d'Alger Bat Au Rythme Du Festival" ("The Heart of Algiers Beats to the Rhythm of the Festival"). On the very same page in that issue of the newspaper, there appeared an article titled "Quand Le Jazz Se Veut Arme De Combat" ("When Jazz Becomes A Weapon of Combat"), which reprinted a poem by the African American poet Ted Joans called "Behind the Smile of Black Jazz." Every day over the period of two weeks, *El Moudjahid* covered, in great detail, the events and happenings in Algiers. There were daily symposia around issues of critical importance to the social, political, cultural, and economic development of Africa as well as musical, theatrical, and dance performances throughout the city.

One of the more popular attractions of the festival was the Afro-American Center, located in the heart of downtown Algiers, a space lent to Cleaver and the Black Panthers by the Algerians. It was part of the support that the Algerian government granted the Panthers, which also included a hilltop villa in which to reside, monthly salaries, and access to telecommunications, among other such amenities. Young Algerians flocked to the center in large numbers and out of great curiosity. The Black Panthers—including Emory Douglas, David Hilliard, and Kathleen and Eldridge Cleaver—staged informational lectures and discussions and handed out plenty of party material and memorabilia, including pamphlets, posters, and pins. With Algeria's support, the Black Panthers were able to publicize their platform, their ideology, and their perspective on the condition of African Americans to a global audience, including revolutionary movements from around the world that were sympathetic to their cause.

American expatriate William Klein was commissioned by the Algerian government to film the festival, but while in the country, he was also able to complete another documentary, titled *Eldridge Cleaver, Black Panther*. In the film, Cleaver expressed what a gathering such as the festival meant for he and the Black Panthers:

When I left the United States, I had no idea that I would end up in Algeria, but I think that I was very fortunate coming to Algeria at the time of the festival and to receive an invitation to participate in the festival, to have the opportunity to establish the Afro-American Center which we opened for the festival, which gave us an opportunity to make ourselves known to the other liberation movements who were brought together by the festival. The stage was set. People came here specifically to check each other out, to see what was going on, and to get some idea as to which movements they could relate to.[43]

At another moment in the film, for which Klein "organized a meeting between Black Panthers and African revolutionary movement leaders at a restaurant"[44] with the intention of them having a substantive exchange of ideas and opinions, a representative of the Zimbabwe African People's Union articulated rather remarkably the degree to which there was understood to be a common struggle between African Americans and Africans. The Zimbabwean declared, "We are following the struggle of our Afro-American brothers in the United States and I am sure they are also following our struggles. The people of Zimbabwe have taken up arms and we are facing a common enemy, and it is this common enemy which we must all crush. If our Afro-American brothers score success in the United States, that success is not only theirs, it is ours too."[45]

The First Pan-African Cultural Festival of 1969 was a moment that embodied both the hopes and desires of the many African and African American radicals gathered there but also demonstrated the great depths of an understood connection between their respective struggles. Host to an incredible array of artists and activists, including for example, Nina Simone, Stokely Carmichael and then-wife Miriam Makeba, Archie Shepp, Maya Angelou, and Ed Bullins, the festival was not without its own contradictions and tensions, but not at the complete expense of progress toward Pan-African unity, strengthened transnational solidarities, substantive exchanges of ideas, and practical political gains. Its impact reverberated out beyond the two weeks of festivities, as those that attended carried their experiences at the festival back with them to their various communities, just as their were a series of practical institutional developments in the domain of the arts and culture for the continent of Africa.[46]

"THE STRUGGLE WAS NOT YET OVER . . . ": THE LEGACY AND SIGNIFICANCE OF THE PAN-AFRICAN FESTIVAL

It is evident that since its staging in July 1969 in Algiers, the First Pan-African Cultural Festival has not yet, one the hand, been the subject of as much interest as it would normally merit, and on the other, been subjected to critical and profound analysis. Outside of publicity articles, brief informational articles, or violent oral responses, no sort of collective or individual position has been taken on the continental or international implications of the Festival or even the problems that were debated there.[47]

—*Souffles: Revue Culturelle Arabe Du Maghreb*

The above excerpt from an editorial in the Moroccan-published journal *Souffles*, which appeared at the beginning of 1970, quite strikingly captured the critical reception of the Pan-African Festival, or lack thereof. In the immediate aftermath of the festival, one could not find a great deal of evidence of meaningful reflection on the festival's significance, just as evidence of its significance cannot be easily found in the secondary historical literature now. And yet it is clear that the First Pan-African Cultural Festival held in Algiers, Algeria, in 1969 represents a watershed moment in the history of linkages between the African American and African freedom movements. The festival gave voice to these important ideas and provided the context for a broad range of African and African diasporic intellectuals, artists, activists, and students come into conversation with one another, in some cases literally, and in others, symbolically. Reflections offered by attendees of the festival provide insight into the unique impact that it had on all who were present.

Dennis Brutus, the South African poet, was in London, England, when he first heard of the Festival. He wrote the Algerian government to notify them of his interest in the event and was subsequently invited as a delegate of South Africa. When asked to reflect on the festival, despite mentioning the elation the moment evoked, Brutus was struck by the degree to which the Algeria he had come to know through hearing of the courageous struggle for independence was very much in crisis even only seven years after its triumph:

> We had this great festival. But, I had a sense that the resistance movement in Algeria, which had been very important for the whole continent—it inspired people from all over the continent—I had a sense that in the society, already there was beginning to develop a division between your middle-class, really affluent Algerians, and the people in the Casbah . . .
>
> When I went into the Casbah, you know, I had that sense of on the one hand you have an elite and on the other hand, people are struggling. So, although Algeria was important for South Africa—many South Africans in the resistance trained with the Algerian army—one had the sense that while there had been a struggle, the struggle was not yet over. And, in fact, the French were returning because their need for Algerian oil . . . Even while they were hating the Algerians, they still wanted the Algerian oil and this conflict developed as a division in this society.[48]

Michele Russell, an African American woman from Washington, D.C., was struck by the apparent failures of the Algerian revolution and the failure of the nation to live up to its image as the country that expelled the colonizer by force, much in the same way Brutus had been disappointed. However, she left the festival while it was still at its "height," allowing her to carry with her a sense of admiration for her ancestral homeland, just as it seems her experience there was a rite of passage of sorts, as it was her first trip to Africa:

> We had seen the film "Battle of Algiers" in the States. Now, wandering the city, each street came upon us with the shock of a double exposure. Neon signs became the flames of bombed cafes. Women in veils became saboteurs. Taxi drivers, the incarnation

of dedicated cadres careening around corners to unknown rendezvous . . . Now, seven years after victory, the liberation struggle has just begun. It goes on in their faces. In almost imperceptible hesitation when they are addressed in Arabic and respond in French. Now, seven years later. The resistance . . .

I left Algeria at the height of the festival. I left her at the point where I would have the most to come back to. And I felt I would come back, if not this land, then to some other part of the Continent that was helping me to return to myself as well. I left for home. I left at the moment I knew that wherever I was I would be, forever, home.[49]

For Barbara Chase-Riboud, the world-renowned African American writer and sculptor, her attendance at the festival helped bring about an artistic breakthrough, for although she was not exhibiting any of her own work there, she had something of an epiphany while amidst the festivities. She explains the very specific developments in her artwork that grew directly out of her experiences in Algiers:

[The festival] was wonderful. It was stupendous. And, at that time, I made a big change in my own work, in my own sculpture . . . I had been, sort of, in a kind of surrealistic mood, with elongated figures and so on. But, they had been getting more and more abstract and I couldn't figure out a way to get them off the legs and off the pedestal. And suddenly, like a light bulb over my head—and it was in Algiers—I realized that I all I had to do was to make them into objects that could be moved by something. And, that something turned out to be silk. That's when I began to do those Malcolm X [sculptures].[50]

Chase-Riboud's series of sculptures in memory of Malcolm X were indelibly shaped by the addition of silk to her artistic repertoire—an idea that came to her in the midst of the fervor of the festival.

Henri Lopes, the prize-winning Congolese novelist and present-day Congolese ambassador to France, also attaches artistic significance to the Pan-African Festival. Prior to the festival, he had only published a couple of poems, but not long after, he began publishing what would become prize-winning novels. Here, he describes the festival's personal and artistic significance:

The importance that the festival had for me, personally, was that after my presentation [at the symposium], a lot of people began thinking of me as a writer. Oh, I had barely written. I had published a few poems in the journal *Présence Africaine*, but no novels. And, I had the impression of being—how does one say it—an impostor with that title [of "writer"]. And, when I returned home, that's when I began writing my first work, my collection of short stories, *Tribaliques*. That was after Algiers. I did two things. I quit smoking and I wrote my first book.[51]

For another writer, Haki Madhubuti, or Don Lee as he was known at the time, the festival held very different meaning. A poet and essayist of the Black Arts Movement, Madhubuti was struck by the sights and sounds of the historic event but it only made him more aware of the work that had yet to be done in Black communities in the United States. He explains,

The level of conversation, the level of dialogue, and the level of interaction was very political, highly charged, and most certainly, our conversations were struggle-driven . . . But, this is a very difficult time because COINTELPRO is coming on strong, the red squads in Chicago, and the Panthers had been driven all out of the country, I mean, Algiers, as well as Cuba and so forth . . . I knew Kwame Ture and Eldridge Cleaver and people like that. So, to go and see them in other spaces and being received royally was very good. But, also, it just spoke to me loudly that the work that we needed to do here [in the U.S.] was just not being done, because if we were doing the proper work here they wouldn't have to leave, they would have protection. And so . . . to my heart, it said, "Go back to Chicago, go back to the States, and just move the work up another volume." And, for me, the volume was, essentially, developing independent Black institutions.[52]

Lastly, for Hoyt Fuller, who had traveled widely on the African continent, the First Pan-African Cultural Festival of 1969 represented the coming into fruition of meaningful transnational solidarities between Africans and African Americans and the advent of a new moment, full of new possibilities and new struggles: "Pan-Africanism is an idea whose time has come. That fact is, for me, the central meaning of Algiers 1969 . . . That was what it meant to me to have Africans from all over the world assembling on the soil of Frantz Fanon's adopted country to consider the direction the peoples of the African continent should take . . . Algiers was the Black World coming of Age."[53] It seems that for Fuller the "coming of age" of the "Black World" meant the collaboration of African peoples across borders of nation and language, just as it meant coming to terms with the complexity inherent in such transnational solidarities. Long had African and African diasporic peoples understood the political and cultural connections they shared, and yet it meant something different to confront one another and learn more intimately the details of the struggles of dispersed revolutionary movements. And therein lies a deeply significant aspect of the presence and official recognition of African American activists in Algeria. The possibility (and reality) of sustained dialogue was transformatory—Barbara Easley, one of the Black Panthers that lived in Algeria, has recounted, "I met women from other liberation movements, and that's where my knowledge of the world became more focused, because when people say, 'I'm from so and so,' you start looking at the world map, whether it's Asia, Africa, South America, and then you start listening to other people's historical battles, and then you realize that you're not the only group of little black people, this select group of African Americans in America that are fighting for that freedom."[54] And, in a 1971 *New York Times Magazine* article, journalist Sanche de Gramont described the status in Algeria of Eldridge Cleaver and his fellow Black Panthers in exile, writing, "in Algiers, the Panthers are respected as one of approximately a dozen liberation movements accredited by the Algerian Government and provided with assistance and support in their task of overthrowing the governments in power in their respective countries . . . They plan to maintain close contact with other liberation movements."[55] He quoted Cleaver as having explained, "This . . . is the first time in the struggle of the black people in America that they have established representation abroad."[56]

HIJACKING FREEDOM, HIJACKING HOPE:
THE "FLEURY FOUR" AND THE END OF AN ERA

Writing from their prison cells in Fleury-Mérogis, France, in 1978, George Brown, Joyce Tillerson, and Melvin and Jean McNair hoped to successfully make a case for their release. As their lawyers explained, their four narratives, published collectively as *Nous, Noirs Americains évadés Du Ghetto* [*We, Black American Escapees From the Ghetto*], were "essentially their defense"[57] against both extradition to the United States and to a prison sentence in France. George, Joyce, Melvin, and Jean did not deny having committed the hijacking of a commercial airplane from Detroit to Algeria in July of 1972 (along with successfully procuring a million-dollar ransom from Delta Air Lines), but they insisted that their actions represented a desperate attempt to escape the oppressive American racism under which they had too long suffered. It was no small accident, however, that they chose Algeria as their destination. They knew full well of the existence of the International Section of the Black Panther Party there as well as the presence of numerous Black Panther members, including Minister of Information Eldridge Cleaver, Communications Secretary Kathleen Cleaver, Field Marshal Don Cox, and, at various other times, fellow Party members Larry Mack, Sekou Odinga, James and Gwen Patterson, and Barbara Easley, among others.[58] In fact, the hijackers intended for the million-dollar ransom to be of aid to the community of exiled Black Panthers that had been living in Algeria since as early as 1969.

In what was the seventeenth American hijacking of 1972, Melvin McNair, George Brown, and George Wright commanded the pilots to land the Delta Air Lines plane in Miami, call in for a million-dollar ransom to be delivered by federal agents wearing only swimsuits (so that they could be sure the agents were not carrying any concealed weapons), and unload the passengers, while their accomplices, Joyce Tillerson and Jean McNair, waited anxiously in their seats with their three children.[59] The hijackers demanded that Delta also provide an international navigator, and since none was available in Miami, the plane took off for Boston, where their last demand was met. Commanding the international navigator to take off without instructions about where to head, it was only after some time in the air that the hijackers informed the pilot that he was to take them to Algeria.[60]

Landing at the airport in Algiers, the capital city of Algeria, the five hijackers and their three children had made it safely to their destination. The United States, responding to what was the most expensive hijacking in American history up until that time,[61] sought the return of the ransom money and "the extradition of the hijackers to the United States, or prosecution of them on air piracy charges in Algeria."[62] But the hijackers were aware that only several months prior, a male and female couple had hijacked a plane from the United States to Algeria (and temporarily made away with a $500,000 ransom from Western Airlines). The couple received conditional asylum in Algeria and ample coverage in the American media.

However, Algeria was becoming less and less patient with such incidents. Having attempted to strengthen economic ties to the United States since the late

1960s, Algeria understood that its ability to do so depended on its standing in the eyes of the U.S. government. An August 19, 1972, a *New York Times* article reported, "[Black] panther officers [in the U.S.] say the government has been putting pressure on Algeria by threatening to cancel a proposed $1-billion plan to import Algerian natural gas to this country. The State Department denied this."[63] The evidence suggests that the Panthers were, in fact, not far from the truth. One memorandum of a conversation between the U.S. Secretary of State under Nixon, William Rogers, and the Algerian Minister of Foreign Affairs, Abdelaziz Bouteflika, held only a couple months after the Detroit-to-Algiers hijacking, contained the following summary of remarks from Secretary of State Rogers: "USG [U.S. Government] appreciates return of planes, crew and money [in the Western and Delta cases], and hopes that if GOA [Government of Algeria] cannot extradite it can at least prosecute the hijackers . . . Algeria has a moderate and reasonable government and is attractive to American investment, but hijacking incidents have damaged Algeria's image."[64] A portion of Bouteflika's remarks were summarized as follows: "GOA [Government of Algeria] would jealously guard its own national independence and would support national liberation movements of peoples deprived of their right by colonial powers. This includes support of liberation movements in Portuguese Africa, South Africa and Rhodesia . . . Bouteflika said GOA is interest in developing its relations with U.S. and added that Black Panthers do not make any effective contribution to Algeria from revolutionary, ideological or moral standpoint."[65] So, as fate would have it, the arrival of the newest contingent of exiled Panthers—George Brown, George Wright, Joyce Tillerson, and Melvin and Jean McNair—although it represented an extraordinary feat, also spelled the beginning of the end of a symbolic and strategic solidarity between African American freedom movements and the country of Algeria. Algeria's economic dependence on American investment and the Algerian administration's pursuit of better diplomatic relations between the two nations essentially required that Algeria cease supporting the exiled Black Panthers. Algeria had not had formal diplomatic relations with the United States since 1967, when the two countries broke ties. But by December 1973, when Secretary of State Kissinger met with Algerian president Boumedienne, "senior American officials said that they expected a rapid intensification of contacts between the two governments and the probable exchange of ambassadors within a couple of months."[66]

Nearly all the exiled Black Panthers left Algeria for Europe around the same time, citing deteriorating relations with the government. While some eventually negotiated deals with the American government to return to the United States, some made lives for themselves in other countries, while still others were caught by the authorities and had to face the possibility of prison time. Black Power activists in the United States were increasingly facing similar fates as local, state, and federal authorities aggressively worked to stamp out movements seen as threats to order and national security. This was the case for the last group of Panther hijackers to arrive in Algiers. On May 26, 1976, four of the five hijackers were arrested in Paris by the French police who were investigating an American request for their extradition.[67] Concerned parties in France quickly formed a committee to publicize the case and generate support for the defendants. The committee was able to get *Nous, Noirs Américains évadés Du Ghetto* published on behalf of the four hijackers. Separated into four parts, each containing the autobiographical narrative of one of the defendants, the book was meant as a testimony to the ugliness and tyranny of racism in America.

Although not exactly acquitted of their crimes, the hijackers escaped extradition to the United States and were sentenced to a few years in French prison. One of their defense attorneys, Louis Labadie, proclaimed, "This is a success . . . It is a condemnation of American racism."[68] The case presented in court "asserted . . . that the hijacking was a 'political act' motivated by racial oppression," and the defense team "produced witnesses to describe instances of police brutality, job discrimination, school segregation, poverty, hunger, and 'armed terrorism' in the United States."[69] And since the defendants had already served about two years of jail time in France as they awaited the completion of the trial, "the women were expected to be released within days and the men in about six months."[70] It was noted that "before a United States court, they would have faced minimum jail terms of 20 years."[71] The hijackers—often called the "Fleury Four" after the well-known prison in the town of the same name—had put American racism on trial and had won. And yet, it was a bittersweet victory. Even as they had hijacked their freedom from American oppression, hope had been hijacked from the possibility for a renewed solidarity between Algeria and African American activists, at least on the same scale as had once been.

In the decade or so following its independence from France, Algeria became a critical node in the constellation of transnational solidarities being forged among revolutionary movements around the world. At the height of the Civil Rights and Black Power eras, just as Algeria looked to Black America as "that part of the Third World situated in the belly of the beast,"[72] so, too, did much of Black America look to Algeria as "the country that fought the enslaver and won."[73] Key figures and factions in the African American freedom movement, often otherwise thought to represent different ideological positions, from King to Malcolm, from the Nation of Islam to the Black Panther Party, as only a few examples among many, took up Algeria and Algeria's revolution in an attempt to imagine and bring into being transnational collaboration. In each case, the implicated parties were looking to leverage the position and resources gained vis-à-vis the other, but that was part and parcel of the work of imagining and creating these transnational connections and also of understanding how they were formed, strengthened, and eventually disintegrated.

Although American Cold War and oil geopolitics ultimately disrupted the transnational linkages between African America and Algeria, the legacy of these connections endure in unique ways in the culture and politics of the twenty-first century. One only has to look to Algerian national life for proof of this fact. At the time of negotiations with American officials for large oil contracts in the mid-1970s (which led, in part, to the demise of the Black Panther community in Algeria), then-Algerian Foreign Affairs Minister Abdelaziz Bouteflika played an important diplomatic role. Three-and-a-half decades later, Bouteflika became president, and oil politics continue to be a defining issue in Algerian national and international affairs. And, although the era of substantive transnational solidarities between African America and Algeria remains a distant memory, few attendees of the festival could probably have imagined that the same downtown streets of Algiers where jazz saxophonist Archie Shepp once collaborated with traditional Algerian musicians are now brimming with Algerian youth who define their cultural lives through the African American cultural form known as hip-hop, in some cases using it to speak out powerfully against the pervasive injustice that plagues their country and the world at large.[74] Looking to seize on this kind of cultural vibrancy and to celebrate the fortieth anniversary of the festival, Algeria is in the midst of planning for the 2nd Pan-African Cultural Festival, scheduled to take place in July 2009.[75] What will be of great interest to historians is how this new festival attempts to make sense of the legacy of the 1969 gathering in Algiers, and of the dynamic fusion of culture and politics that defined that bygone era.

NOTES

1. Richard B. Moore, "Africa Conscious Harlem," *Americans from Africa: Old Memories, New Moods*, ed. Peter I. Rose (Chicago: Atherton, 1970), 399.
2. Michele Russell, "Algerian Journey," *Freedomways* (4th quarter, 1969), 357.
3. Hoyt W. Fuller, "First Pan-African Cultural Festival: Algiers Journal," *Negro Digest* (October 1969), 74.
4. Boutkhil Alla, "Festival Culturel Panafricain," *Jeune Afrique* 444 (July 7–13, 1969): 58.
5. Ted Joans, "The Pan African Pow Wow," *Journal of Black Poetry* 1, no. 13 (Winter/Spring 1970): 4.
6. James Meriwether, *Proudly We Can Be Africans: Black Americans and Africa, 1935–1961* (Chapel Hill, NC: University of North Carolina Press, 2002), 242.
7. Meriwether, *Proudly We Can Be Africans*, 242.
8. Jonathan Scott Holloway, "What America Means to Me? Defining Black Life Through the Motherland," *Reviews in American History* 31 (2003): 93–100.
9. James Meriwether, "African Americans and the Mau Mau Rebellion: Militancy, Violence, and the Struggle for Freedom," *Journal of American Ethnic History* 17, no. 4 (Summer 1998): 63–86.
10. Roland Wolseley, *The Black Press, U.S.A.* (Ames: Iowa State University Press, 1990), 90.
11. Meriwether, *Proudly We Can Be Africans*, 8.
12. Thomas Ronan, "Ben Bella Links Two 'Injustices,'" *New York Times*, October 14, 1962, 20.
13. Karl Evanzz, *The Judas Factor: The Plot to Kill Malcolm X* (New York: Thunder's Mouth Press, 1992), 128.
14. Ronan, "Ben Bella Links Two 'Injustices,'" 20.
15. Ibid.
16. Ibid.
17. Ibid.
18. Martin Luther King, Jr., "My Talk With Ben Bella," *New York Amsterdam News*, October 27, 1962, 12.
19. King, Jr., "My Talk With Ben Bella," 12.

20. Evanzz, *The Judas Factor*, 128.

21. George Breitman, ed., *Malcolm X Speaks* (New York: Grove Press, 1965), 66.

22. Francee Covington, "Are the Revolutionary Techniques Employed in *The Battle of Algiers* Applicable to Harlem?" in *The Black Woman: An Anthology*, ed. Toni Cade Bambara (New York: Penguin, 1970), 245.

23. Covington, "Revolutionary Techniques," 244.

24. Ibid., 251.

25. Ronald Walters, "The Impact of Frantz Fanon on the Black Liberation Movement in the United States," in *Mémorial International Frantz Fanon: Interventions et Communications Prononcées à l'Occasion du Mémorial International Frantz Fanon de Fort-de-France (Martinique) du 31 mars–3 avril 1982* (Paris and Dakar: Présence Africaine, 1984), 210.

26. Walters, "The Impact of Frantz Fanon," 210.

27. Robert Scheer, ed., *Eldridge Cleaver: Post-Prison Writings and Speeches* (New York: Ramparts/Random House, 1969), 18.

28. *Black Panther Party: Part 1, Hearings Before The Committee on Internal Security, House of Representatives, Ninety-First Congress, Second Session* (Washington, DC: U.S. Government Printing Office, 1970–71), 2807.

29. Stokely Carmichael, *Black Power: An Address by Stokely Carmichael at the Dialectics of Liberation Congress, London, England, 1967.* LP record. Liberation, 1967.

30. William Van Deburg, *New Day In Babylon: The Black Power Movement and American Culture, 1965–1975* (Chicago: University of Chicago Press, 1992), 61.

31. Aristide Zolberg and Vera Zolberg, "The Americanization of Frantz Fanon," in *Americans From Africa: Old Memories, New Moods*, ed. Peter I. Rose (Chicago: Atherton, 1970), 198.

32. Nathan Hare, "Algiers 1969: A Report on the Pan-African Cultural Festival," *Black Scholar* 1, no. 1 (November 1969): 3.

33. Robert Scheer, *Eldridge Cleaver*, 74–75.

34. Kathleen Cleaver, "Back to Africa: The Evolution of the International Section of the Black Panther Party (1969–1972)," in *The Black Panther Party Reconsidered*, ed. Charles E. Jones (Baltimore: Black Classic Press, 1998), 217.

35. *All-African Cultural Festival Handbook* (Algiers, Algeria: Organization of African Unity General Secretariat, 1969), 1.

36. *All-African Cultural Festival Handbook, 1.*

37. "La Presse Africaine et Le Festival," *Alger 1969: Bulletin D'Information du 1er Festival Culturel Panafricain* 4 (May 15, 1969): 40–43.

38. Kenneth C. Landry, "Algerian Festival to Spotlight Africa's Vast Cultural Heritage," *Muhammad Speaks*, July 13, 1969, 31.

39. Copy of original in author's possession.

40. Dave Burrell, interview by James G. Spady, date unknown. Unpublished manuscript in possession of James G. Spady.

41. Henri Lopes, interview by author, Paris, France, June 13, 2003. Unpublished manuscript in author's possession. Translated from French by the author.

42. Houari Boumedienne, "The Algerian Festival: Inaugural Address," in *New African Literature and the Arts: Vol. 3*, ed. Joseph Okpaku (New York: The Third Press, 1973), 1.

43. *Eldridge Cleaver, Black Panther*, 35mm film, directed and filmed by William Klein (Algiers, Algeria: ONCIC (Algerian National Film Board), 1970).

44. *Eldridge Cleaver, Black Panther.*

45. Ibid.

46. *Manifeste Culturel Panafricain: Premier Festival Culturel Panafricain, Alger, 21 Juillet 1 Août 1969,* microfilm (Addis Abeba: Service de Presse & d'Information du Secrétariat Général de l'O.U.A., 1969).

47. "Le Festival Culturel Panafricain D'Alger 1969," *Souffles* (4th trimester 1969/January–February 1970).

48. Dennis Brutus, interview by author, Philadelphia, PA, February 14, 2004. Unpublished manuscript in author's possession.

49. Russell, "Algerian Journey," 355–64.

50. Barbara Chase-Riboud, interview by James G. Spady, Philadelphia, November 14, 2003. Unpublished manuscript in possession of James G. Spady.

51. Lopes, interview by author.

52. Haki Madhubuti, interview by author, Philadelphia, February 14, 2004. Unpublished manuscript in author's possession.

53. Hoyt Fuller, *Journey to Africa* (Chicago: Third World Press, 1971), 94–95.

54. Barbara Easley, quoted in Robin J. Hayes, "'A Free Black Mind is a Concealed Weapon': Institutions and Social Movements in the African Diaspora," *Souls* 9, no. 3 (July 2007): 231.

55. Sanche de Gramont, "Our Other Man In Algiers," *New York Times Magazine*, November 1, 1970, 30.

56. de Gramont, "Our Other Man In Algiers," 30.

57. Melvin McNair, Joyce Tillerson, George Brown, and Jean McNair, *Nous, Noirs Américains évadés Du Ghetto* (Paris: éditions Du Seuil, 1978), 314.

58. Cleaver, "Back to Africa"; McNair et al., *Nous, Noirs Américains.*

59. McNair et al., *Nous, Noirs Américains*; Richard Witkin, "3 Hijack Jet, Collect $1-Million and Fly to Algeria," *New York Times*, August 1, 1972, 1, 69; Robert Lindsey, "Once More Into The Breach . . . " *New York Times*, August 6, 1972, E2; "No-Profit Hijack," *New York Times*, August 3, 1972, 32.

60. McNair et al., *Nous, Noirs Américains.*

61. "The ransom impounded in Algiers was the largest ever paid to hijackers in this country," reported Robert Lindsey, "Algerians Seize 1-Million Ransom," *New York Times*, August 2, 1972, 1.

62. Lindsey, "Algerians Seize 1-Million Ransom," 1.

63. "Natural Gas Deal Strengthens U.S. Economic Ties to Algeria," *New York Times*, July 26, 1969, 2.

64. Memorandum of conversation between Algerian Minister of Foreign Affairs Bouteflika and Secretary of State Rogers. Telegram 188030 From the Department of State to the Mission to the United Nations and the Interests Section in Algeria, October 14, 1972, 1933Z. National Archives, RG 59, Central Files 1970–73, POL 7 ALG.

65. Memorandum of conversation between Algerian Minister of Foreign Affairs Bouteflika and Secretary of State Rogers.

66. Bernard Gwertzman, "Kissinger Arrives in Cairo After 2-Hour Algeria Talk," *New York Times*, December 14, 1973, 14.

67. McNair et al., *Nous, Noirs Américains évadés Du Ghetto*, 22.

68. "France Sentences 4 Americans Convicted in Hijacking of Airliner," *New York Times*, November 25, 1978, 8.

69. "France Sentences 4 Americans," 8.

70. Ibid.

71. Ibid.

72. Larry Neal, "The Black Writer's Role," *Liberator* 6, no. 6 (June 1966): 8.

73. Joans, "The Pan-African Pow-Wow," 4.
74. See James G. Spady, H. Samy Alim, and Samir Meghelli, *Tha Global Cipha: Hip Hop Culture and Consciousness* (Philadelphia: Black History Museum Press, 2006).
75. As of the time of submission of this chapter for publication, the 2nd Pan-African Festival (July 2009) had not yet occurred.

CHAPTER 7

LET US BE MOORS

RACE, ISLAM, AND "CONNECTED HISTORIES"

HISHAAM D. AIDI

"¡SEAMOS MOROS!" WROTE THE CUBAN POET AND NATIONALIST JOSE MARTI IN 1893, in support of the Berber uprising against Spanish rule in northern Morocco. "Let us be Moors . . . the revolt in the Rif is not an isolated incident, but an outbreak of the change and realignment that have entered the world. Let us be Moors . . . we [Cubans] who will probably die by the hand of Spain."[1] Writing at a time when the scramble for Africa and Asia was at full throttle, Marti was accentuating connections between those great power forays and Spanish depredations in Cuba, even as the rebellion of 1895 germinated on his island.

Throughout the past century, particularly during the Cold War Latin American leaders from Cuba's Fidel Castro to Argentina's Juan Perón would express support for Arab political causes and call for Arab-Latin solidarity in the face of imperial domination, often highlighting cultural links to the Arab world through Moorish Spain. Castro, in particular, made a philo-Arab Pan-Africanism central to his regime's ideology and policy initiatives. In his famous 1959 speech on race, the *jefe maximo* underlined Cuba's African and Moorish origins: "We all have lighter or darker skin. Lighter skin implies descent from Spaniards who themselves were colonized by the Moors that came from Africa. Those who are more or less dark-skinned came directly from Africa. Moreover, nobody can consider himself as being of pure, much less superior, race."[2]

With the launching of the "war on terror," and particularly with the invasion of Iraq, political leaders and activists in Latin America have been warning of a new imperial age and again declaring solidarity with the Arab world. Some refer rather

This chapter originally appeared in *Middle East Report*, Issue 229, Winter 2003.

quixotically to a Moorish past. Linking the war on Iraq to Plan Colombia and to the Bush administration's alleged support for a coup against him, the erratic Venezuelan strongman Hugo Chavez has repeatedly urged his countrymen to "return to their Arab roots" and has attempted to mobilize the country's *mestizo* and black majority against white supremacy. "They call me the monkey or Black," Chavez says of his domestic and international opponents. "They can't stand that someone like me was elected."[3]

In less contentious terms, Brazil's left-leaning President Lula da Silva visited the Middle East in early December 2003 to seek "more objective" relations with the Arab world, to call for an "independent, democratic Palestinian state," and to launch a common market with the Arab world as an alternative to the North American market (particularly with many in Arab countries boycotting American products).[4] Brazil's largest trade union federation strongly denounced post-September 11 U.S. intervention in Colombia, Venezuela, and the Middle East, praising the protest movements that have appeared against U.S. and Israeli "militarism" and calling on Brazilian workers to join in the struggle "against Sharon's Nazi-Zionist aggression against the Palestinian people" and in support of the intifada.[5]

THE OTHER SEPTEMBER 11 EFFECT

In the age of the "war on terror," such expressions from the Western world of affinity with the Arab world are not confined to statements of political solidarity. In Latin America, Europe, and the United States, for example, there has been a sharp increase in conversion to Islam. At the first world congress of Spanish-speaking Muslims held in Seville in April 2003, the scholar Mansur Escudero, citing "globalization," said that there were 10 to 12 million Spanish speakers among the world's 1.2 billion Muslims.[6] In the United States, researchers note that usually 25,000 people a year become Muslim but, by several accounts, that number has quadrupled since September 11.[7] In Europe, an Islamic center in Holland reported a tenfold increase and the New Muslims Project in England reported a "steady stream" of new converts.[8] Several analysts have noted that in the United Kingdom, many converts are coming from middle-class and professional backgrounds, not simply through the prison system or ghetto mosques, as is commonly believed.[9] The Muslim population in Spain is also growing due to conversion as well as immigration and intermarriage.[10]

Different explanations have been advanced to account for this intriguing phenomenon, known as "the other September 11 effect," the primary effects being anti-Muslim and anti-immigrant backlash and infringements upon civil liberties. Commenting on how the accused "dirty bomber" Jose Padilla and the shoe bomber Richard Reid converted to Islam, French scholar Olivier Roy observes, "Twenty years ago such individuals would have joined radical leftist movements, which have now disappeared or become 'bourgeois' . . . Now only two Western movements of radical protest claim to be 'internationalist': the anti-globalization movement and radical Islamists. To convert to Islam today is a way for a European rebel to find a cause; it has little to do with theology."[11] This portrayal of Islam as an outlet for the West's political malcontents ignores the powerful allure of certain aspects of Islamic theology and begs the question of why, for at least a century, even when communism was still in vogue, minorities in the West have seen Islam as a particularly attractive alternative. Roy's formulation also neglects the critical elements of racism and racialization. At least since Malcolm X, internationalist Islam has been seen as a response to Western racism and imperialism.

Though Westerners of different social and ethnic backgrounds are gravitating toward Islam, it is mostly the ethnically marginalized of the West historically, mostly black but, nowadays, also Latino, Native American, Arab, and South Asian minorities who, often attracted by the purported universalism and color-blindness of Islamic history and theology, are asserting membership in a transnational *umma* and thereby challenging or "exiting" the white West. Even for white converts like John Walker Lindh, becoming Muslim involves a process of racialization renouncing their whiteness because while the West stands for racism and white supremacy on a global scale, Islam is seen to represent tolerance and anti-imperialism. This process of racialization is also occurring in diasporic Muslim communities in the West, which are growing increasingly race-conscious and "black" as anti-Muslim racism increases. To cope, Muslims in the diaspora are absorbing lessons from the African American freedom movement, including from strains of African American Islam.

Over the past two years, Islam has provided an anti-imperial idiom and imaginary community of belonging for many subordinate groups in the West, as Islamic culture and art stream into the West through minority and diaspora communities and, often in fusion with African American art forms, slowly seep into the cultural mainstream. Subsequently, many of the cultural and protest movements—antiglobalization, anti-imperialist, antiracist—in the West today have Islamic or African-American undercurrents. At a time of military conflict and extreme ideological polarization between the West and the Muslim world, Islamic culture is permeating political and cultural currents, remaking identities and creating cultural linkages between Westerners and the Muslim world.

LATINO BACK CHANNELS

Recent journalistic accounts have noted the growing rate of conversion to Islam in the southern Mexican state of Chiapas and the often violent clashes between Christian and Spanish Muslim missionaries proselytizing among the indigenous Mayan community. The Muslim campaign in Chiapas is led by a Spaniard from Granada, Aureliano Perez, member of an international Sufi order called al-Murabitun, though he is contending with a rival missionary, Omar Weston, the Nation of Islam's local representative. Particularly interesting about the several hundred Mayan Muslims is the view of some of the converts that although some of the missionaries are Spanish like the conquistadors, their embrace of Islam is a historic remedy for the Spanish conquest and the consequent oppression. "Five hundred years ago, they came to destroy us," said Anastasio Gomez Gomez, age twenty-one, who now goes by Ibrahim. "Five hundred years later, other Spaniards came to return a knowledge that was taken away from us."[12]

The view of 1492 as a tragic date signaling the end of a glorious era and the related idea that conversion to Islam entails a reclaiming of that past is common among the Latino Muslim community in the United States. That community, estimated in 2000 at 30,000 to 40,000 members, has grown in the past two years, with Latino Muslim centers and *da'wa* (proselytizing) organizations in New York, Los Angeles, Miami, Fresno, and Houston.[13] The banner hanging at the Alianza Islamica center in the South Bronx celebrates the African and Islamic roots of Latin America: against a red, white, and blue backdrop stands a sword-wielding Moor, flanked by a Taino Indian and a black African. The Spanish conquistador is conspicuously absent. Imam (Omar Abduraheem) Ocasio of the Alianza Islamica speaks passionately about the continuity between Moorish Spain and Latin America: "Most of the people who came to Latin America and the Spanish Caribbean were from southern Spain, Andalusia; they were Moriscos, Moors forcefully converted to Christianity. The leaders, army generals, *curas* [priests] were white men from northern Spain . . . *sangre azul*, as they were called. The southerners, who did the menial jobs . . . servants, artisans, foot soldiers . . . were of mixed Arab and African descent. They were stripped of their religion, culture, brought to the so-called New World where they were enslaved with African slaves . . . But the Moriscos never lost their culture . . . we are the cultural descendants of the Moors."[14] The Puerto Rican imam writes, "Islamically inspired values were conveyed ever so subtly in the Trojan horse of Spanish heritage throughout the centuries and, after 500 years, Latinos were now ready to return."[15]

In the past two years, Islam and the Arab-Muslim world seem to have entered even more poignantly into the Latin American imagination, gaining a presence in political discourse and strongly influencing Hispanic popular culture. This Arab cultural invasion of Latin America, which has reverberated in mainstream American culture, is often attributed to the Brazilian telenovela El Clon and Lebanese-Colombian pop icon Shakira.

El Clon, the highest-rated soap opera ever shown on Telemundo, a U.S. Spanish-language channel, reportedly reaches 2.8 million Hispanic households in the United States as well as 85 million people in Brazil and tens of millions across Latin America. The series, which began broadcasting shortly after September 11, 2001, tells the story of Jade, a young Brazilian Muslim who returns to her mother's homeland of Morocco after her mother's death in Brazil. There she falls in love and settles down with Lucas, a Christian Brazilian, and adapts to life in an extended family setting in the old city of Fez. Filmed in Rio de Janeiro and Fez, the telenovela offers a profusion of Orientalist imagery from veiled belly dancers swaying seductively behind ornate latticework to dazzling shots of Marrakesh and Fez spliced with footage of scantily clad women on Rio's beaches and, of course, incessant supplications of "Ay, por favor, Allah!" from Jade's neighbors in the medina. The Moroccan ambassador to Brazil, in a letter to a Sao Paolo newspaper, criticized the series for its egregious "cultural errors," "gross falsification," and "mediocre images" promoting stereotypes of Muslim women as submissive and men as polygamists leading lives of "luxury and indolence."

Despite the kitsch, *El Clon* has triggered what *Latin Trade* called "Mideast fever" across Latin America. Belly dancing and "Middle Eastern-style jewelry" became "the rage in Rio and Sao Paolo," Brazilians began throwing "A Thousand and One Nights" parties, "Talk to a Sheikh" chat rooms cropped up online, and two new agencies opened up to offer package tours to North Africa. (In his letter, the Moroccan ambassador acknowledged that Brazilian tourism to Morocco had increased by 300 percent thanks to *El Clon*.) A journalist visiting Quito, Ecuador, found viewers of the series "wide-eyed and drop-jawed for all things Arab."[16] Even in the United States, where *El Clon*'s broadcast was almost blocked due to alleged potential controversy, it has exerted cultural influence upon the Latino community and others. In New York, observers note the *El Clon*–triggered fashion for Arab jewelry and hip scarves, the overflowing belly dancing classes, and a recently opened beauty parlor called *El Clon* in Queens.[17]

Through the Latino back channel, the impact of Shakira in bringing Arab culture to the MTV audience has also been considerable. The Lebanese-Colombian singer was bombarded with questions by the media about her views "as an Arab" on the September 11 attacks and advised to drop the belly dancing and the Arabic riffs from her music because it could hurt her album sales, but she refused. "I would have to rip out my heart or my insides in order to be able to please them," said the songstress, and expressed horror at hate crimes against "everything that's Arab, or seems Arab."[18] During the run-up to the Iraq war, Shakira's performances took on an explicitly political tone, with her dancers wearing masks of Tony Blair, George W. Bush, and Fidel Castro. Backdrop screens flashed images of Bush and Saddam Hussein as two puppets playing a sinister game of chess, with the Grim Reaper as the puppeteer. She also undertook a highly publicized tour of the Middle East (though her concerts in Casablanca, Tunis, and Beirut were postponed), during which she visited her father's ancestral village in the Bekaa Valley. Viewers across the region were delighted when Shakira appeared on Egyptian television singing the tunes of Fairuz. In Europe, the United States, South America, and even the Middle East, the belly-dancing star has fostered a reported mania for hip scarves with coins and tassels. In a random check of Cairo nightclubs, Egyptian government officials confiscated twenty-six Shakira outfits "weighing no more than 150 grams [5 ounces]" and deemed "scandalous,"[19] but local filmmakers are currently negotiating with government officials over rights to a film project called Shakira fi al-Munira, about a young Egyptian girl infatuated with the Colombian chanteuse.

While the craze for Arab culture has occurred in the wake of September 11 and the ensuing war on terrorism, it is not necessarily political. Commenting on the popularity of shawarma and hookahs in Quito, one journalist observes that "the new fascination with Arabia comes at a time when there are new reasons for anti-American sentiment"—the recent policy of currency dollarization—but adds reassuringly that "El Clon's following surely won't produce a new sect of Islamic fundamentalist terrorists in Latin America." It is also not clear that conversion to Islam necessarily constitutes political or cultural resistance. Referring to the vogue for Islam and Arabic among Spanish youth, one Catalan journalist wryly observes, "It will take more than teenagers converting to an Islam-lite to stop [Spanish Prime Minister Jose Maria] Aznar's Christian nationalism and Castilian imperialism. We need a civil dialogue about our relations with the Orient."[20] Belly dancing and learning elementary Arabic may not be acts of resistance but such activities create important, albeit imaginary, cultural linkages that can be activated for political purposes. As Miles Copeland, head of the Mondo Melodia label, who plans to release a film on the American belly dancing craze PR Newswire, "Belly dancing is about art, not politics, but in experiencing the art, you also experience the culture, and that becomes political in and of itself."[21] Interest in Arab culture and conversions are bringing Islam into the imagination of Western youth, feeding powerful movements and cultures of protest.

FROM HARLEM TO THE CASBAH

In *No Name in the Street*, James Baldwin reflects on the "uneasy" reaction he would get when, while in France in 1948, he would "claim kinship" with the Algerians living there. "The fact that I had never seen the Algerian casbah was of no more relevance . . . than the fact that the Algerians had never seen Harlem. The Algerian and I were both, alike, victims of this history [of Europe in Africa], and I was still a part of Africa, even though I had been carried out of it nearly four hundred years before."[22] Most French-born Arabs have never been to Harlem but "claim kinship" with African Americans as they draw inspiration from the black freedom struggle. Numerous French-Arab (Beur) intellectuals and activists have noted their indebtedness to African American liberation thought, and the secular prointegration Beur movement of the early 1980s organized campaigns and marches modeled on the U.S. civil rights struggle. But in the early 1990s, as the impoverished, ethnically segregated *banlieues* mushroomed around French cities, the discourse of integration began to give way to talk of self-imposed exclusion and warnings that the children of immigrants "had gone in a separate direction." The region of Lyons, where 100,000 gathered for the famous march for integration in 1983, is today cited by commentators as evidence of the failure of assimilation. Lyons, by one account, has become a "ghetto of Arabs" and has fallen to Islamist influence, boasting six neighborhood boys in the U.S. military detention center at Guantanamo Bay.[23]

The generation of Black and Arab Muslim youth that came of age in crime-ridden *banlieues* that periodically explode into car-burning riots, and that are monitored by a heavy-handed police force, is in no mood for integration. By some estimates, 50 to 60 percent of the French prison population is Muslim.[24] French commentators are increasingly wondering if they have developed a "race problem" like that of the United States, with the attendant pathologies of ethnic ghettoes, family breakdown, drugs, violence, and, of particular concern these days, Islamism. As in the American ghetto, disintegrating family units have been replaced by new organizations: gangs, posses, and religious associations, particularly Islamic groups, which provide services and patrol the *cités*, the housing projects where most immigrants live.[25]

The confluence of Islam and urban marginality in France was displayed in a consummately postcolonial moment on October 6, 2001, when France and Algeria met in their first soccer match since the Algerian war of independence. The match was stopped prematurely when thousands of French-born Arab youth, seeing Algeria losing, raided the field chanting "Bin Laden! Bin Laden!" and hurled bottles at two female French ministers.[26] The ill-fated match, coming on the heels of September 11, led to hysterical warnings of an intifada simmering in the heart of France, an Islamic fifth column, the "unassimilability" of certain immigrants, and, again, an American-style "race problem." Like American pundits, the French are concerned about whether Islamic and Muslim organizations that have emerged in the *banlieues* will keep youths out of trouble or radicalize them. An American writing for the *Weekly Standard* notes, "It's the Farrakhan problem. Mosques do rescue youths from delinquency, idleness and all sorts of other ills. But in so doing, they become power brokers in areas where almost all disputes are resolved by violence and the most tribal kind of *woospeh* [respect, in a French accent, supposedly]. And it is that mastery of a violent environment—not the social service record—that these groups call on when they make demands on the larger society."[27]

The French media has shown a keen interest in the rising conversion to Islam in the United States and Europe, particularly in the overlap of Islam and race, or, more specifically, ethnic awareness, mobilization, and self-segregation. An expose in an April 2003 edition of the magazine *L'Express* opened with the following statement: "Blacks, whites, Latinos, Asians . . . every year, 50,000 to 80,000 [Americans] convert to Islam. Internal enemies, members of the 'axis of evil'?" The French government's attempts to control Islamic mobilization in the *banlieues* through elections for a national Islamic council (aimed, in the words of the interior minister, at taking Islam out of "cellars and garages") backfired when the conservative Union of Islamic Organizations, inspired by Egypt's banned Muslim Brotherhood, won 14 out of 41 seats.

Zacarias Moussaoui, the "twentieth hijacker" awaiting trial in the United States, in many ways embodies the story of Islam and racial exclusion in France. Although he did not grow up impoverished in the cities, by all accounts, the French-Moroccan harbored a deep racial rage. In his youth, Moussaoui was often ridiculed because of his dark skin and frizzy hair, and repeatedly called *négre* (nigger), but it was after the 1991 Gulf War that he became politicized. He began to consider himself "black," joining the "Kid Brothers," a university group modeled after the Egyptian Muslim Brotherhood, and came back from a stint in London deeply hostile toward whites. "He became a racist, a black racist, and he would use the pejorative African word *toubab* to describe white people," said his brother.[28] Moussaoui raged against Western permissiveness and imperialism in Algeria, Palestine, and Chechnya.[29]

Richard Reid, the "shoe bomber," who became radicalized in the same Brixton mosque as Moussaoui, embodies the similarly distressing urban and racial situation in Britain. West Indian and South Asian youth live in benighted "mill and mosque" towns, devastated by capital flight in the late 1980s and 1990s, where the anti-immigrant British National Party is making inroads and race riots erupt frequently. Many of these youth have drifted toward radical Islamist groups. By all accounts, the petty thief and graffiti artist known as ENROL embraced Islam while in Feltham Young Offenders Institution, to seek solace from racism. His father Robin tried to explain Reid's odyssey to Islam as a result of the difficulty of being of mixed race. "Islam accepts you for who you are," the father told CNN talk show host Larry King. "Even I was a Muslim for a little bit . . . because I was fed up with racial discrimination." In an interview with the *Guardian*, Robin continued, "About ten years ago, I met up with Richard after not seeing him for a few years. He was a little bit downhearted. I suggested to him, 'Why don't you become a Muslim? They treated me all right.'"

The mixing of Islam and racial awareness in Europe is also leading to political mobilization. The Arab European League (AEL), headed by the fiery Lebanese-born Dyab Abou Jahjah, is explicitly modeled on the American civil rights movement, borrowing slogans ("By Any Means Necessary!") and protest techniques from the Black Panthers and the Nation of Islam and aiming to mobilize Arab and Muslim youth across Europe to lobby European governments to make Arabic one of the official languages of the European Union and to gain state funding for Islamic schools. Based in Brussels, but with chapters opening in France and Holland, the AEL has launched a cross-border Arab pride movement and organized marches against the U.S. war in Iraq and in solidarity with the Palestinian intifada. Known as the "Arab Malcolm X," Abou Jahjah, who says he finds the ideas of integration "degrading," admits being inspired by the slain African American civil rights leader, who "was also against assimilation . . . fought for civil rights and was also inspired by Islam."[30] "We're a civil rights movement, not a club of fundamentalist fanatics who want to blow things up," he told the *New York Times* on March 1, 2003. "In Europe, the immigrant organizations are Uncle Toms. We want to polarize people, to sharpen the discussion, to unmask the myth that the system is democratic for us." The AEL has also organized Black Panther-style "Arab patrols" to "police the police." Groups of unarmed Arab youths dressed in black follow the police around, carrying video cameras and flyers that read, "Bad cops: the AEL is watching you." Fusing African American, Islamic, and Arab elements in its style and rhetoric, the AEL has become a political force to be reckoned with, even prompting the Belgian government to attempt to ban its patrols on the basis of a 1930s law that proscribes private militias.

"LE RESPECT" AND "LES PITBULLS"

Seul le beat aujourd-hui nous lie et nous unit.
[Today only the beat links and unites us.]

—Saliha, "Danse le Beat"

Hip-hop has emerged as the idiom for the urban activism of minority youth in Europe. For Muslim youth experiencing the crackdown on immigrants, as well as state withdrawal and welfare cuts, hip-hop offers a chance to express critiques, vent rage, declare solidarity with other marginalized youth (particularly African Americans), and display cultural pride—to show, as New York rapper DMX says, "who we be."[31]

If American rap has been criticized for its materialism, nihilism, and political nonchalance, French hip-hop offers trenchant critiques of racism, globalization, and imperialism. Numerous groups such as Yazid and La Fonky Family deal explicitly with the challenges of being Arab and Muslim in the West and relations between Islam and the West. In their hit single, "Je Suis Si Triste" ("I'm So Sad"), the Marseilles-based rap crew 3ème Oeil (Third Eye), made up of the Comorian-born Boss One (Mohammed), Jo Popo (Mohammed), and Said, offer biting social commentary over an infectious, looping bass line. Decrying hate crimes against veiled Muslim women in France, condemning police brutality and mass incarceration (with a special shout out to Mumia Abu Jamal), the rappers focus their lyrical fire on the West's "stranglehold" (*la main-mise*) on the East.

In addition to verbal release, hip-hop is also used to combat racism and to promote black-white-Arab relations, as in the Urban Peace Festivals and spoken-word poetry events (*les slameurs*) organized by SOS Racisme. Hip-hop, interestingly, is also being used to counter Islamist influence in the *banlieues*. The Beurette leader Fadela Amara, who organized the march "Ni putes ni soumises" ("Neither whores nor submissive"), a march that has now developed into a women's rights organization affiliated with SOS Racisme, often invites Muslim female rappers to spread a feminist message. "Ni putes, ni soumises" aims to mobilize youth against ghettoes and for equality, but also to counter the Islamist organizations such as the powerful Union of Islamic Organizations, which delivers services in the cites in exchange for veiling. Amara says discrimination and unemployment make many young men feel "excluded from the French project." These youths, she says, often return to Islamic traditions, opposing gender mixing and women's education and sometimes assaulting women who do not dress according to their idea of modesty.[32] French Muslim rappers and R&B singers publicly and collectively condemned the September 11 attacks, saying the terrorists were, in the words of Ideal J, a Franco-Haitian convert to Islam, "dishonoring the faith." Abd Al Malik of the New African Poets, a Congolose convert to Islam, noted the importance of rap and Islam to young ghetto dwellers: "Rap has opened a world to us, empowering us young men, and Islam has allowed us to flourish by teaching us respect for 'the other.' [But] the Taliban are instrumentalizing the religion."[33]

Attempts by some French Islamists to boycott American products and market products like Mecca Cola are failing since *banlieusards* remain loyal to American streetwear labels like Fubu and Phat Farm, often claiming that such clothing is an anti-American, but pro-black, statement. More recently, local *banlieue* streetwear clothing lines have appeared with names like Bullrot (a combination of pitbull and rottweiler) and Adedi (an acronym for *Association de differences*), the latter founded by a Moroccan, a Gabonese, and a Senegalese to combat racism and extremism and to celebrate difference.[34]

French commentators associate hip-hop with Islam, claiming that rap, like Islam, often brings rage, pathology, and dysfunction. The anti-immigrant National Front of Jean Le Pen and its splinter, the National Republican Movement, have historically denounced hip-hop. In March 2001, both Far Right parties opposed the use of public funds to finance the first Hip-Hop Dance World Cup in Villepinte, stating that "hip-hop is a movement belonging to immigrants of African origin installed in France and which constitutes a call to sedition against our institutions."[35] More recently, however, the National Front has begun to use hip-hop as a way to spread its political message, "win back" French youth, and counter Arab and American influence in French culture. The white supremacist rap crew Basic Celto, affiliated with the National Republican Movement, has as its objective to break "immigrants' monopoly" over hip-hop "which diffuses the immigrants' complaints." Basic Celto aims to promote a "national revolutionary" rap with a "Christian identity" and to draw "*francais d'origine*" away from immigrant influence.[36]

But the allure of Islam and Islam-inflected cultures like hip-hop and rai to French youth continues to grow, prompting much editorial pondering. *Le Monde* ran a story on how Ramadan is increasingly observed in French schools, even by non-Muslims, and there have also been reports of many non-Muslim girls wearing headscarves in solidarity with Muslim schoolgirls sent home for wearing le foulard. Commenting on *Le Pen*'s remark that hip-hop is a dangerous musical genre that originated in the casbahs of Algeria, rapper Boss One (Mohammed) of 3eme Oeil said, "For *Le Pen*, everything bad—rap, crime, AID—comes from Algeria or Islam . . . The more Bush and Chirac attack Islam and say it's bad, the more young people will think it's good, and the more the oppressed will go to Islam and radical preachers. Especially here in America. Because life is hard in France, but we have a social safety net."[37]

Commentators have also blamed hip-hop for bringing social ills associated with the American ghetto to France. "[French-Arab youth] intentionally imitate belligerent Afro-American lifestyles, down to 'in-your-face' lyrics for booming rap music," moaned one observer.[38] Some have pointed to the "African-Americanization" of the speech patterns of French youth, noting that their verbal jousting is similar to that of "American rappers from black ghettoes."[39] Indeed, the culture of France's suburban ghettoes is heavily influenced by the trends of the American inner city—the urban argot, street codes of conduct, and "honor system" are strikingly similar.[40] In January 2000, a law was passed creating a police unit to monitor the behavior of pit bulls and Rottweilers in housing projects where, as in the United States, such dogs had become very popular during the 1990s among urban youth.[41] The slurs used against blacks (*negres*) and Arabs (in France, *bougnoles*, in Spain, *Moros*, and in Belgium, *makkak*, which means white ape) have become commonly used terms of endearment among Muslim youth, as with the term "nigger" in the United States. But clearly, Muslim European youth have not learned misogyny and rage from hip-hop or from African Americans. The fact that hip-hop is being used by secular urban movements to counter Islamism and racism is an illustration of the growing racial consciousness of Muslim youth in Europe, the deep resonance of the African American experience, and how imagination can help construct a cultural world to resist state oppression and religious fanaticism.

KEEPIN' IT HALAL

The hip-hop movement has a powerful oppositional streak that makes it both attractive and troubling to political actors. Hip-hop's ability to jangle the hegemonic discourse was recently seen with Jay-Z's "Leave Iraq Alone" verse and Outkast's antiwar hit "Bombs Over Baghdad," denouncing the first Gulf War, which was yanked off the air by MTV and Clear Channel when bombs began raining on Baghdad in March 2003.[42] Hip-hop artists have strongly opposed the war without fear of the social opprobrium visited upon the Dixie Chicks and other white pop stars. As hip-hop mogul Russell Simmons put it, "Rappers don't have to worry about anything. No one likes what they have to say anyway, so they're not afraid to speak up." But when hip-hop is infused with Islamic themes and political allusions, the establishment press has found it particularly unsettling. Hence, the outrage over rapper Paris's recently released and rapidly selling *Sonic Jihad*, the cover of which features an airplane flying toward the White House, and the alleged purging of Arabic terms and references to Hussein from Tupac Shakur's recently released Better Dayz (though the slain rapper was referring not to the late Iraqi dictator, but to Hussein Fatal, a member of his Outlawz posse, which also includes Khadafi, Kastro, Komani, and Idi Amin).[43]

In the fall of 2002, accused sniper John Muhammad, formerly of the Nation of Islam, sent notes to the police that referenced lyrics from rappers who are "Five Percenters," a heterodox Black Muslim sect. The subsequent media frenzy triggered a soul-searching conversation within the Islamic hip-hop community that was rendered particularly urgent when Muslim hip-hoppers found themselves linked to the war on terror by Niger Innis, chairman of the conservative Congress of Racial Equality. Shortly after the arrest of John Muhammad, Innis met with Department of Justice officials to express concern over "domestic Black Muslims as a national security issue" and launched a campaign to counter Islamic recruitment efforts in the nation's prisons and colleges.[44] (A similar uproar occurred more recently in the United Kingdom when a hip-hop group named Shaikh Terra and the Soul Salah crew released a video "Dirty Kuffar" ["Dirty Unbelievers"], in which they salute Hamas and Hizbullah and praise Osama bin Laden. The "hate video" drew the attention of Labor MP Andrew Dismore, who described the video as "disgust[ing]" and "inexcusable" and launched a police investigation into the radical Muslim group.[45]) Muslim rappers are asking themselves, should we be expected to "represent" Islam positively and avoid the misogynist and material-istic excesses of mainstream hip-hop artists? Or should the aim be to "get paid" and gain wide success even if it means "playing with the haram (illicit)"? Of the U.S.-based Muslim hip-hop crews, Native Deen and Sons of Hagar have been praised for their positive political and religious messages. Native Deen, made up of three African American rappers who will not perform in venues that allow mixed dancing or serve alcohol, have been profiled in the *New Yorker* and even received praise from the State Department, but the group has yet to garner airtime on mainstream radio stations. The Des Moines–based Sons of Hagar, made up of Allahz Sword (Ahmad) and Ramadan Conchus (Abdul), both Arab Americans, and Keen Intellect (Kareem) and Musa, Irish American and Korean American converts to Islam, respectively, have also been praised for socially conscious lyrics. Their poignant single "Insurrection" ("It's the Arab hunting season, and I ain't leavin' / I'm pushin' the conscience button on you people / Where is the reason?"), and their track "Sisterssss," in support of polygamy,[46] are popular in the under-ground Muslim-Arab hip-hop scene. But Sons of Hagar has also not achieved mainstream exposure.

The Muslim rap crew that is gaining worldwide notoriety for its lyrical dex-terity, stylistic appeal, and explicitly positive portrayal of Islam is the Denmark-based trio Outlandish. Made up of a Moroccan, a Pakistani, and a Honduran, Outlandish has topped the charts with hits including "Guantanamo" (the chorus: "And I got all my Moros here, Guantanamo")[47] and "Aicha," a remake of Cheb Khaled's 1995 hit. The latter track, which saw heavy rotation on MTV Europe and climbed to fourth on the charts in Germany, has been hailed as the most positive depiction of Muslim women in a music video, with shots of preprayer ablution and veiled and unveiled Arab, South Asian, and African women. Rather than playing with the haram, Outlandish is about "keepin' it halal" (licit).

American hip-hop commentators note that political, cerebral rap may be popular in Europe but if it cannot be "bling-blinged," or sexed up, it will not sell in the United States. A recent dispute between Simmons and a segment of the African American Sunni community is illustrative. Though not a Muslim, Simmons has frequently declared his respect for Islam and the Nation of Islam (NOI) in particular. "I grew up on Farrakhan," he said in one interview. "Where I grew up, there were dope fiends and black Muslims. If Muslims came by, you stood up straight."[48] He also tried to broker talks between the NOI and American Jewish organizations, denounced the invasion of Iraq, helped organize Musicians United to Win Without War, and is currently planning a Middle East youth peace summit. But when a recent issue of his *OneWorld* magazine ran a cover with female rapper Li'l Kim wearing a "burka-like garment over her face" and "lingerie from the neck down" and in the same issue said, "Fuck Afghanistan," Najee Ali, director of the civil rights group Project Islamic Hope, demanded an apology to America's Muslims. As someone active in brokering truces in the hip-hop world, Ali cited his Islamic duty "to the people of hip-hop and humanity" and called on Simmons to apologize for the magazine cover and for the "pornographic female rapper" Foxy Brown, who, in her song "Hot Spot," produced by the Simmons-founded Def Jam, says "MCs wanna eat me but it's Ramadan."

The Li'l Kim incident instigated a discussion over other not-so-halal trends in Islamic hip-hop. The cover of *XXL* magazine showing rapper Nas holding a glass of cognac and wearing prayer beads around his neck outraged many Muslims. "Why he imitatin' the kufar (unbelievers, in Arabic) with the Hail Mary beads?!" fumed one blogger. Many Sunni Muslims have also criticized the style of some female Muslim hip-hoppers of wearing a headscarf (hijab) with midriff tops and the low-riding jeans popularized by Jennifer Lopez. These sartorially adventurous young Muslim women, known variously as "noochies" (Nubian hoochies), "halal honies," and "bodacious bints" ("girls," in Arabic) have provoked heated cyber debates about freedom of expression, female modesty, and the future of Islam in America. "Our *deen* ("religion," in Arabic) is not meant to be rocked!" says hip-hop journalist Adisa Banjoko, author of the forthcoming book *The Light From the East* on Islamic influence in hip-hop. "I see these so-called Muslim sistas wearing a hijab and then a bustier, or a hijab with their belly button sticking out. You don't put on a hijab and try to rock it! Or these brothers wearing Allah tattoos, or big medallions with Allah's name—Allah is not to be bling-blinged!"[49]

Just as controversial are the Arabic calligraphy tattoos that women, even outside the hip-hop community, have taken to wearing. The words "halal," "haram," and "sharmuta" ("whore" in Arabic but a term of endearment in certain circles these days) are tattooed on shoulders, thighs, or lower backs, and worn with bathing suit tops or hip-hugging jeans. Some of these haram trends in Islamic hip-hop are deliberate responses to orthodox or fundamentalist Islamic dress, like the "high-water pants" or "total hijabs" seen in some inner-city areas. Among young Muslim males, equally provocative are black T-shirts worn by some Shiite youth, which read, in crimson, "Every Day Is Ashura, Every Day Is Karbala"—references to Shiite rituals commemorating the death of Imam Hussein in the seventh century and the Iraqi plain where he died in battle. Also troubling to some is the growing popularity of martial arts among urban Muslim youth, who say self-defense skills are necessary against gangsters and violent police. If many Black Muslims in the 1960s were practicing syncretic forms of martial arts like "Kushite boxing," many of today's young male hip-hoppers are learning "Islamic wrestling." "The Prophet was a grappler," explains one enthusiast. "The hadith (saying of the Prophet) teaches us to never hit the face of our opponent and that [Islamic] grappling allows you to win over an opponent without punching them and risking brain damage."

Russell Simmons has said that "the coolest stuff about American culture, be it language, dress, or attitude, comes from the underclass—always has and always will."[50] If so, then as Islam seeps into the American underclass and as

Muslims populate the underclass in Europe, Islamic cultural elements will per-colate upward into mainstream culture and society. For many American youth, Islamic hip-hop is their first encounter with Islam and often leads them to strug-gle with issues of race, identity, and Western imperialism. As Bakari Kitwana has noted, "If asked about a specific political issue many hip-hop generationers can easily recall the first time their awareness on that issue was raised by rap music."[51] In Europe, many North African youth are rediscovering Islam and becoming race conscious through Five Percenter and NOI rap lyrics. For many white hip-hop-pers in the United States, the sought-after "ghetto pass" acceptance in the hip-hop community comes only with conversion to Islam, which is seen as a rejection of being white. The white rapper Everlast, formerly Eric Schrody of House of Pain, claims that conversion to Islam and mosque attendance allow him to visit ghetto neighborhoods he could never enter as a non-Muslim white.[52] Curiously, Ever-last's espousal of Islam caused static with the white rapper Eminem, who accused him of becoming Muslim to deny that he is a "homosexual white rappin' Irish." One young white Latino youth explained the link between Islam and his street credibility as follows: "In the Bronx, looking like me, you don't get much respect. When I took the shihada (professed Islam), the brothers gave me respect, the white folk got nervous, even the police paid attention."[53]

Efforts are being made to direct the energy of Islamic hip-hop. In late July 2003, the First Annual Islamic Family Reunion and Muslims in Hip-Hop Con-ference and Concert was held in Orlando, Florida, with prominent imams from across the country leading three days of workshops on Muslim youth and stress-ing the importance of deen, family, schooling, and organizing. Activities included Islamic spelling bees, Islamic knowledge competitions, and performances by "pos-itive lyricists" like Native Deen. The conference also established Hallal Entertain-ment, Inc., and helped launch the Islamic Crisis Emergency Response System, a Philadelphia-based organization which provides services to needy Muslim and non-Muslim families.[54] Fusing Islamic themes with the preeminent global youth culture, Islamic hip-hop has emerged as a powerful internationalist subculture for disaffected youth around the world.

ROARING FROM THE EAST

"The specter of a storm is haunting the Western world," wrote the Black Power poet Askia Muhammad Touré in 1965. "The Great Storm, the coming Black Revolution, is rolling like a tornado; roaring from the East; shaking the moorings of the earth as it passes through countries ruled by oppressive regimes . . . Yes, all over this sullen planet, the multi-colored 'hordes' of undernourished millions are on the move like never before in human history."[55] Toure was pondering the appeal of "the East" to African American youth in the aftermath of the 1955 Bandung conference. There, President Sukarno of Indonesia had told the repre-sentatives of twenty-nine African and Asian nations that they were united "by a common detestation of colonialism in whatever form it appears. We are united by a common detestation of racialism." Those were the days when Malcolm X met with Fidel Castro at the famed Teresa Hotel in Harlem, and when Malcolm, from his perspective of "Islamic internationalism," came to understand the civil rights movement as an instance of the struggle against imperialism, seeing the Vietnam War and the Mau Mau rebellion in Kenya as uprisings of the "darker races" and, like the African American struggle, part of the "tidal wave" against Western imperialism.

Some commentators, pointing to the current antiwar and antiglobalization movement, have suggested that a new era of Afro-Asian-Latin solidarity may be in the offing. In the United States, the past two years has seen a political ferment and coalition-building between progressive groups—in particular between Arab American and Muslim American groups and African American groups—not seen since the 1960s when the Black Panthers and the Student Non-Violent Coordinating Committee declared solidarity with the Palestine Liberation Organization, which in turn declared solidarity with Native Americans. September 11 and the subsequent backlash has led many African American leaders to stand with Muslim and Arab Americans, not least because African American Muslims are also targeted in the post-September 11 profiling and detention campaigns. Activists like Al Sharpton are mobilizing against the USA PATRIOT Act "because it is used to profile people of color" and is "impacting Muslims everywhere, including Brooklyn and Harlem."[56]

Given the centrality of Islam and the Arab world to the war on terror and the presence of *kaffiyyas* and (regrettably) bin Laden T-shirts at protests from Porto Alegre to Barcelona, it appears that the new Bandung may have a distinct Arab or Islamic cast. In the past two years, a number of Latin American leaders have called for "concrete action" to establish a Palestinian state. Castro has signed agreements of bilateral cooperation with Algeria and the United Arab Emirates and continues to rail against "global apartheid" in general and "Israeli apartheid" in particular. Castro has also been accused of building ties with Iran and selling biotechnology in exchange for cheap oil. When he visited Iran in 2001, Castro spoke of his rapport with President Mohammad Khatami and reported that he "had the longest sleep of his life in Tehran." Most recently, he has been accused by the United States of jamming the satellite broadcasts of U.S.-based Iranian opposition groups.[57] Recent articles in right-leaning American newsmagazines claim to have discovered evidence that Venezuela is providing identity papers to suspicious numbers of people from Arab and South Asian "countries of interest" (as well as Colombians and Cubans). One article also features the claim of the former Venezuelan ambassador to Libya, Julio Cesar Pineda, to possess correspondence from Hugo Chavez, stating his desire to "solidify" ties between Latin America and the Middle East, including use of the oil weapon.[58] Chavez challenged the reporters in question to produce "one single shred of evidence" for their claims.[59]

These stories of Cuban and Venezuelan ties to Middle Eastern radicals may be little more than partisan puffery, and Chavez's repeated calls for solidarity with the Arab world may be nothing more than petroleum diplomacy or an embattled leader's desperate plea for allies. Yet the Venezuelan leader's appeal to "Arab roots" is indicative of a trend in the West. Among Western subordinate groups and opposition movements that feel victimized or neglected by globalization, the Arabs are seen as bearing the brunt of the worldwide imperial assault in the era of the war on terror. As Western nationalists portray Islam as a threat to freedom and security, and launch wars to bring democracy to the Muslim world, "the multi-colored hordes" of the West are reaching for teachings and precedents (like Moorish Spain) in Islam that they hope will make the West more compassionate and free.

Islam is leaking into the West through conversion, migration, and media-driven cultural flows, and, to many, the Islamic world is presenting a repertoire of alternative identities. As marginalized Westerners are finding inspiration in Islam, Muslims in the diaspora are inspired by the African American experience. The cross-fertilization taking place between Islamic, black, and Latin cultures is creating fascinating trends and art forms. Many would argue that the fashion for Arabic tattoos, Allah chains, Orientalist soap operas, belly dancing, and hip scarves is just that—fashion. But as the Arab pride movement in Europe and Islamic hip-hop demonstrate, the vibrant cultural intermingling can have significant political implications. Cultural flows can spark forceful challenges to state policies, state-imposed identities, and the claims of Western nationalism. For many of the minority convert communities and the diaspora Muslim communities, Islamic Spain has emerged as an anchor for their identity. Moorish Spain was a place where Islam was in and of the West and inhabited a golden age before the rise of the genocidal, imperial West, a historical moment that disenchanted Westerners can share with Muslims. Neither Muslim nostalgia for nor Western Orientalist romanticism about Andalusia is new, but it is new for different subordinate groups in the West to be yearning for "return" to Moorish Spain's multiracialism. In this worldview, the year 1492 is a historical turning point. On Columbus Day in October, Chavez urged Latin Americans to boycott celebrations of the "discovery," saying that Columbus was "worse than Hitler." That the longing for pre-1492 history is shared by many minorities throughout the West is an indication of their lasting exclusion and how the stridency of Western nationalism since September 11 has revived memories of centuries-old trauma. As one African American activist put it recently, "The profiling and brutalizing of African-Americans didn't begin after September 11. It began in 1492."[60] In a similar spirit, after Moussaoui was arrested in the United States and granted the right to represent himself in court, one of his first demands was "the return of Spain to the Moors."

With African American and Latino converts speaking of the tragedy of 1492 and with Muslim minorities in the West becoming increasingly race-conscious and inspired by black America, the world is witnessing a new fusion between Islam and Pan-Africanism. Today, however, this racialized Islamic internationalism contains elements of other cultures and diasporas as well. Islam is at the heart of an emerging global antihegemonic culture, which postcolonial critic Robert

Young would say incarnates a "tricontinental counter-modernity" that combines diasporic and local cultural elements, and blends Arab, Islamic, black, and Hispanic factors to generate "a revolutionary black, Asian and Hispanic globalization, with its own dynamic counter-modernity . . . constructed in order to fight global imperialism."[61]

NOTES

1. Jose Marti, "España en Melilla," in *Cuba: Letras*, vol. 2 (Havana: Edicion Tropico, 1938), 201.

2. Quoted in Rene Depestre, "Carta de Cuba sobre el imperialismo de la mala fe," *Por la revolucion, por la poesia* (Havana: Institute del Libro, 1969), 93.

3. *El Pais*, April 17, 2002.

4. *Latin American Weekly Report*, October 4, 2003.

5. CUT National Plenary, Conjuntura Internacional e Nacional, Resolution 10, "Cresce a polaizacao politica a social em todo o mundo." Accessible online at http://cutnac-web.cut .org.br/10plencut/conjtex5.htm.

6. *Deutsche Presse-Agentur*, April 3, 2003.

7. *New York Times*, October 22, 2001; *The Economist*, October 26, 2001. Imams and converts also made this claim in interviews carried out by Columbia University's Muslim Communities in New York Project on June 4 and June 16, 2003.

8. *Times* (London), January 7, 2002.

9. *Evening Standard*, March 15, 2002.

10. *Christian Science Monitor*, October 2, 2002. See also Yusuf Fernandez, "Spain Returning to Islam," *Islamic Horizons* (July-August 2002).

11. Olivier Roy, "Euro-Islam: The Jihad Within?" *The National Interest* (Spring 2003).

12. *Cox News Service*, August 11, 2002; see also *Knight-Ridder News Service*, June 20, 2003.

13. *El Diario-La Prensa*, October 6, 2001. See also *Islamic Horizons* (July–August 2002).

14. Author's interview with Rahim Ocasio, April 16, 1999.

15. Rahim Ocasio, "Latinos, The Invisible: Islam's Forgotten Multitude," *The Message*, August 1997.

16. Kimi Eisele, "The Multicultural Power of Soap Operas," *Pacific News Service*, November 25, 2002.

17. Interview with Rosa Margarita of *El Diario-La Prensa*, August 8, 2003. El Clon-inspired fashion can be viewed online at http://www.laoriginal.com/especiales.htm.

18. *Independent*, July 19, 2002.

19. *Agence France-Presse*, May 28, 2003.

20. Interview with Fernando Casado Caneque, September 8, 2003. Casado was referring to the conservative Aznar's effort to insert a reference to Europe's Christian roots in the EU's constitution, a measure that has provoked the Spanish Left and the regions of Andalusia and Cataluna who resent how the Aznar government has made Catholicism so central to the state's identity. See *El Pais*, July 28, 2003.

21. James Baldwin, *No Name in the Street* (New York: Dial, 1972), 41.

22. See, for instance, the interview with Ferida Belghoul in Alec Hargreaves, *Voices from the North American Community in France: Immigration and Identity in Beur Fiction* (Providence, RI: Berg Publishers, 1991), 126.

23. *Le Monde*, February 12, 2003.

24. *Jerusalem Report*, May 6, 2002.

25. See Loïc Wacquant, "Red Belt, Black Belt: Racial Division, Class Inequality and the State in the French Urban Periphery and the American Ghetto," in *Urban Poverty and the Underclass*, ed. Enzo Mingione (London: Blackwell, 1996).

26. *New York Times*, October 16, 2001.

27. *Weekly Standard*, July 15, 2002.

28. *Times* (London), September 29, 2001.

29. Abd al-Samad Moussaoui, *Zacarias, My Brother: The Making of a Terrorist* (New York: Seven Stories Press, 2003), 129.

30. *Independent*, April 3, 2003.

31. See Paul Silverstein, "Why Are We Waiting to Start the Fire? French Gangsta Rap and the Critique of State Capitalism," in *Black, Blanc, Beur: Rap Music and Hip-Hop Culture in the Francophone World*, ed. Alain-Philippe Durand (Lanham, MD: Scarecrow, 2002).

32. *Le Figaro*, June 17, 2003; *Le Monde*, March 11, 2003.

33. *Le Monde*, September 27, 2001.

34. *L'Expansion*, June 11, 2003.

35. *Independent Race and Refugee News Network*, April 1, 2001.

36. The group's manifesto is online at http://infosuds.free.fr/082001/enquete_bc.htm. I am grateful to Paul Silverstein for this point.

37. Interview with 3ᵉᵐᵉ Oeil and DJ Rebel, Bronx, New York, July 24, 2003.

38. *Jerusalem Report*, May 6, 2002.

39. *L'Express*, March 27, 2003.

40. David Lepoutre, *Coeur de banlieue: Codes, rites et languages* (Paris: O. Jacob, 1997).

41. *Le Figaro*, June 3, 2000.

42. I am grateful to Zaheer Ali for this point.

43. Interview with Napoleon, March 22, 2004, New York. Tupac Shakur's former companion, Napoleon, a Muslim convert who will be releasing a CD titled "Have Mercy," featuring a collaboration with the Pakistani-American crew The Aman Brothers, speaks about this allegation in an interview with the Tupac fan site HitEmUp.com, published on April 16, 2003. Accessible online at http://www.hitemup.com/interviews/napoleon-part1 .html#Bush.

44. *Washington Times*, November 13, 2002.

45. "Islamic rappers' message of terror," *The Observer*, February 8, 2004.

46. When told that polygamy is illegal in the United States, Allahz Sword responded, "A lot of rappers out there talk about pimpin'—is that good? I'm just talking about part of my religion." *Seattle Post-Intelligencer*, February 17, 2003.

47. The Spanish slur Moro has long been a term of endearment in Morocco and in the Moroccan diaspora the Arabic adaptation is *moro khal al-ras* (black-headed Moor).

48. Hishaam Aidi, "'Building A New America': A Conversation with Russell Simmons," *Africana.com*, February 5, 2002.

49. Personal communication with author, August 4, 2003.

50. Quoted in John McWhorter, "How Hip-Hop Holds Blacks Back," *City Journal*, Summer 2003.

51. Interview with Adisa Banjoko, "Everlast: Taking Islam One Day at a Time," July 12, 1999. The interview is accessible online at http://thetruereligion.org/everlast.htm.

52. Interview with Adisa Banjoko.

53. Interview with Columbia University's Muslim Communities of New York Project, June 16, 2003 (Converts Focus Group).

54. Sister Kalima A-Quddus, "Verily This Is a Single Ummah," *MuslimsInHipHop Newsletter*, August 7, 2003.

55. Quoted in Robin Kelley, *Freedom Dreams: The Black Radical Imagination* (Boston: Beacon, 2002), 60.
56. *Village Voice*, December 24, 2002.
57. *Financial Times*, July 21, 2003.
58. See Martin Arostegui, "From Venezuela, a Counterplot," *Insight on the News*, March 4, 2003 and "Terror Close to Home," *US News and World Report*, October 6, 2003.
59. Agence France-Presse, October 2, 2003.
60. Interview with Columbia's Muslim Communities of New York Project, July 21, 2003 (focus group for Muslims in NYPD and Fire Department).
61. Robert Young, *Postcolonialism: An Historical Introduction* (London: Blackwell, 2001), 2.

CHAPTER 8

CONSTRUCTING MASCULINITY

INTERACTIONS BETWEEN ISLAM AND AFRICAN AMERICAN YOUTH SINCE C. ERIC LINCOLN'S *THE BLACK MUSLIMS IN AMERICA*

RICHARD BRENT TURNER

THE BLACK MUSLIMS IN AMERICA WAS A GROUNDBREAKING WORK—THE FIRST monograph that presented a detailed sociological study of a twentieth-century African American Muslim community. C. Eric Lincoln demonstrated a profound understanding of the dynamic changes in black American identity that developed from the convergence of the civil rights movement, Black nationalism, and the independence of African nations from colonialism after World War II; simultaneously, he provided evidence of African American Islam's success in redefining black manhood in America's inner-city communities and how the Nation of Islam's "emphasis on youth and masculinity . . . generated a strong appeal" to converts.[1] One of his book's major themes is that the Nation of Islam in the 1960s was a black nationalist "united front of Black men . . . predominantly lower class . . . trying to reach all Black men . . . in the colleges . . . [and]in the jails."[2]

In retrospect, contemporary scholars might criticize the above characterization of African American Islam for its almost single-minded focus on a proto-Islamic community and the theme of black manhood. My article shifts the discourse of black masculinity in African American Islam away from the patriarchal models of racial uplift that may have influenced the work of scholars in Lincoln's time.[3] Instead, I examine how young African American male converts and second-generation Sunni Muslims in a university community have critically constructed an understanding of manhood through the interaction between Islam and hip-hop culture in their daily lives. The young men interviewed for this study chose to dedicate themselves to mainstream Islam, and various aspects of hip-hop culture reflect the social justice values and political consciousness of African American Muslim communities. Their orientation to Islam and hip-hop culture also reflects a new contemporary discourse in academic and popular culture on black masculinity that encourages African American men "to rethink who and what we are"[4] and to deconstruct the "ideological traps" that prevent "a progressive engagement" with social justice in the United States and the world.[5]

Notwithstanding the above critique of "male-centered"[6] studies of African American Islam, *The Mosque Study Project 2000*, which is the "most comprehensive survey of American mosques," confirms the contemporary numerical significance of Black Muslim men in mainstream Islam. In that survey, African Americans comprised 63 percent of the converts to Sunni Islam in 2000 and 68 percent of those African American converts were male.[7] Presently, the religious identities of African American Muslim men remain one of the central themes in the contemporary story of Islam in the United States.

This study is one small step using qualitative research that attempts to answer the question of how the religious, political, and demographic dynamics for African American male converts have changed since the publication of Lincoln's book. Using oral history interviews conducted at the University of Iowa after September 11, 2001,[8] black autobiographies, and recent social-scientific and cultural studies of African American masculinity, I have evaluated, from contemporary and historical perspectives, identity formation among Sunni Muslim converts and a second generation Muslim involved in hip-hop culture. I also discuss visual representations in television and cinema that frame youth conversion experiences.

HIP-HOP CULTURE, ISLAM, AND BLACK MASCULINITY

Rap has always been here in history. They say when God talked to the prophets, he was rappin' to them.

—Afrika Bambaataa,
Yes Yes Y'all—the Experience Music Project Oral History of Hip-Hop's First Decade

Grandmaster Flash defined hip-hop as "the only genre of music that allows us to talk about almost anything. Musically, it allows us to sample and play and create poetry to the beat of the music. It's highly controversial, but that's the way the game is."[9] Hip-hop culture began in the 1970s in the Bronx among African American, Latino, and Caribbean youth who created intercultural crews ("new kinds of families") of MCs, DJs, break dancers, and graffiti artists whose art and style expressed resistance to "postindustrial conditions" and the destruction of jobs, affordable housing, and support systems in working class black and Latino neighborhoods in New York City at the beginning of the Reagan-Bush era.[10] These conditions became the norm for black America in the 1980s and 1990s. The rise of a new prison-industrial complex eventually jailed nearly one-third of all African American men in their twenties. Drawing on subjects such as racial profiling, a new underground economy generated by widespread drug addiction in America, youth deaths from police brutality and torture, AIDS, and new assaults on affirmative action, rap music became the primary medium for documenting and protesting the hardness of life for black youth in the "hood"[11] and beyond. The music, poetry, style, clothing, language, and life experiences of hip-hop artists reflected black people's encounters with Islam, Christianity, prison life, death, violence, drugs, love, sex, hope, and despair in their urban communities. Soon, rap poetry was being composed by young people across the globe even though they did not live in the same conditions.

In the 1980s, hip-hop culture became an important signifier of black masculinity at the very same time that the mainstream print and visual media began to explore the "endangered" status of black men in the United States. Kobena Mercer writes, "Overrepresented in statistics on homicide and suicide, misrepresented in the media as the personification of drugs, disease and crime, such invisible men . . . suggest that Black masculinity is not merely a social identity in crisis. It is also a key site upon which the nation's crisis comes to be dramatized, demonized and dealt with."[12]

The music industry decided to focus on the sensationalist lifestyles and rhymes of gangsta rap in the 1990s, first popularized by Ice Cube, Snoop Doggy Dog, and Dr. Dre and the rap group N.W.A., which highlighted black-on-black violence, police brutality, drug dealing, homophobia, and sexism in southern California hip-hop culture. Of course, these themes validated the stereotype of black men that have dominated the music and film industry since the early twentieth century and provided yet another rationale for a criminal justice system that now imprisons more people per capita than any other nation in the world. Nonetheless, the music and poetry of hip-hop culture contains, in Tricia Rose's useful phrase, complex "hidden transcripts." Rap music presents serious social, political, and spiritual critiques of systematic racism and classism.[13]

Charise L. Cheney analyzes the transformation of gangsta rap from the oppositional culture of the "underground street"[14] reporting in the inner city to "contemporary minstrelsy"[15] in which "caricatures became the norm,"[16] and she emphasizes "that many observers mistakenly assume that this genre represents all rap music."[17] However, the young Muslim men in this study are inspired by the genre of rap music that attained its greatest commercial popularity in the "golden age of rap nationalism"[18] in the late 1980s and early 1990s, with groups such as Public Enemy that reflected new expressions of Pan-African, Black Nationalist, and Black Atlantic identities in their music, politics, and spirituality.[19] As we shall see, the culture of this "golden age" of hip-hop continues to influence select contemporary rap artists and their African American Muslim fans.

If one of the important functions of rap music, according to Anthony B. Pinn, is to answer "the ultimate questions of life," certainly Islam has significantly influenced the construction of black masculinity in hip-hop culture since its beginning as a local New York City phenomenon in the 1970s.[20] Afrika Bambaataa, one of the most important originators of rap music, put African American Islam at the center of the political and religious influences that created hip-hop in the Bronx, New York:

> I grew up in the southeast Bronx . . . an area where back in the . . . early '70s there was broken glass everywhere . . . But it was also an area where there was a lot of unity and a lot of social awareness going on . . . Seeing all the violence that was going on with the Vietnam War . . . in Attica and Kent State and being aware of . . . the late '60s, with Woodstock and the Flower Power, the Love Power movement . . . put a lot of consciousness in my mind to get up and do something . . . to say "we've got to stop the violence with the street gangs . . . *Hearing the teachings of the Nation of Islam made a lot of people get up and try to get the drugs out of their community*; and seeing a lot of the struggles that was going on all around the world through television gave a lot of hope to this area to do something for itself . . . and when we started this music called hip hop, which didn't have a name at the time, it brought a lot of the elements of these different movements together.[21]

Afrika Islam and Richard Sisco, two of Bambaataa's associates in their rap group, called the Zulu Nation, contemplated the significance of this early political and religious message of hip-hop for young African American men, "Bam always used to talk about positive things for the Black race . . . It was good for those wild kids that didn't have no direction. Bam gave them . . . some self-respect . . . something to keep their head up—self esteem—because in this world we're living in, there's not too much for a Black man out here to have self-esteem about."[22]

As rap music developed its commercial and cultural power in the post–civil rights era, hip-hop culture began to explicitly engage aspects of the religious tradition of Islam in important discourses about black masculinity and how young people could redefine blackness and its place in American popular culture. Thus, in the poetry of many of the early message rappers such as Public Enemy, Paris, Poor Righteous Teachers, Eric B. and Rakim, KRS-One, Ice Cube, Brand Nubian, Big Daddy Kane, Super Jay, Stetasonic, Defiant Giants, Q-Tip, Sister Souljah, Prince Akeem, Flavor Flav, and others, Malcolm X, Elijah Muhammad, Louis Farrakhan, and Muhammad Ali were highlighted as the masculine heroes of a new political musical vibe that reflected a dramatic shift in Islam's place in American society and culture.[23]

In the 1960s, the numbers of Sunni Muslims in Western Europe and North America began to expand because of the U.S. Immigration Act of 1965 and global Muslim immigration to Great Britain, France, and the United States. Eventually, African Americans became "the largest convert community to [Sunni Islam] in the West."[24] This new black American vanguard in mainstream Islam was influenced by Malcolm X's departure from the Nation of Islam and his *hajj* to Mecca in 1964. As a result of these factors, in the late 1980s, Islam became the second largest religious tradition after Christianity in the United States. Thus, the "religious sensibilities" of hip-hop culture reflected the new life experiences of African American youth who were born in the aftermath of the black power era and saw the growing influence of Islam in their local urban communities.[25]

Perhaps the most useful way to see the religion's influence on black youth in hip-hop culture may be as the counterconception to the hegemonic discourse of oppressive majority media and music industry images that have demonized and exoticized urban life, rap music, and young black people in the late twentieth century. In this context, Islam identifies an ideological fulcrum that has enabled politically and spiritually conscious youth in the hip-hop community to achieve independence from the race, gender, and class stereotypes disseminated by the music industry and select political groups in the dominant culture.

In the late 1990s, numerous rappers with Islamic messages started important conversations about mainstream Islam in the hip-hop community that were influential in the conversion of African American youth to Islam. These messages ranged from a simple awareness of Sunni Muslims in the black community, expressed in Jay Z's rhyme about body tattoos: "I never read the Quran or Islamic scriptures, the only psalms I read were on the arms of my niggas"[26] to the more complex poetry about the construction of black American identity from the Muslim hip-hop artists Mos Def and Talib Kweli in their album masterpiece *Black Star*, which is sampled below, "We feel we have a responsibility to shine a light into the darkness . . . Black like the veil the Muslim Aminah wears . . . Black like the slave ships that brought us here . . . Black like the Creator loves all creatures who are knowledge and truth seekers."[27]

Hip-hop culture's significant impact on the production of black cinema and television in the 1990s also influenced youth conversions to Sunni Islam. Prison conversions and the positive influence of Islam among African American inmates were the themes in several high-profile television series and films that appeared in this era in which there was an alarming increase in black men's involvement with the criminal justice system. Obie Clayton and Joan Moore's analysis of a 1997 Urban Institute report estimated

> that six more African-Americans are incarcerated than their White counterparts. If we disaggregate the figures we would find that one third of African-American men between the ages of twenty and twenty-nine are either in prison or in jail . . . Many legal and extralegal factors have been cited for the high incarceration rates of African-Americans. Racial profiling, mandatory minimum sentences, and especially the disparities in drug laws have had a dramatic effect on the incarceration rates of young males especially in urban inner-city neighborhoods.[28]

These unfortunate statistics helped to create a critical hip-hop youth audience for television such as the HBO prison drama *Oz* (in which Minister Said is the inspiring inmate leader of a large community of African American Sunni Muslim men who have reconstructed their masculinity in prison through conversion to Islam) and films such as Spike Lee's 1992 *Malcolm X* (with Denzel Washington as the young Malcolm who converts to the Nation of Islam in prison)[29] and the 1992 *South Central* (in which Carl Lumbly as Ali, the Muslim mentor who teaches a black inmate, Glenn Plummer as Bobby Johnson, how to reclaim his manhood by renouncing his gang affiliation and assuming his responsibility as a strong father to his young son).

Ali's advice to Bobby Johnson in the following prison scene from *South Central* echoes the psychological and social reality of many young African American men in the hip-hop community:

Ali: "Black man is in prison or dodging prison and his kids suffer. It's the anger in the Black man the cycle of hate; you're in that cycle little brother, did you grow up with your daddy?"

Bobby Johnson: "No."

Ali: "I see inside you man, you still got time before it eats you up. But you got to break your cycle, hate is broken by giving, I give to you and I break my own hate cycle . . . me and you, we brothers and we got to be here for our children. For me it's too late, not for you. [Ali reaches and picks up a book] You owe me fifteen boxes of cigarettes, you repay me by reading [Ali hands Bobby Johnson a book] two hours every day."[30]

Certainly, the above film and television productions continue to resonate with the lives of many young black men who desperately need to reconstruct their masculinity during and after incarceration. Some of these youth have been influenced by scholars' stories about the dedicated work of Muslim chaplains in state and federal penitentiaries and have been inspired by hip-hop era autobiographies like Nathan McCall's *Makes Me Wanna Holler*, Sanyika Shakur's *Monster: The Autobiography of a L.A. Gang Member*, and the classic Alex Haley and Malcolm X's *Autobiography of Malcolm X*, which also underline the significance of Islam for black men in prison.[31] Finally, all of the above information suggests that the future of African American Muslim communities in the twenty-first century may be significantly influenced by the conversion experiences and social-political perspectives of young hip-hop artists and their fans.

HIP-HOP CULTURE AND AFRICAN AMERICAN MUSLIM MEN ON CAMPUS

One of the main things about Islam is the establishment of African American manhood.

—Khalid, interview, October 8, 2004

Hishaam Aidi believes that the Islamic renaissance in contemporary Black America is related to some of the conditions that initially produced hip-hop culture: the withdrawal of state support, capital, social institutions, and welfare agencies from inner cities and the expansion of a powerful race-based system that left gaps of support for youth in black neighborhoods. Muslim organizations filled these gaps during the Reagan-Bush era. Aidi traces black youth conversion experiences to the breakdown of traditional family values, searches for identity and spirituality, and rejection of "permissive consumerist culture"—issues for which the faith of Islam provides answers. Overall, his study provides some provocative reasons for Islamic conversions among African American youth who are leading "nihilistic and chaotic lives" in the urban ghettoes. In this respect, Aidi's work resonates with most of the previous research on the interaction between Islam and hip-hop youth culture which focuses on the influence of the Nation of Islam and the Five Percenters (a splinter group of the Nation of Islam) on African Americans in large cities.[32]

However, the interviews that I conducted with African American Sunni Muslim men at the University of Iowa suggest a broader range of attraction to Islam among black youth in suburbs and college campuses and from middle- and upper-class households. In these diverse contexts, perhaps the attraction of and access to hip-hop's "consumerist culture" also has a lot to do with the acquisition of information about Islam that is central to their spiritual quests.

I first met the five young Muslim men who constitute the African American nucleus of the University of Iowa Muslim community as students in my course, "African American Islam in International Perspective" in September 2002.[33] As I taught my course, the American government was advancing its plans to invade Iraq despite worldwide grassroots opposition to war in the Middle East. The excitement and contemporary relevance of the discussions, the combination of graduate and advanced undergraduate students, and their excellent research papers fostered an environment that proved to be one of the most rewarding teaching experiences in my academic career. Our subject matter took us outside the classroom to the Iowa City Mosque and to an evening meal sponsored by the Muslim Student Association in which our class broke the fast with the Muslim community during Ramadan.[34]

In class and in the oral history interviews conducted in my office, it became clear that the African American Muslim men are a distinct community within the Muslim Student Association (with an estimated 130 members), which is dominated numerically by Sunni Muslim immigrants and Muslim Americans of South Asian and Arab descent. My interviewees are primarily recent converts to Islam who influenced each other's conversion experiences on the campus, with the exception of one second-generation Muslim who is affiliated with Imam Warith Deen Muhammad's American Society of Muslims. They are African American men in their twenties from middle-class households, and their understanding of race reflects the multiracial ethos of Islam—one member of their group is a white American convert who is married to an African American Muslim woman. Although their faith is based on striving to adherence to the general beliefs, practices, and legal and social issues of Islam,[35] their "masculine identity," that is, their religious and political consciousness, is mediated through the interaction between Islam and hip-hop culture. Hip-hop as an interpretive prism for these young men's spirituality is an important theme in the excerpts from their narratives that follow.

One of the Muslim converts on campus was spiritually inspired by "Islam's long intricate history in hip hop." He said that

> when the DJ culture first started to explode in the South Bronx, the Islamic identity was already part of the culture . . . Afrika Bambaataa . . . was known for his

Afrocentric style of dress. He wore Muslim prayer caps, kufis, and prayer beads around his neck . . . Artists like Rakim, Public Enemy, Queen Latifah, Stetasonic, Big Daddy Kane, and De La Soul [brought] their Islamic identity to the center of the music. While not all of these artists presented a Sunni version of Islam in their music, the images and words used in their music conveyed an Islamic theme . . . Beginning with the Afrocentric revolution in hip hop . . . The music has been a forum for Muslims to gain exposure and recognition of their beliefs in the public eye.[36]

For another young Muslim man, there was a direct connection between hip-hop culture and the conversions to Sunni Islam among his friends on campus, "Probably for a lot of us hip hop was an entry way to Islam . . . We gravitated toward the message of not only social consciousness but also of God consciousness . . . Me and Musa had a hip hop group . . . I ended up joining Musa's group because his message was [spiritually] conscious. We openly rapped about Islam before I accepted Islam."[37]

Also, Muslim rappers such as "common, Talib Kweli, and Hitek, the Roots, and Wu Tang Clan" are strong role models for politically and religiously conscious black men, according to one of my Muslim interviewees on campus.[38] He emphasized the influence of Mos Def, an orthodox Muslim MC, who, like Malcolm X, made the pilgrimage to Mecca. Mos Def's CD *Black on Both Sides* begins with the liner notes "Allah, the Lord of the worlds,"[39] dedicates his creative work to political prisoners such as Assata and Fred Hampton, Jr., and says "Free the Land!"[40] Nas's poetry about racism as a manifestation of "domestic terrorism"[41] in *Stillmatic* is also a significant influence for the aforementioned interviewee. Nas raps, "Been blessed with Allah's vision, strength, and beauty"[42] but the domestic terrorism of racism makes him "feel foreign"[43] although he is a citizen of the United States.

The reconstruction of black masculinity was a central theme in Khalid's conversion to Islam:

> I believe that the common thread for all of us was that we were at the university trying to find our places as Black men in Iowa . . . I received racially motivated e-mails . . . so I started getting political because in Chicago, being Black was not a big thing, I was one of a million Blacks. Here, I came to understand how I was different as a Black man. From the Black Panthers to Malcolm X, I read. It was an evolution . . . I was searching for my own personal identity. Who am I, what do I believe in? Until that time I was a Christian because that's what I was born in. I was searching to see what was true.[44]

Many young African American men reportedly have not found in some black churches a critical mass of youth programs that relate to the social and political issues that are important to them.[45] Instead, some black youth are seeking out Islam. Khalid's search for answers to the ultimate questions in his life first took him to a Protestant church near campus, a church that did not allow its members to question the African American pastor's interpretation of the Bible. Eventually, he found his manhood and spiritual answers in Islam, a religion that he first encountered in his rap group:

> This friend of mine, Amir, used to hang out with Musa who is a Muslim. Then Amir accepted Islam . . . He was also a rapper . . . Musa had a [rap] group . . . One day he invited me to mosque . . . I started going to Jumah. Then Ramadan came around. I was reading Malcolm X's autobiography—the part where he went on hajj . . . There was that point where I was real radical and had that hate going on for what my environment was. But then when Malcolm got to hajj, he eventually shed that hatred and I was seeing the same thing in front of my eyes. Seeing these different brothers greeting each other, asking for God's peace to be on you, feeding each other, doing acts of humility to each other, I could relate to Malcolm X's sincerity . . . Everything just seemed like this is it. This is internally what I was looking for . . . I eventually accepted Islam at the end of 2000.[46]

The above narrative provides evidence that the biography, the spirituality, and the political insights and courage of Malcolm X continue to inspire young people in hip-hop culture and to influence new converts to mainstream Islam. For some youth, Malcolm X is a powerful model that black men "who accept Islam in prison are not condemned but praised for accepting the religion . . . [and he is] the essence of Black manhood,"[47] according to an African American Muslim man on campus. Moreover, the current revival of serious research on Malcolm X in the academy and his numerous images in popular culture underline the fact that young black Americans in the hip-hop community currently receive more critical information about Islam through written and visual media and technology than any previous generation in American history.[48]

For one second-generation African American Muslim man on campus, the definitive experience in the construction of his masculinity was his participation in a Muslim funeral ritual in Boston with his father:

> When we went to the beginning of the Jinazah service, my father said . . . We have to clean his body from head to toe. We will be real delicate with the body . . . We used a bucket of soap and warm water. After we got finished cleaning him up he had a smile on his face. As we were doing it we said short prayers in Arabic, then we put some sandalwood musk oil on his head. We had incense burning in the background. There was no maliciousness, everyone's intentions were clean. Then we had three white sheets. In the end, we wrapped him in the sheets . . . then we placed him in his coffin. We bury him in his grave so his head faces east toward Mecca . . . Everyday I have no taqwa-God consciousness since I got back to Iowa City, I think about this.[49]

Although the brunt of the 9/11 backlash was borne by America's Middle Eastern population, the entire Islamic community felt some of the sting. With much of the post-September 11 media coverage focusing on the immigrant communities, the social and political perspectives of African American Muslims, who constitute roughly 30 percent of regular mosque participants,[50] have largely been

ignored. That is unfortunate because the ethnic identity of this group is shaped by
its long history of contributions to the American experience.

The African American Muslim students on Iowa's campus were inspired by the
poetry of Muslim and politically conscious rappers who offer critical analysis of
September 11. Nas, in *Stillmatic*, raps about the "domestic terrorism" of racism
and the external terrorism of 9/11 and war. He sees 9/11 as a global wake-up call
to end senseless hatred and killing.[51] The DPZ view the current state of affairs as
a domestic war on African American citizens, which is as serious as the international war on terrorism, a view expressed in their CD *Turn Off the Radio*. Their
repetitive chant—"That's war"—suggests ongoing terrorism against blacks that
has many sinister manifestations in the American system.[52] Chuck D., the MC of
Public Enemy, presents an argument against war in the CD *Revolverlution*, which
echoed the global antiwar protests against the American government's plans to
invade Iraq: "Seen four planes kill everyday folks, I guess 9/11 ain't no joke."
Chuck D then urges the rulers of the world and the terrorists to stop murdering
innocent people and to give the power to ordinary people to avoid a "sequel"
to war.[53]

The following excerpts from my interviews with African American Muslim
students reflect hip-hop's emphasis on the "hood"—the black urban landscape as
a site for political and social discourse and the tensions between young African
American Muslims and new Arab and Asian immigrants in the country, especially
in the wake of 9/11.[54]

A college student saw the difference between African American and Arab Muslim immigrants as a matter of devotion. "I've met a lot of Muslims in the U.S.
from Muslim countries ruled by the *sharia* who do not practice Islam, but their
parents do," he said. "Among immigrant Muslims, enthusiasm for Islam is at an
all-time low." Nevertheless, he felt that "it is a major priority and duty for the
African American Muslims to sustain and build the foundations for a strong community in America; not to separate from the Arabs but unite with them. Imam W.
D. Muhammad is trying to implement that now . . . The Muslim community in
America can be a stronger community when Arab and African-American Muslims
start to respect each other."[55]

The cultural differences between the two groups can be deep. "I think I have a
better picture of the political and cultural landscape of America than immigrant
Muslims," one young man said. "Most immigrants who come here don't go in the
hood. They don't see the inequalities between the races."[56] "I am not anti-Arab,"
said another. "I am just letting you know the divisions we have as a Muslim community." Then he added, "But there are also divisions between African-American
Muslim communities."[57]

Paradoxically, there may be something dark in American culture that could
help bind these communities together. "The reaction by the American public to
September 11 did not surprise me at all," said one respondent. "I don't know if
Muslim immigrants realized that they are looked down upon in America but they
are. America is the most racist country in the world."[58]

Finally, many young African American Muslims felt that the best response to the 9/11 backlash was to proclaim their faith. "I saw more African-Americans trying to physically express their religion by wearing *hijab* in the roughest neighborhoods such as the West Side of Chicago," said a young student whose introduction to Islam came from two hip-hop musicians. "African-Americans stood on Islam. I took a hard stance on how I felt about Islam; my father and his girlfriend were offended. They thought I had flipped [but] I never felt like I was going to be attacked. I wasn't afraid to proclaim Islam. I am a Muslim; it made me stronger."[59]

CONCLUSION

The information presented in this article suggests that the space and place of Islam's messages to African American youth has expanded from inner-city urban black communities in C. Eric Lincoln's time in the 1960s to America's urban and suburban university campuses in the twenty-first century. Hip-hop, North America's most important youth culture, is utilized as an interpretive prism through which some young Muslims construct meaning as African American men and believers. The insights of this research provide evidence that for some black youth on college campuses, rap music is more than just an "oppositional subcultural music" of resistance and pleasure.[60] Hip-hop culture is also a powerful medium through which students search for the answers to the ultimate spiritual and political concerns in their lives and their identities are paradigms for global Muslim youth.[61]

NOTES

1. C. Eric Lincoln, *The Black Muslims in America*, 3rd ed. (Trenton, NJ: Africa World Press, 1994), 25. An earlier version of this article was presented in the Afro-American History Group at the American Academy of Religion Annual Meeting in San Antonio, Texas, November 22, 2004.

2. Lincoln, *The Black Muslims in America*, 79–80.

3. Carolyn Moxley Rouse, *Engaged Surrender: African-American Women and Islam* (Berkeley: University of California Press, 2004), 5; Paula Giddings, *When and Where I Enter* (New York: Bantam Books, 1984), 317–18; Kevin K. Gaines, *Uplifting the Race* (Chapel Hill: University of North Carolina Press, 1996).

4. Ellis Cose, *The Envy of the World: On Being a Black Man in America* (New York: Washington Square Press, 2002), 15. There is a blossoming literature on twentieth-century and contemporary African American men. See the following select works: Alfred A. Young, Jr., *The Minds of Marginalized Black Men* (Princeton: Princeton University Press, 2004); Steven Estes, *I Am A Man: Race, Manhood, and the Civil Rights Movement* (Chapel Hill: University of North Carolina Press, 2005); Maurice O. Wallace, *Constructing the Black Masculine* (Durham: Duke University Press, 2002); Marlon Ross, *Manning the Race* (New York: New York University Press, 2004); Phillip Brian Harper, *Are We Not Men* (New York: Oxford University Press, 1996); and Herb Boyd and Robert Allen, eds., *Brotherman* (New York: One World, Ballantine Books, 1995).

5. Rudolph P. Byrd and Beverly Guy-Steftall, eds., *Traps: African-American Men on Gender and Sexuality* (Bloomington: Indiana University Press, 2001), xiii and xv.

6. Rouse, *Engaged Surrender*, 5.

7. Ishan Bagby, Paul M. Perl, and Bryan T. Froehle, *The Mosque in America: A National Portrait: A Report from the Mosque Study Project* (Washington, DC: Council on American Islamic Relations, 2001), 21. This report does not count African American participation in proto-Islamic communities such as the Nation of Islam.

8. The information in this article is part of a larger oral history project of interviews that I have conducted with African American Muslim men in California, New Orleans, Chicago, Boston, and Iowa since the early 1990s.

9. Grandmaster Flash, "Foreword," *The Vibe History of Hip Hop*, ed. Allan Light (New York: Three Rivers, 1999), viii.

10. Tricia Rose, *Black Noise: Rap Music and Black Culture in Contemporary America* (Hanover, CT: Wesleyan University Press, 1994), 34–35.

11. Marc Mauer, *Race to Incarcerate* (New York: The New Press, 1999); David Cole, *No Equal Justice* (New York: The New Press, 1999). Also see the articles in *Souls: A Critical Journal of Black Politics, Culture and Society* 2, no. 1 (Winter 2002), which focus on the theme "Race-ing Justice: Black America vs. the Prison Industrial Complex."

12. Kobena Mercer, "Endangered Species: Danny Tisdale and Keith Piper," *Artforum* 30 (Summer 1992): 75, and Thelma Golden, *Black Male: Representations of Masculinity in Contemporary American Art* (New York: Harry N. Abrams and the Whitney Museum of American Art, 1994), 19. See also Haki R. Madhubuti, *Black Men: Obsolete, Single, Dangerous?* (Chicago: Third World Press, 1990); Richard Majors and Janet Mancini Billson, *Cool Pose: The Dilemmas of Black Manhood in America* (New York: Lexington Books, 1992), and Michelle Wallace, *Black Macho and the Myth of Superwoman* (New York: Dial, 1979).

13. Eithne Quinn, *Nuthin' but a "G" Thang: The Culture and Commence of Gangsta Rap* (New York: Columbia University Press, 2005); Tricia Rose, *Black Noise*, 100; Michael Eric Dyson, *Holler if You Hear Me: Searching for Tupac Shakur* (New York: Basic Civitas Books, 2001); Nelson George, *Hip Hop America* (New York: Penguin, 1998); Imani Perry, *Prophets of the Hood: Politics and Poetics in Hip Hop* (Durham, NC: Duke University Press, 2004); Cheryl L. Keyes, *Rap Music and Street Consciousness* (Urbana: University of Illinois Press, 2002).

14. Charise L. Cheney, *Brothers Gonna Work It Out: Sexual Politics in the Golden Age of Rap Nationalism* (New York: New York University Press, 2005), 6.

15. Ibid.

16. Ibid.

17. Ibid

18. Ibid., 21.

19. Ibid., 19.

20. Anthony B. Pinn, ed., *Noise and Sprit: The Religious and Spiritual Sensibilities of Rap Music* (New York: New York University Press, 2003), 14.

21. Fricke and Ahearn, *Yes Yes Y'all*, 44; emphasis added.

22. Ibid., 55.

23. Joseph D. Eure and James G. Spady, eds., *Nation Conscious Rap* (New York: PC International Press, 1991); Todd Boyd, *The New H.N.I.C. The Death of Civil Rights and the Reign of Hip Hop* (New York: New York University Press, 2002), 150–51; Juan M. Floyd-Thomas, "A Jihad of Words: The Evolution of African American Islam and Contemporary Hip Hop," in *Noise and Spirit*, ed. Pinn, ed., 51–61; Ernest Allen, Jr., "Making the Strong Survive: The Contours and Contradictions of Message Rap," in *Droppin' Science: Critical*

Essays on Rap and Hip Hop Culture, ed. William Eric Perkins (Philadelphia: Temple University Press, 1996).

24. See Richard Brent Turner, *Islam in the African-American Experience*, 2nd ed. (Bloomington: Indiana University Press, 2003), 234–35 for details about U.S. immigration laws and Muslim immigrants in the late twentieth century. Yvonne Haddad, "Muslim Communities and the New West." *Middle East Affairs Journal* 5, nos. 3–4 (Summer–Fall, 1999): 14.

25. Haddad, "Muslim Communities and the New West," 17.

26. Jay Z's CDs include *Reasonable Doubt* (Roc-A-Fella, 1996), *In My Lifetime, Vol. 1* (Roc-A-Fella, 1997), *Hard Knock Life, Vol. 2* (Roc-A-Fella, 1998), *The Dynasty: Roc La Familia* (Roc-A-Fella, 2000), *The Life and Times of Sean Carter, Vol. 3* (Roc-A-Fella, 2001), *MTV Unplugged* (Roc-A-Fella, 2001), and *The Blueprint 2* (Roc-A-Fella, 2003).

27. *Mos Def and Talib Kweli are Black Star* (Rawkus Records, 1998, 2002).

28. Obie Clayton and Joan W. Moore, "The Effects of Crime and Imprisonment on Family Formation," in *Black Fathers in Contemporary American Society*, ed. Obie Clayton, Ronald B. Mincy, and David Blankenhorn (New York: Russell Sage Foundation, 2003), 85–86; S. Craig Watkins, *Representing: Hip Hop Culture and the Production of Black Cinema* (Chicago: University of Chicago Press, 1998).

29. Spike Lee with Ralph Wiley, *By Any Means Necessary* (New York: Hyperion, 1992); "The Inmates of 'Oz' Move into a New Emerald City," *New York Times*, July 15, 2001, pp. 24, 33.

30. *South Central* (Warner Brothers, 1992).

31. Nathan McCall, *Makes Me Wanna Holler: A Young Black Man in America* (New York: Random House, 1994); Sonyika Shakur, *Monster: The Autobiography of an L.A. Gang Member* (New York: Atlantic Monthly, 1993); Alex Haley and Malcolm X, *The Autobiography of Malcolm X* (New York: Ballantine Books, 1965); Robert Dannin, *Black Pilgrimage to Islam* (New York: Oxford University Press, 2002); Aminah McCloud and Frederick Thaufeer al-Deen, *A Question of Faith for Muslim Inmates* (Chicago: ABC International Group, 1999).

32. Hishaam Aidi, "Jihadis in the Hood: Race, Urban Islam, and the War on Terror," *Middle East Report* 224 (Fall 2002): 38–40; Floyd-Thomas, "A Jihad of Words" 49–70; Yusuf Nuriddin, "The Five Percenters: A Teenage Nation of Gods and Earths," in *Muslim Communities in North America*, ed. Yvonne Y. Haddad and Jane I. Smith (Albany: State University of New York Press, 1994), 109–32; Brett Johnson and Malik Russell, "Time to Build," *Source: The Magazine of Hip Hop Music, culture, and Politics* 158 (November 2002): 118–22. The Five Percent Nation (The Nation of Gods and Earths) was founded in Harlem by Clarence 13X in 1964. They focus on the esoteric literature of W. D. Fard and believe that all black men are gods. They see themselves in the 5 percent of African Americans who are the "poor righteous teachers" of their race. In their beliefs, Harlem is Mecca and Brooklyn is Medina. Felicia M. Miyakawa, *Five Percent Rap: God's Hop's Music, Message, and Black Muslim Mission* (Bloomington: Indiana University Press, 2005).

33. See Garbi Schmidt, *Islam in Urban America: Sunni Muslims in Chicago* (Philadelphia: Temple University Press, 2004), chap. 4 on the Muslin Student Associations. Currently, there are an estimated 15–20 African-descended Muslims on the University of Iowa campus.

34. See Marcus K. Hermansen, "Teaching About Muslims in America," in *Teaching Islam*, ed. Brannon M. Wheeler (New York: Oxford University Press, 2003).

35. McCloud and Thaufeer al-Deen, *A Question of Faith for Muslim Inmates*, iii.

36. Interview with Musa. "Islamic Influence and Identities in Hip Hop," Musa's research paper, December 2002, n.p.

37. Interview with Khalid, October 8, 2004.
38. Interview with A.J., February 26, 2002.
39. Mos Def, *Black On Both Sides* (Rawkus, 1999).
40. Ibid.
41. Nas, *Stillmatic* (Ill Will Records, 2001).
42. Ibid.
43. Ibid.
44. Interview with Khalid, October 8, 2004.
45. For studies of the African American Church in the post–civil rights era, see Anthony B. Pinn, *The Black Church in the Post-Civil Rights Era* (Maryknoll, NY: Orbis Books, 2003); R. Drew Smith, ed., *New Day Begun: African-American Churches and Civic Culture in Post Civil Rights America* (Durham: Duke University Press, 2003); and Smith, *Long March Ahead: African-American Churches and Public Policy in Post Civil Rights America* (Durham: Duke University Press, 2004).
46. Interview with Khalid, October 8, 2004.
47. Ibid.
48. Edward Rothstein, "The Personal Evolution of a Civil Rights Giant," *The New York Times*, May 19, 2005, B1, B7; Gerald Fraser, "Manning Marable's Project to Restore the Legacy of Malcolm X," *Journal of Blacks in Higher Education* (Summer 2001); Lewis V. Baldwin and Amiri YaSin Al-Hadid, *Between Cross and Crescent: Christian and Muslim Perspectives on Malcolm and Martin* (Gainesville: University Press of Florida, 2002).
49. Interview with Dirul, October 7, 2002.
50. Bagby, Perl, and Froehle, *The Mosque in America*, 17, 21.
51. Nas, *Stillmatic* (Ill Will Records, 2001).
52. DPZ, *Turn Off the Radio*, the Mix Tape, Vol. 1 (Holla Black, 2002).
53. Public Enemy, *Revolverlution* (Koch Records, 2002).
54. See Jamillah Karim, "Between Immigrant Islam and Black Liberation: Young Muslims Inherit Global Muslim and African-American Legacies," *The Muslim World* 95, no. 4 (October 2005): 497–513.
55. Interview with Dirul, October 7, 2002.
56. Interview with A. J., April 2 and 9, 2002.
57. Interview with Dirul, October 7, 2002.
58. Interview with A. J., April 2 and 9, 2002.
59. Interview with Khalid, October 3, 2002.
60. Yasue Kuwahara, "Power to the People Y'All: Rap Music, Resistance, and Black College Students," *Humanity and Society* 16, no. 1 (1992): 56.
61. See Manning Marable, *The Great Wells of Democracy: The Meaning of Race in American Life* (New York: Basic Civitas Books, 2002), chap. 10; Cornel West, *Democracy Matters: Winning the Fight Against Imperialism* (New York: Penguin, 2004), chap. 6; Hishaam Aidi, "Let Us Be Moors: Islam, Race and 'Connected Histories,'" *Souls: A Critical Journal of Black Politics, Culture, and Society*, 7, no. 1 (Winter 2005): 36–51.

CHAPTER 9

THROUGH SUNNI WOMEN'S EYES

BLACK FEMINISM AND THE NATION OF ISLAM

JAMILLAH KARIM

"ISLAM HELPS MEN A GREAT DEAL BECAUSE IT TEACHES THEM HOW TO TREAT THEIR women," stated Sister Levinia X.[1] E. U. Essien-Udom cited Sister Levinia in his book published in 1962, *Black Nationalism: A Search for an Identity in America.* Over forty years later, Sunni African American Muslim women continue to make this claim about Islam—that it demands the fair treatment of women—despite the fact that a large percentage of Americans see Islam as a religion that oppresses women. Not only do African American Muslim women assert this positive claim about Islam but also a growing number call themselves feminists. Most Muslim women, however, find the label unnecessary because they believe that identifying themselves as Muslim inherently assumes a commitment to gender justice. As Carolyn Rouse writes in her study of Sunni African American Muslim women, "Muslim women recognize Islam to be the first 'feminist' monotheistic religion," that is, they see it as the first religion to bring systematic social reform and rights to women in a patriarchal society.[2]

Rouse effectively argues that black feminists should see African American Muslim women converts as feminists. This article seeks to do the same through a method absent in most studies on the Nation of Islam (hereafter, "the Nation"). Historical accounts and analyses of the Nation, including Sherman Jackson's *Islam and the Blackamerican* and Michael A. Gomez's *Black Crescent,* construct how we should remember the Honorable Elijah Muhammad's Nation of Islam. Few studies, however, tell us how Sunni African American Muslims who are former followers of Elijah Muhammad remember the Nation's legacy. How do Elijah Muhammad's past followers, now followers of his son Imam W.D. Mohammed, remember and teach their children how to remember the Nation? This is possibly the most important way that the Nation is remembered, that is, by the community of people who experienced it firsthand and who pass down their Nation stories to later generations. These ex-member accounts are likely to have the greatest impact on how third-, fourth-, and fifth-generation Muslims will imagine their American Muslim past, which includes how African American Muslim women will imagine their long legacy of feminism. I argue that their feminist consciousness as Sunni Muslim women manifests in compelling ways as they reconstruct their Nation past. How Sunni African American Muslim women remember the Nation reveals not only their place within black feminist tradition but also the enduring contribution that they bring to this tradition.

I was raised in a Sunni African American Muslim community in Atlanta, Georgia, under the leadership of Imam W. D. Mohammed. My parents joined the Nation of Islam in 1971 but turned toward Sunni Islam in 1975 under the direction of Imam W. D. Mohammed. I grew up hearing Nation stories not only at home but also at the mosque. Sometimes they were mentioned in religious lectures during the Friday congregational prayer. Although my parents and the other ex-Nation members rejected the race theology of the Nation, they recalled and celebrated their Nation history because they saw it as an important step to Islam as they practice it now. As my mother described, "I had been a Christian, then I joined the Nation, and now came Islam. And it was beautiful. It was like another shackle was being lifted from my mind and soul. This was it," she said. "I loved the idea that there was this One God. It was wonderful. Whatever the prayers of our ancestors were when they were taken from their homes and lands, I believe Allah has answered their prayers through us."

FEMINIST NARRATIVES OF THE NATION

Many scholars, including Gomez, have described how the Nation reduced women to their biological capacities. In the Nation, "the black woman's body," Gomez writes, "is seen primarily as a tool of procreation and reproduction." Elijah Muhammad called black men to protect black women "not really" for the benefit of women but to preserve black men's "proprietary rights" over black women's bodies.[3] While this analysis cannot be ignored, it does not represent how most Sunni African American Muslim women talk about their Nation experiences. Since Muslim women often recall Nation stories to honor it as their doorway to Islam, they tend to highlight positive aspects of the Nation more than they do negative ones. This does not mean that they never identify the problems that they experienced. Indeed, describing the Nation's problems helps to shed light on why they ultimately chose Sunni Islam over the Nation. Similarly, the favorable memories that they share of the Nation usually indicate Nation practices or ideas that are compatible with Sunni Islam.

In this article, I analyze the Nation stories of two middle-aged Sunni women, Lynda and Marjorie. Both joined the Nation in 1971 but now follow the Sunni

leadership of Imam W. D. Mohammed. Lynda was raised in Woburn, Massachu-
setts. Now married with four children, she currently lives in Marlton, New Jersey.
She owns her own day care in Lawnside, New Jersey. Marjorie was born and raised
in Washington, D.C. Divorced with four children, she works as a project manager
in an internationally renowned consulting firm. She currently lives in Atlanta,
Georgia. Both women are exceptionally active in their mosque communities.
Marjorie is one of the founding members of the Muslim women's organization,
Sisters United in Human Service (founded in 1998). In 1999, Lynda and her
family spearheaded the building of a mosque in Lawnside that they named Masjid
Freehaven to commemorate the Underground Railroad. I conducted interviews
with both Marjorie and Lynda in March and April of 1999. Neither of the two
women identified herself to me as a feminist; however, I analyze their consciousness and acts of agency as feminist.

As Gomez indicates, protecting the black woman's body is an important part
of Nation discourse. In Sunni Islam, the most emphasized reference to the female
body is the duty for women to protect it through self-covering. Covering includes
not only the hair but also the chest, legs, and arms. Some people find Muslim
women's covering oppressive; however, the covering transfers the obligation to
protect women's bodies from men to women themselves. In other words, Muslim
women take it upon themselves to protect their bodies from the male gaze. I call
this a feminist move since agency is transferred to women, on the condition, of
course, that Muslim women are covering what they wish as a personal choice,
which is the case for most Muslim women in the United States. Certainly, the
covering, specifically the headscarf, can and does function as a symbol of oppression when women are forced to cover. In contrast, however, the headscarf functions as a symbol of piety and Muslim identity when it reflects women's choice.
For Muslim women, dress highlights "the role the body plays in the making of the
self, one in which the outward behavior of the body constitutes both the poten-
tiality and the means through which" one cultivates her inner spirituality and/or
projects her religious and cultural loyalties.[4] Still, it is possible that some women
who choose to cover do so out of subtle, or not so subtle, community or family
pressures to conform to traditional gender expectations. Notions of male control
may inform some of these expectations but even in such cases, Muslim women
discover and articulate benefits of covering. As anthropologist Saba Mahmood
and others have argued, it is possible for women to practice agency within accommodation to established gender expectations. In the Nation narratives below, the
women describe how the requirement to cover attracted them to the Nation. I
propose that their current Sunni position, that is, to cover their bodies, explains
why covering is a constant theme celebrated in their Nation narratives. Other
practices consistent between the Nation and Sunni Islam also emerge.

Lynda highlights woman's covering when she tells her Nation stories. Instead
of protection, she describes another benefit of covering: "Coming into the Nation
where you cover up the hair and everything else, take off any makeup, it was *all*
right. I didn't have a problem with it because that wasn't really me anyway." The

requirement to cover her hair in the Nation relieved Lynda of the notion that physical appearance determined her worth and value. "Short hair, dark skin, that was me. Skinny, skinny legs." As a little girl, Lynda felt less beautiful than the white girls at her school in Woburn, Massachusetts. When she was older, she took up modeling because "I had to feel that I looked just as good as them." Lynda gained confidence from her modeling because she learned how to wear makeup and fix her hair. But when the Nation taught her to cover her hair and remove the makeup, Lynda embraced this as what was naturally and truly her. For Lynda, the Nation signified honoring one's natural beauty as a black woman.

In addition to covering, another Sunni practice that Muslim women experienced in the Nation before coming to Sunni Islam is separation between men and women for some activities. In Sunni Islam, men and women separate during the prayer to prevent physical contact, and depending on how conservative the mosque community, men and women separate during social events as well. Lynda remembers this separation in the Nation and speaks favorably of it as helping to create sisterhood. The separation of men and women in temple activities and services gave the women "our own situation," Lynda described. "I saw a lot of women that were in here and they were like welfare people, struggling sisters, uneducated. I enjoyed that we were able to force them to read and all kinds of things to do to make them better." Lynda feels that coed activities would have served as a distraction at the expense of progress. "You are competing for the opposite gender to notice you, whereas if you are separate, you are able to progress a lot better [and] become a more well-rounded, self-fulfilled person. And then you enter into the arena" to deal and compete with both men and women.

Lynda's representation of the Nation not only teaches us about her past but also her present. As Sarah Lawrence-Lightfoot asserts, our reflection and perspective in the present render a "deliberate and imaginative" reconstruction of our past.[5] In other words, the past becomes colored by the present. Therefore, Lynda's Nation narrative reflects her current Sunni Muslim perspective that women should have autonomy over their bodies through dress and some level of separation from men. This perspective certainly represents a feminist position when it functions to empower women.

From both a Sunni and Nation perspective, however, this autonomy does not refer to women's interactions with their husbands. Sunni Islam instructs both women and men to be sexually intimate only with their spouses. Islamic law gives husbands the right to their wives' bodies upon marriage and expects for women to be intimate with their husbands except when menstruating or ill. This expectation generally does not create controversy. The marital expectation that does, however, is the Qur'anic statement that "husbands should take full care of their wives, with [the bounties] God has given to some more than others and with what they spend out of their own money" (4:34 Abdel Haleem translation). The more popular translation of this verse is Yusuf 'Ali's translation, "men are the protectors and maintainers of women, because Allah has given the one more (strength) than the other, and because they support them from their means." Sunni Muslims debate both how this verse should be translated and how it should be interpreted. Does it mean that men are providers and women are not? Does it mean that women are inherently inferior to men and therefore unable to adequately take care of themselves? Or is it simply a recommendation for men to take care of their wives? If so, why does the Qur'an make this recommendation?

Both Amina Wadud and Carolyn Rouse argue that this verse could not possibly mean that men are the providers of women at all times, in all cases. This simply is not reality. On the contrary, Wadud argues in *Qur'an and Woman* that the Qur'an is reminding men of their responsibility to take care of women in the context of "the responsibility and right of women to bear children." The Qur'an specifies the man's responsibility to provide "material maintenance" to see "to it that the woman is not burdened with additional responsibilities which jeopardize that primary demanding responsibility that only she can fulfil."[6] African American Sunni women tend to interpret this verse in ways that converge and diverge with Wadud's. Unlike Wadud, many Sunni women do essentialize men as natural providers (even though they may not live up to their nature). But, like Wadud, they do not take the verse to mean that women are naturally unable to provide for themselves. Instead, they see this verse as a challenge for men to live up to their God-given role and act as responsible men who take care of their families. "The rhetoric of patriarchy," that is, that men are the maintainers of women, "may be deployed [by Sunni African American women] not to make women submissive," Rouse writes, "but to instill in men a sense of responsibility."[7] It is this emphasis on responsibility that comes across in Marjorie's reconstruction of the Nation. Here I tell her story with her words in quotations.

Until the age of five, Marjorie had a father in the home. He provided for his family of seven as a chauffeur, with another job on the side. This livelihood landed them in the projects of Washington, D.C. After her parents separated, Marjorie's mother "had to go on welfare because she believed in staying home with her five children." Due to her mother's example, Marjorie envisioned the ideal woman as one having "a husband to take care of her." Hence, her ideal man was one who could provide.

With the separation of her parents, however, Marjorie did not see this ideal fulfilled in her childhood. Instead, she witnessed her mother's live-in boyfriend abuse her mother after many evenings of drinking. Marjorie explained, "I didn't like the boyfriend-girlfriend thing, especially in that class of folks. A lot of them would escape from their lifestyles and would indulge in drinking." Although her mother participated in this lifestyle, she still emerged as a positive role model for Marjorie. She recalls how her mother went back to school while raising her children, entered a profession, and moved her family off of welfare by the time Marjorie was a teenager. However, no positive images of men emerge in Marjorie's memory. From the men in her childhood, she knew only what she did not want in a husband. With a bitter voice, she expressed, "I didn't want a man that drank. I had an uncle that drank. I saw how he acted, and I didn't want that."

What piqued Marjorie's early interest in the Nation, however, relates more to her image of herself than to the image of the men and women in her life. Marjorie believed that the blacker you were, the less beautiful you were. As a little girl, she would spend hours in the tub after playing under the hot sun hoping to wash away the sun's mark. Thus, when she first heard the phrase "black is beautiful" at age fifteen, she said, "it was one of the most beautiful and refreshing ideas that ever came to my mind. It was almost like my mind was being unshackled—[to know] that black was beautiful, that I'm beautiful, that brown was beautiful, that kinky hair could be beautiful, thick lips could be beautiful, and it was wonderful because this was me." Likewise, the Nation taught Marjorie to love her blackness. Marjorie said, "I wanted to go into something that would make me feel good about myself, and in the Nation they said, 'The black man is God.'" With a big hearty laugh, Marjorie remembers her response as "Wow! He is?!"

The discipline of the Nation men also appealed to Marjorie. She says, "They were in the forefront selling their papers, well-dressed and well-mannered . . . They respected you . . . If they wanted to be with you, they wanted to marry you. They weren't into the drugs. If they were, they stopped." This "very clean, decent, and disciplined life" appealed to Marjorie "because it was closer to being pious." The Nation women's dress appealed to her for the same reasons. Marjorie loved the head covering because "when I was Catholic, I wanted to be a nun. All the nuns that we knew were African American, and they just seemed so pious." Although her childhood fantasy was to become a nun, she also wanted to have children. As a Black Muslim, she could have it all. As a Muslim, "you got to wear all the different colors . . . and still look modest like [the nuns], and then of course I could have babies." The headpiece and modest uniform made Marjorie feel distinguished. "Not only were we black and proud . . . we were the supreme race of people on the earth . . . Because we dressed differently, we definitely stood out back then."

For Marjorie, the Nation taught that "the best woman was a modest woman. A woman was someone who took care of her home and family. I wanted to do that. She was good to her husband, submissive to her husband. I had no problem with that." Because of this ideal of strong dedication and commitment to the husband that a Muslim woman expected to find reciprocated in her husband's treatment of her, Marjorie saw that the Nation set high, comparable standards for both men and women. The Nation "pumped both of us up. When they would pump up one, they were pumping up the other. I didn't see any difference from the other." As far as staying at home with the children, Marjorie celebrated this aspect of womanhood. "We were taught *even then* that Paradise lies at the foot of the mother," Marjorie said, referring to a statement attributed to the Prophet Muhammad.

Analyzing Marjorie's narrative, we can see why she found the Nation's ideology appealing. Foremost, it ennobled her perception of herself and her race. Also, the male-provider role in the Nation reflected for her the foundation on which to restore strong family life. This way of seeing and embracing the Nation can be described as a black feminist perspective. As bell hooks has argued, a black feminist perspective develops from the everyday experiences of black women as they face race, class and other forms of discrimination.[8] Recognizing multiple forms of discrimination is essential to understanding black feminism. African American women experience gender injustices within broader struggles of racism and economic exploitation. Their feminism does not privilege "gender over race or class," which means that black feminism broadens its scope from the individual to the family and community.[9]

In the spirit of the collective good, Nation women embraced gender roles that placed men as breadwinners and women as homemakers. As Sister Levinia X stated, the Nation "teaches women how to raise their children, how to take care of their husbands, how to sew and cook, and several domestic things which are necessary for a family."[10] And this was empowering for women who wanted better families. At the same time, there were aspects of the Nation that women did not embrace. But this did not cause them to reject the Nation. They believed that the economic and moral progress of African Americans depended on the progress of men and women; therefore, they absolutely maintained alliances with their men to combat racism. Like black women in the church, Nation women positioned themselves within "a tradition of protest and cooperation," what Cheryl Townsend Gilkes calls "a dialectical tradition."[11] Evelyn Higginbotham's fascinating book *Righteous Discontent* also describes how black churchwomen maintained alliances with black men while simultaneously challenging male hegemony in the church. They broadened racial uplift work beyond the confines of the church and formed women's organizations to expand their reach in the broader community: "For women, race consciousness did not subsume or negate their empowerment as women but rather encouraged it."[12] In their eyes, commitment to the race required that black women challenge the limits placed on them by black men.

Similarly, Nation women accommodated their roles at the same time that they resisted them. Marjorie remembers how the Nation's mandate that men could not work for the white man left her family surviving on her husband's sales of bean pies and fish. "It was like living from day to day." Marjorie helped by making bow ties, sandwiches and money pouches for sale. "I told him that I was making it so that I could buy myself a dress, but it always had to go to pay the bills." Marjorie was honoring Nation ideology by living the role of a domestic but at the same time challenging the idea that only men provided. Whether the extra money went to purchase her personal items or to pay the bills, she understood the value of her work to her family. By announcing to her husband how she wanted the money to be used, she recognized the personal rights and benefits due to her for her hard work in the home. She deserved at least a new dress.

Other Nation women, like Lynda, did work outside the home. Lynda worked partly out of necessity but also because "I just wanted to have self-worth." Lynda did not receive any criticism for working because she was the lieutenant over the females at her temple; hence, "they listened to what I said." Lynda resisted the fancies of Nation men who "wanted you to be barefoot and pregnant" and expanded the scope of motherhood. And she did this while remaining loyal to the Nation. Ironically, the very structure of the Nation gave women the agency to reconstruct defined gender roles to work in their favor. The Muslim Girls Training and Civilization Class (MGT and CC), a core activity for Nation women, created an autonomous working space for them. Lynda described the intensity of their work: "We were at the temple all the time—three, four, or five days a week. Coming out to the classes, you were learning your lessons, you were drilling, you were learning how to cook. I was responsible for going over the Honorable Elijah Muhammad's lecture. I would prepare tests. We would have discussions. We were just busy."

The purpose of MGT and CC was to train Nation women to be good wives and mothers. In reality, however, it became the space for broadening the scope of motherhood to community activism. This ability to transform potentially limiting roles for women into ones that empower women dates back to slavery. In the living quarters, enslaved women were responsible for the domestic duties such as cooking, sewing, cleaning, and rearing the children. While domestic work limited white women's influence and power in the community, enslaved women were "performing the *only* labor of the slave community that could not be directly and immediately claimed by the oppressor":

> Precisely through performing the drudgery that has long been a central oppression of the socially conditioned inferiority of women, the black woman in chains could help to lay the foundation for some degree of autonomy, both for herself and her men. Even as she was suffering under her unique oppression as female, she was thrust by the force of circumstances into the center of the slave community. She was, therefore, essential to the *survival* of the community. Not all people have survived enslavement; hence her survival-oriented activities were themselves a form of resistance. Survival, moreover, was the prerequisite of all higher levels of struggle.[13]

Similarly, Lynda interprets the Nation's gender roles as a strategy of survival, not a mechanism intended to oppress women: "Really when they were saying that the woman should be at home, taking care of their kids, basically, they were trying to say, 'Woman your role is to nurture,' . . . though not realizing that a woman can nurture in all aspects of society that she's involved in." Hence, Lynda expands the significance of motherhood in the same moment that she defends the efforts of the Nation. She is defending her men even as she argues for women's expanded social roles. Justifying the Nation's ideology, Lynda said, "The Nation, probably in a more simplistic way, broke down and defined the role of male and female at its basic level. The Nation was like a seed and a sign of putting things back in its form."

For Lynda, "putting things back in its form" means for African American men and women to reclaim the "different but equal roles that God gave originally" in order to restore stable families. During slavery, traditional gender roles were undermined as slave women were often expected to produce as much as men while still having to reproduce and endure critical phases of motherhood such as pregnancy and nursing. Unfortunately, the legacy of struggle that has made

African American women capable of both providing and mothering has been betrayed as many African American men have left the financial and emotional support of children to women alone. Thus, Lynda's notion of the Nation "putting things back in its form" means that the Nation worked to make both men and women accountable for their families. Thus, Nation women embraced gendered roles because they returned responsibility to men.

In this way, Nation women were engaging race, class, and women's liberation simultaneously. Mutual responsibility would mitigate the burdens of African American women who too often live in poverty and raise their children alone and, at the same time, uplift the entire race. However, Nation women were not willing to surrender the valuable role outside the home that their legacy of survival had afforded them. Tynnetta Deanar, a regular female columnist for the Nation newspaper in the 1960s, showed this commitment to work outside the home when she stated, "The Muslim woman is able to coordinate family responsibilities and community obligations as she educates on a level not only local and national but international in scope."[14] In one statement, Deanar has transformed herself from a mother, teaching children in the home, to an international activist in the name of education. And this statement was made not behind closed doors but published in the paper.

The Nation was empowering for women because it gave them the most basic seeds for community building—positive self-image and the ability to both take on and insist responsibility. Marjorie said, "Whether the Nation was liberating depends on what you consider liberating. The fact that we weren't doing all kinds of crazy stuff out there in the street, and we were serious about taking care of our families and communities, yes, it liberated in that sense." Lynda said, "The Nation gave me a place to develop the confidence that I needed. It was a womb that got me ready to come out into the world." Marjorie's and Lynda's voices, though distinct, reflect the spectrum of voices of Sunni women, once Nation women.

FEMINIST CONSCIOUSNESS IN SUNNI ISLAM

In her study of Sunni African American Muslim women in Los Angeles, Carolyn Rouse demonstrates how Sunni Muslim women continue to emphasize men's responsibility to support women. She places them within black feminist tradition because they claim gender roles to improve African American families and communities. However, she sets Muslim women apart from black feminists. "Muslim women are unabashedly clear about the methods they use to arrive at truth."[15] They rely on the Qur'an and the Sunnah (prophetic example) to form their understanding of gender roles and practices. Black feminists, on the other hand, are not confined to any textual sources to define their feminist theory. Instead, black women's experiences of multiple forms of discrimination define their feminism. While this flexibility presents an advantage, the religious foundation of Muslim women's feminist strategies also indicates a unique contribution.

For African American Muslim women, the need to insist men's responsibility arises from everyday experiences, for example, their experiences as single mothers. Again, everyday struggle is what situates Muslim women within black feminist tradition. Marjorie's narrative demonstrates this very clearly. As Sunni Muslims, however, this preexisting need becomes sanctioned by the Qur'an through the verse "husbands should take full care of their wives." Now that they assume Qur'anic sanction, their preexisting black feminist strategies to ensure men's accountability appear even more persuasive and essential. Sunni Islam further instills the idea of men's responsibility through the Islamic ruling that whatever income a wife brings in or whatever property she owns, all of it belongs to her and her alone. On the other hand, a husband is expected to maintain his wife and children with his earnings. Of course, this comes with rights that men have over women, including women's obedience to men according to most traditional and modern interpretations. (There are also feminist interpretations of what obedience means.)

As they did in the Nation, many Sunni women continue to work despite this Qur'anic expectation for men. Lynda, for example, continues to work, and she remains married to the same responsible man to whom she was married in the Nation. However, Lynda emphatically defends why the Nation idealized women's work in the home and men's work outside. Her Sunni understanding, I argue, explains her commitment to the fundamental position that men should be pushed to work. Having established this basic expectation, women's roles can be interpreted in multiple ways, for example, nurturing in the home and in the community.

I imagine that the experiences of African American women, Muslim and non-Muslim, will make male accountability a critical goal in black feminist agendas. However, one alternative to focusing on male accountability in African American communities is encouraging nontraditional family networks. Most nontraditional families tend to be female-centered extended kin networks. They function without a consistent male presence. Instead of men, women, including mothers, grandmothers, aunts, and sisters, assume primary responsibility for themselves and their children. In Carol Stack's study *All Our Kin*, published in 1974 when Lynda and Marjorie were still in the Nation, she described such networks: "Commonly, the mother's personal domestic network includes the personal networks of her children, who are half siblings with different fathers . . . Mothers expect little from the father; they just hope that he will help out. But they do expect something from his kin, especially his mother and sisters."[16]

By 2000, female-centered kin networks, or female households, characterized many African American families. Census figures show that in 1999, 30 percent of black households were maintained by women without a spouse present. This is three times the percentage of total U.S. households maintained by women (11.8 percent). Only 5.7 percent of black families were maintained by men without a wife present. If we take Central Harlem as one example, studies in the 1990s found that "69 percent of all families are headed by women" and "54.3 percent of all households headed by women that include children under eighteen have incomes below the poverty line."[17] In 1999, the median household income for all black families was $33,300, compared to $50,000 for all American families. Households headed by African American women significantly influenced this disparity. African American married-couple families had a median income of $50,700, whereas African American families maintained by women had an annual income of $20,600.[18]

These numbers clearly indicate that female-centered kin networks are not a viable alternative. Sunni African American Muslim women know by experience that they are not. And because the Qur'an clearly asserts that men are expected to support women, they know it with a firm conviction. Muslim women's commitment to this expectation as they struggle to improve African American families marks their contribution to black feminist tradition. In other words, the Qur'anic expectation for men poses a unique set of questions that African American Muslim women bring to black feminist thought: is the male-provider role an ideal for promoting balanced responsibility between men and women in the family unit? Never does the Qur'an make a general statement about women akin to its assertion about men. Does the Qur'an's greater emphasis on the role of men anticipate social realities that demonstrate that men are more likely to threaten mutual accountability? The economic consequences of racism make providing difficult for African American men and women. Is the Qur'an's concern with men a divine acknowledgement that men are more likely to abandon their families? Childbearing motivates women to provide. What motivates men? Is it healthy sexual and spiritual relations with women? This would mean that men's incentive to provide is contingent on, and, therefore, threatened by, external challenges. Carol Stack's study supports this idea as she shows that men are less likely to provide once they end relations with the mothers of their children.

In cases when men are committed through marriage, do men still need the reminder to support women? Are men likely to do their share in the family without defined roles? Or are gendered roles necessary for basic community building, to ensure that men will do their part? What then happens when quality community life is established? Is it still necessary to promote some form of gendered roles to ensure that mutual responsibility will be maintained? Or can black men ever be expected to adequately support women in a society in which they are demonized? Does the male-provider role maintain patriarchy since it assumes women's dependency? If so, is this accommodation of patriarchy worth the benefit that Muslim women claim, that is, restoration of the family? As African American Sunni women engage these complex questions, they influence the direction and broaden the scope of black feminist thought. Since they are included among black women, the experiences, ideas, and questions of Sunni African American women represent part of black feminist tradition. Imagining Muslim women outside this tradition denies their experiences as black women.

NOTES

1. Essien Udosen Essien-Udom, *Black Nationalism: The Rise of the Black Muslims in the U.S.A* (Harmondsworth: Penguin, 1967), 86.

2. Carolyn Moxley Rouse, *Engaged Surrender: African American Women and Islam* (Berkeley: University of California Press, 2004), 150.

3. Michael Angelo Gomez, *Black Crescent: The Experience and Legacy of African Muslims in the Americas* (Cambridge: Cambridge University Press, 2005), 324.

4. Saba Mahmood, *Politics of Piety: The Islamic Revival and the Feminist Subject* (Princeton: Princeton University Press, 2005), 159.

5. Sara Lawrence-Lightfoot, *I've Known Rivers: Lives of Loss and Liberation* (Reading: Addison-Wesley, 1994), 11.

6. Amina Wadud, *Qur'an and Woman: Rereading the Sacred Text from a Woman's Perspective*, 2nd ed. (New York: Oxford University Press, 1999), 72, 73.

7. Rouse, *Engaged Surrender*, 16.

8. bell hooks, "Black Women: Shaping Feminist Theory," in *Words of Fire: An Anthology of African-American Feminist Thought*, ed. Beverly Guy-Sheftall (New York: New Press, 1995), 278.

9. Rouse, *Engaged Surrender*, 143.

10. Essien-Udom, *Black Nationalism*, 86.

11. Cheryl Townsend Gilkes, "'Together and in Harness': Women's Traditions in the Sanctified Church," *Signs* 10, no. 4 (1985): 697.

12. Evelyn Brooks Higginbotham, *Righteous Discontent: The Women's Movement in the Black Baptist Church, 1880–1920* (Cambridge, MA: Harvard University Press, 1993), 166.

13. Angela Davis, "Reflections on the Black Woman's Role in the Community of Slaves," in *Words of Fire: An Anthology of African-American Feminist Thought*, ed. Beverly Guy-Sheftall (New York: New Press, 1985), 205.

14. Tynnetta Deanar, "Muslim Woman Is Model Personality," *Muhammad Speaks*, June 1962, 15.

15. Rouse, *Engaged Surrender*, 148.

16. Carol B. Stack, *All Our Kin: Strategies for Survival in a Black Community* (New York: Harper & Row, 1974), 53.

17. Leith Mullings, "Households Headed by Women: The Politics of Race, Class, and Gender" in *Conceiving the New World Order*, ed. Faye D. Ginsburg and Rayna Rapp (Berkeley, Los Angeles, and London: University of California Press, 1995), 125, 126.

18. Jesse D. McKinnon and Claudette E. Bennett, "We the People: Blacks in the United States," Census 2000 Special Reports, August 2005, http://www.census.gov/prod/2005pubs/censr-25.pdf (accessed March 30, 2006).

CHAPTER 10

BLACK ARABIC

SOME NOTES ON AFRICAN AMERICAN MUSLIMS AND THE ARABIC LANGUAGE

SU'AD ABDUL KHABEER

DESPITE THE DECEMBER COLD OUTSIDE, IT WAS TOASTY INSIDE THE QUEENS apartment. This was thanks to what some New Yorkers jokingly nickname "project heat" that blasts from the apartment radiator whose thermostat a renter cannot control. A group of about seven African American Muslim women had come together for an *iftar*, a meal to break the daily Ramadan fast. After dinner, they gathered in conversation on topics ranging from the criminal justice system to an upcoming fashion show. At one point Saleemah, a middle-aged woman, was asked to share her *shahadah* story, the account of how she converted to Islam. Although she converted in 1975, she traced the origin of her journey to Islam to the late 1960s and her activism within the Black Panther Party and Pan-African movements. She told the group:

> Allah had me go up one day to a prison, I was looking for my girlfriend's uncle who was um, unjustly, um, sentenced and I wanted to speak to him but his block didn't come out, but this [other] block came with all these brothers with white kufis, and I always had an eye for nice looking men so I said [to myself]: "well, who are all these nice looking brothers with these white kufis on their heads?!" I knew they were Sunni Muslims, people tried to give me dawah in '72 but I wasn't ready to hear it 'cause I felt the Arabs put black people in slavery, so I didn't want to hear nothing about Arabs.

Saleemah's story ends with conversion, yet her initial aversion to Islam because of Arab participation in the enslavement of Africans is notable. It is illustrative of the complexity and contestation around the meaning and role of the Arabic language for African American Muslims. Arabic is the language of the Qur'an and daily prayers, and the Prophet Muhammad was an Arab—are these merely coincidences of history or do they carry a deeper significance?

African American Muslims, both converts and those raised in Muslim families, join the majority of the world's Muslims (most of whom are non-Arabs) who habitually use the Arabic language. The prevalence of Arabic use among non-Arab Muslims can be attributed in part to the religiously required use of Arabic in prayer. The language of the Qur'an and the early Muslim community, Arabic has also historically been the lingua franca of the Islamic intellectual tradition. Some Muslims argue that the relationship between Arabic and the Islamic community is merely coincidental: Arabic was the language of Muhammad and his community, which necessitated an Arabic Qur'an. However, for many Muslims, both scholars and laity, Arabic assumes preeminence over other languages. Arabic is not merely a language used by God to communicate to the world; rather, it is God's language and carries all the divinity and authority of its speaker.

In African American Muslim communities, Arabic enters everyday language through the practice of ritual. Arabic appears in the speech of African American Muslims when talking about religion, in greetings and other customary expressions, and in the use of words borrowed from the Arabic language that replace English equivalents, such as *umi* for mother, *bayt* for house, and so on. Importantly, these borrowed or loan words are of specific types: terms of endearment, titles to address family and community members, specific places, clothing, things and events, civilities, and theoretical concepts.[1] The introduction of the Arabic language in speech is also a conversational tool used to produce specific reactions in a discussion and direct the course of conversation.[2] Therefore, any analysis of the use of the Arabic language among African American Muslims must be understood in the context of who is speaking, to whom is she or he speaking, and what is he or she speaking about.

Most interesting about the use of Arabic in African American Muslim speech is that word choice and how often words are used index competing beliefs and attitudes toward the use of the Arabic language among African American Muslims. These beliefs and attitudes, also known as language ideologies, further index beliefs about what it means to be an "authentic" African American Muslim. In this debate, African American Muslims confront questions of identity. How does a community construct an identity that is distinctive yet not artificially so? What determines that choices are sincerely attuned to the African American Muslim's cultural, religious, and racial here and now? And why is that important? In this chapter, I examine Arabic language use and the role of competing language ideologies in authenticity debates among African American Muslim communities in the United States.[3]

I argue that the religious significance of the Arabic language functions beyond ritual religious practices. The religious origin of Arabic, for African American Muslims, imbues the language with a power that travels with Arabic as it is used in everyday speech. Therefore, even outside religious events, the Arabic language is given a semiholy character and value among groups of African American Muslims. I will show how the authority imbued in Arabic allows for more than the simple substitution of Arabic words for English equivalents but the effective *replacement* of certain English words and phrases in the speech of African American Muslims. Despite the fact that English is the native language of African American Muslims, in certain contexts, English becomes inadequate—unable to meaningfully describe social realities. Moreover, because of the cultural frameworks African American Muslims bring to Islamic practice, the sociolinguistic meanings of Arabic words can be related to both a religious (Islamic) context and parallel meanings found within the context of broader African American culture. Further, I will unpack different ideas about the Arabic language by exploring its role in relationships of power among African American Muslims. By examining opposing language ideologies among African American Muslims, I will seek to determine beliefs regarding which ideas and persons are seen as "authentically" Islamic as well as which ideas and persons are seen as "authentically" *African American* Muslim.

My arguments are drawn from an analysis of the general African American Muslim linguistic context and specific incidences of Arabic use in interactions among African American Muslims. In the recorded ethnographic moments I present, I was a participant as well as observer; an anthropologist whose position in these social spaces was marked by ivy league pedigree, overseas study, gender, and religious and racial identity. Focusing on a community that is only beginning to be seriously studied ethnographically is both exciting and challenging. This chapter will not exhaust the possible analytical angles through which to approach this topic. However, what follows offers some notes—a series of questions and some thoughts—which will hopefully open up a conversation or stand as an invitation to carry out more research on African American Muslim sociality.

AN AFRICAN AMERICAN MUSLIM SPEECH COMMUNITY?

As a black woman of Caribbean and Latin American descent, a fourth-generation (mother's side) and second-generation (father's side) American, and a second-generation Muslim who speaks Arabic fairly decently, I certainly do not describe African American Muslims as *a* speech community to ignore the diversity that is characteristic of this community. African American Muslims are defined in this chapter as Muslim Americans descended from Africans enslaved in the western hemisphere, and those who self-identify as such. African American Muslims—poor, wealthy, working, and middle-class—live in America's inner cities, suburban enclaves, and rural communities. African American Muslims are thus, by definition, a heterogeneous speech community.

Therefore, my argument that African American Muslims constitute *a* speech community is not related to cultural or linguistic homogeneity. As Linguistic anthropological scholarship has shown, speech communities are not homogeneous entities where all members have the same knowledge of grammar and culture, as well as the same ability to use language.[4] Rather, African American Muslims constitute a speech community because they share the same context within which Arabic enters and functions in their linguistic worlds, although its manifestations, incorporations, and rejections are not strictly formulaic.

The Arabic linguistic context shared by African American Muslims is marked by two characteristics. Firstly, Arabic is not their first language; thus, for the overwhelming majority of African American Muslims, the encounter with Arabic is one that occurs through a process of religious conversion, be it direct (personal conversion) or inherited (conversion of a child's parents).[5] Religious conversion, as Susan Harding suggests, "is the process of acquiring a specific religious language."[6] Yet, if we understand language to be "a complex inventory of all the ideas, interests and occupations that take up the attention of the community," then conversion yields more than a new way to talk about and to God.[7] The inclusion of religious terminology in ritual speech and, subsequently, in everyday talk, can be described as acquiring a means of expressing newer and rearticulating older worldviews and forms of sociality.

The second characteristic that shapes the shared Arabic linguistic context of African American Muslims is the racial politics of Islam in America. A major fault line in the American Muslim community is the question of who speaks and interprets for American Muslims—*indigenous* or *immigrant* Muslims. Within the American Muslim community, the term *indigenous* is used to describe Muslims whose ancestors were enslaved in the Americas, that is, African Americans and Latina/os as well as Native American Muslims and Muslims whose ancestors are Europeans but are not recent immigrants, that is, white Americans. The term *immigrant* is used to refer to Muslims who have immigrated to the United States within the last century and their children. This distinction is not nuanced to describe, for example, the Arab American Muslim family that has resided in the United States since the nineteenth century. Therefore, the distinction appears to be connected to a particular relationship to the United States, a historic relationship that African American, Latina/o, white American, and Native American Muslims share that American Muslims of other descent are presumed not to have.[8]

Outside of the American Muslim community, the term indigenous, particularly in human rights discourse, is meant to refer to *native* peoples or the original inhabitants of a particular territory who, based on their identity, claim rights to land and sovereignty against the hegemonic power of *settler* peoples. At this point, I have been unable to determine when this term began to be used in the American Muslim context; however, its usage is neither coincidental nor incidental but derives its power from this more common definition.[9] By claiming to be indigenous, the *indigenous* Muslim makes similar claims: to be native and therefore have proprietary rights that predate those of the newly arrived, yet more socioeconomically powerful party, the *immigrant* Muslim.

The *indigenous-immigrant* divide is one concerning symbolic power, defined by Bourdieu as "[the] power of constituting the given through utterances, of making people see and believe, of confirming or transforming the vision of the world and, thereby action on the world and thus the world itself."[10] Specifically, the issue at hand is of legitimate religious authority within the American Muslim commu-nity—who has the legitimate authority to interpret for the Muslim community in the United States.[11] This issue of legitimate religious authority is tied to the cultural capital of the Arabic language in the global Muslim community. The abil-ity to wield Arabic in social spaces invokes the authority of the Divine and thereby increases symbolic power—the power to name and define reality.

As noted earlier, Arabic is the lingua franca of the intellectual Islamic tradition from which follows the reasonable expectation that a technical competency in the language is one criterion to be considered in determining legitimate religious authority. A leap taken, which *indigenous* Muslims object to, is that technical competency is necessarily the equivalent of deep religious understanding. More-over, as the majority of *immigrant* Muslims are not native Arabic speakers, *indig-enous* Muslims also object to a related attitude that Sherman Jackson identifies as *Immigrant Islam.* [12] The idea that *immigrant* Muslims are, by virtue of national origin, technically competent and thus endowed with the deep understanding that confers religious authority.[13] It is important to note that because of the cul-tural capital of Arabic among Muslims worldwide, Arabic will always be a trump in negotiations of symbolic power, even in contexts that do not include Arabs, like those of African American Muslims whose main *immigrant* interlocutors are South Asian.

The *indigenous-immigrant* divide is also marked by, and reflective of, the perfor-mance of race and racism in the United States. Particularly for African American Muslims, interactions between themselves and *immigrant* Muslims has proven that *immigrant* Muslims typically replicate and embody white supremacists attitudes toward African American history and culture. Indeed, the residential patterns and acculturation choices of the majority of the *immigrant* Muslim community seem to reflect this pursuit of whiteness and the prejudiced attitudes toward African American Muslims, spiritual brethren or not.[14] This *immigrant* performance of race is particularly an affront as African American claims of indigeneity are further buttressed by the fundamental role of the African American Muslim community in the establishment of Islam as a legitimate form of American religiosity.[15]

With the patterns of social life within the American Muslim community reflective of the racial segregation that is commonplace in American society, Afri-can American Muslims tend to have limited incidences of face-to-face interac-tions with *immigrant* Muslims. Nevertheless, this tension is well known, felt to be shared experientially, and is the backdrop of interethnic relations within the American Muslim community. It is within this shared linguistic context, deter-mined by conversion and racial politics, that language ideologies about the Arabic language have emerged within the African American Muslim community—ide-ologies that respond to the questions why Arabic should be used, how Arabic should be used, and by whom.

WHEN ENGLISH IS NOT ENOUGH

During the weekly study circle, Leticia, who has only recently released her first amateur rap album, informed the group that she had chosen the titles for the songs on her next CD: "one is gonna be called *sista fitna*." Leticia then went on to describe an incident at a local Brooklyn mosque that inspired the title. Briefly, a woman brought food to the daily Ramadan break fast held at the mosque and refused to share the food except under certain strange conditions. The other members of the study circle were shocked by the woman's action, particularly because it contradicts the overly generous behavior customary during the month of Ramadan. Upon hearing the details of this story, another participant, Jamillah, exclaimed, "Oh she [the woman] must be *majnoon*!" The other women nodded their heads and voiced their agreement to affirm Jamillah's assessment. This scene concluded with a question by twelve-year-old Maisha, "What is *majnoon*?" I responded, "Girl, that just means crazy."

Leticia, Jamillah, and the other participants in this story met every Wednesday evening in Queens, New York. This group of African American Muslim women gathered for the purpose of religious education where, during each meeting, the informal group leader shared information and led a discussion on a religious topic. I frequently attended these meetings and, despite the goal of religious education, socializing was a major element of these gatherings. The group's size generally ranged between seven and twenty women and consistently began approximately two hours after the scheduled start time. The youngest participant in this group was twelve years old and the oldest was seventy but most of the participants were in their early forties and fifties. Most of the women in the group converted to Islam and had been Muslim for an average of fifteen years. The younger women, mostly teenagers, had been raised Muslim since birth.

The word *fitna*, used by Leticia in her future CD title, *sista fitna*, is defined as temptation, trial, and discord in Arabic. In addition to the literal meanings, the term is also generally associated with two historic concerns in the Muslim community: communal chaos and the temptation of women in (male) society. Among African American Muslims, I have observed that *fitna* is used to describe a variety of disappointing experiences in private and public life. For example, *fitna* may be used to describe the occurrence of an argument, losing a job, or a marital dispute. Therefore, Leticia's use of *fitna* impresses upon her listeners the reprehensibility of the woman's behavior because it evokes not only the literal meanings of *fitna* and but the sociohistorical understandings of the word as well. By evoking the meanings of *fitna*, the woman's behavior moves from a matter of poor manners to a much more significant social ill.

In the light of associations between meanings of *fitna* and women, Leticia's word choice raises the question if she created the title *sista fitna* because she interpreted the incident as another example of the *fitna* caused by women. Alternatively, did Leticia rename the woman *fitna* because her behavior is another example of the trials facing the Muslim community? The group's conversation in reference to this story does not lend any evidence to support the first supposition. Moreover, *fitna* has more than a gendered meaning, which seems to support the idea that Leticia would have used *fitna* to describe similar behavior by a male. Leticia confirmed the latter and said that she renamed the woman *fitna* because she "causes trouble wherever she goes."

As with any language, knowledge of the possible meanings of Arabic words and phrases and when to use them is not an innate ability but is learned through a process of socialization—a process which includes both socialization through language and socialization to the use of language.[16] Although she was being raised in a Muslim family and was a student at a local Islamic school, Maisha was not familiar with any of the meanings of *majnoon*. In Maisha's question "what is *majnoon*?" we can trace this socialization process. Maisha, like the other adolescents at this study circle, attends at the behest (or perhaps command) of her mother. She is made to participate in the study circle as a means of introduction, clarification, and affirmation of what she *should* believe as a Muslim. She is also made to attend in order to be socialized to the concepts and practices that *should* be exhibited by a member of the *ummah*. The potential answers to her question (including my own) are meant to teach her what the word means, thereby building her "vocabulary" and ability to function with competence in a community in which Arabic is a common linguistic resource and tool. The different definitions of *majnoon* given in response to her question are also exemplary of the fact that shared or dominant meanings of Arabic terminology within the community do not go uncontested— they do not mean the same thing to everyone. Maisha is an adolescent but other African American Muslims, for example, converts, would be on the same learning curve and would encounter the meaning and use of Arabic words and phrases through a variety of spaces and practices, including study groups, Friday sermons, pamphlets and religious books, and other everyday events.

The discourse that emerges from these practices produces a lexicon that circulates within the African American Muslim community. The words that make up this African American Muslim Arabic lexicon are almost all nouns. This is a characteristic that Jane Hill identifies as "a sign of very restricted bilingualism, since it has long been recognized that nouns are the earliest borrowing in an incipient bilingual context."[17] Hill notes further that the category of words used also indicates the realms of interaction between the carriers of the loan language and the borrowers.[18] Due to the ethnic segregation common within the American Muslim community, it is doubtful that the introduction of Arabic in the everyday speech of African American Muslims was the result of intimate contact between African American Muslims and Arabic speakers. The major Islamic movements among African American Muslims from the early to late twentieth century—the Nation of Islam, the community of Imam W. D. Mohamed, the Darul Islam, and others—were not tied to Arab communities neither in terms of spiritual authority, community leadership, nor social life. Furthermore, any relationships they did have were often tenuous at best.

In a separate interview with Saleemah, she noted that as a new Muslim in the mid 1970s, she attended a variety of Islamic meetings and conventions in which her interactions with Arabic speakers, particularly men, were limited. Where gender might have facilitated more intimate contact between African American women and Arabic-speaking women, cultural distance reigned supreme, as even interactions between women never seemed to transcend the weekend conference milieu. Therefore, it is more likely that Arabic was introduced into African American Muslim speech through religious study and the elementary study of the Arabic language that was particularly common among African American converts in the mid- to late twentieth century. In the educational contexts of religious lectures and Arabic language classes, words and phrases drawn directly from the Qur'an, hadith texts and religious practice manuals, as well as simple everyday terms, were taught and retained by students. The ability of these students to engage in more advanced language study, due to financial and familial constraints, was often limited.[19]

Two words that exemplify this type of learning, retention, and incorporation are *umi* and *khimar*. It is easy to find many second-generation African American Muslims who were raised to call their mothers *umi*. *Umi* is an Arabic phrase that means "my mother," yet in the Arabic speaking world, and among Arab Americans, mothers are almost exclusively referred to as *mama*.[20] Similarly, in many African American Muslim communities, the headscarf that is worn by Muslim women is referred to as *khimar*, whereas among Arab Americans and in the Arab world, a Muslim woman's head covering is called *hijab*. Both terms are Arabic and are featured in the Qur'an; however, *khimar*, in its plural form *khumur*, is the precise term used in the Qur'anic verse describing female modesty.[21] *Hijab*, in contrast, is used in the Qur'an in several instances that have "a common semantic theme of separation, albeit not primarily between the sexes."[22] Thus, there is a distinctive adoption of the Arabic language in a way that is not tied to the norms of Arab culture, to which access is limited. These words and others, such as *zawj* (spouse) and *wali* (guardian), also follow this pattern of incorporation as a result of intensive interaction with religious and linguistic texts rather than an intensity of interaction with Arab Muslims. These words are subsequently used outside scholastic contexts and become commonplace in African American Muslim speech.

In an article that traces the use of broadcast radio parlance in everyday Zambian speech, Debra Spitulnik writes that "certain institutions provide common linguistic reference points" such that they function as the "source and reference point for phrases and tropes which circulate across communities."[23] These phrases and tropes are *public words*, defined by Spitulnik as "standard phrases such as proverbs, slogans, clichés, and idiomatic expressions that are remembered, repeated, and quoted long after their first utterance."[24] The concept of public words is particularly useful when analyzing Arabic language use among African American Muslims. Arabic terms are used in institutions such as study circles, hip-hop, and *dawah* (invitation to Islam) pamphlets, these institutions serve as shared linguistic reference points through which *Arabic public words* circulate among African American Muslims.

In the exchange transcribed above, Jamillah interjected a public word, *majnoon*, into the conversation to further embellish Leticia's description of *fitna* by the woman. By using *majnoon* she implied that the woman's behavior is not simply an error in judgment but a fundamental problem with her psychological constitution and perhaps even more than that. *Majnoon* is derived from the root word *janna*, which means to become crazy and to be possessed. It is not clear to what extent the etymology of the word is known by all members of the group, yet it is the act of choosing an Arabic word as opposed to a potential English equivalent that is key—it illustrates the power of meaning that Arabic wields in this speech community. In essence, the English equivalents to these words—*fitna* and *majnoon* and others—are inadequate because they do not express the value and meaning this Islamic community has invested in Arabic terms and language. Moreover, although the English language may be inadequate, choosing Arabic terminology over English can also be an act that engages the broader contexts the African American Muslim performs within: American Muslim *and* African American.

The root word *janna* is also the etymological root for the word *jinn*, which describes created beings who, according to Islamic belief, have been given free will and live along side humanity but are unseen to the human eye. Like human beings, there are righteous *jinn* and there are *jinn* who are thought to work alongside Satan in attempting to thwart believers from righteousness. These types of beings are similar to the spirits that are part of African American folk traditions.[25] As Braziel Robinson, a former slave who claimed to be able to see spirits described, "my two spirits are good spirits, and have power over evils spirits, and unless my mind is evil, can keep me from harm. If my mind is evil my two spirits try to win me, if I won't listen to them, then they leave me and make room for evil spirits and then I'm lost forever."[26]

Spirits, their existence, and their potential to do harm in human life is related to different practices such as conjure, hoodoo, and Christian spirit possession, which are undergirded by a general belief that the supernatural plays an intimate role in human life. The existence of the concept of *jinn* in African American folk tradition is important because it is through familiarity with the supernatural through these traditions that African American Muslims can make an experiential connection between a religious concept and their profane reality. The concept of *jinn* makes sense, it is familiar, and therefore resonates culturally because it reflects and reaffirms a preconversion cultural belief. This is particularly important because racial divides within the American Muslim community preclude African American Muslims access to an experience of the supernatural through the stories that circulate amongst their *immigrant* brethren. Thus, certain Arabic words, such as *majnoon*, should be seen as *doubly* meaningful because they reflect both a particular Islamic cosmology as well as a spiritual orientation that has roots in the Christian and folk belief traditions of African Americans.

One notable point in the discussion was my interjection, "girl, that [*majnoon*] just means crazy." As a participant in the conversation, I injected a definition of *majnoon* devoid of any sociohistorical meaning. This was motivated (admittedly, heavy-handedly) by my personal concern with practices that might elevate Arab culture to the detriment of African American Muslims, under the auspices of being more "Islamic." By claiming that *majnoon* "just means crazy," I attempted to interfere with the socialization process by reducing the definition to only its simplest literal meaning. Yet it is clear that *majnoon* does *not* just mean crazy; embedded in its meaning are community beliefs on the supernatural and the authority of the Arabic language. In this moment and in the encounter that I recount below, my attempt at linguistic intervention is no match for a language firmly rooted in the discourse on what it means to be an African American Muslim.

SPEAKING IN GOD'S NAME

Recently, I was in the company of another group of African American Muslim women who had gathered for a Ramadan break fast at a home in Queens, New York. Arabic words and phrases were used in many different ways throughout the evening's conversations; here, I focus on one participant, Naimah, and her use of Arabic as a means of asserting authority. I have divided the evening's discussions into two parts, distinguished by two different patterns of speaking. During the first part of the discussions, the following pattern developed. One woman would tell an anecdote or give a personal opinion related to the topic at hand. Naimah would then interject by summarizing the speaker's narrative and providing her Islamic commentary on the issue. By Islamic commentary, I am referring to her habit of couching each topic in the context of the challenges (as she understood them) facing the world Muslim community. Whether using English exclusively or using Arabic as well, through the course of the evening's conversations, she asserted herself as a religious commentator on each topic of discussion.

Naimah was a powerful speaker because she used a variety of linguistic tools to make her points. Overall, Naimah used Arabic significantly more than any other speakers in the discussion. She used Arabic in two specific ways: (1) she used words commonly known to African American Muslims while emphasizing an Arabic accent in her pronunciation of these words, and (2) she recited Qur'anic verses in Arabic, with and without a subsequent English translation. Naimah was also a powerful speaker because her speaking style evoked the elocutionary techniques of African American Christian preachers and Muslim Imams (mosque leader) in terms of cadence and intonation. In the tradition of Imams, she used Arabic as a stylistic feature of her speech and thus the authority she wielded in the conversation was in part drawn from her ability, through performance, to embody the authority of an Imam. Through this technique, we find Arabic as doubly meaningful again, drawing significance from its religious authority as the language of the Qur'an and from the cultural authority inscribed in the style of African American clergy.

Through her "Islamic" reframing of the conversation, she was attempting to establish herself as the representative of the one correct Islamic authority. Her interjections did not simply move her into the position of speaker vis-à-vis the listeners; rather, she promoted herself into the position of authority vis-à-vis those who must listen to, obey, and accept authority. Moreover, rather than embody the position of a speaker whose role is "animator, author and principal as one," I believe Naimah did not perceive herself to be the author of her speech.[27] She rarely used "I" statements when speaking; instead, she would consistently state that "we [Muslims]" need to "realize" or "understand" that Allah wants or says "X" and the *sunnah* (tradition) of the Prophet directs us to do "Y." Therefore it appeared that she perceived herself as playing the role of God's representative in the conversation, and her authority was derived from being the orator of God's speech. This perception—and the authority it afforded—appeared to be supported by the other women in the group, as in most cases she was able to deliver her commentary-monologues without interruption, interjection, or challenge by the other participants.

Her authority as God's representative drew on the preeminence of Arabic in the African American Muslim community. Wielding Arabic terminology is a common conversational tool among African American Muslims, especially in debates, because it implies a religious knowledge base; if you speak the *'Arabiya* you must know the religion.[28] This is, of course, the same claim implicitly made by *immigrant* Muslims as noted above, although in these speech contexts, there are no immigrants present. Much like the white normative gaze articulated by Cornel West, we can understand the language ideologies of African American Muslims to function under an *immigrant* normative gaze.[29] The contestations over authority and representation that define the relations of power between the *immigrant* and *indigenous* communities shape behavior, in this case, language use, even when the *immigrant* is not present. Yet although the *immigrant* Muslim is not present, his claims of normativity still function and are often hegemonic within interactions between African American Muslims. Therefore, Arabic language use is designed to shift the power to be authoritative in the discussion or debate to the speaker who has the ability to supplement their argument with the incorporation of Arabic words and phrases, whether drawn directly from religious texts or not.

Just as was the case in the *sista fitna* story, my participation in the conversation was particularly significant. My participation marks the second half of the discussions during which I consistently challenged Naimah's commentary-monologues. The first incidence of this followed Saleemah's narration of her conversion to Islam during a visit to a prison inmate:

Naimah: You know what I realized . . . for many people, for many people going to prison can be a life saving movement.
[In the background: Oh yeah. Sure.]
Suad: For most people, it's not.
[Silent pause]
Naimah (low voice): True . . .
Naimah (raised voice): But um, I have, I have that hope because right now I have two brothers.

As in the pattern of speaking that had developed in earlier conversations, Naimah interjects in order to provide her commentary, "You know I realized . . . ," to which some of the other participants affirmed her comments with "Oh yeah" and "Sure." My interjection was followed by a notable pause and it appeared as if Naimah was caught off guard. Her "True . . . " response to my statement was stated in a lowered voice and it appeared that, during the pause, she was processing this challenge to the pattern of conversation. However, she quickly regained her composure, conceded my comments, and resumed her commentary-monologue ending with a story of the positive impact of imprisonment on the behavior of her two brothers.

Yet from that point until the end of the conversations that evening, the pattern of speaking changed dramatically. I would consistently interrupt Naimah's monologues and challenge her Islamic commentary with my own personal opinions on what is religiously desirable. As a result of my resistance to the hegemonic religious impulse and in (I hope) deference to Naimah's right to her own opinion, I prefaced many of my comments with "I" statements. In contrast, Naimah continued to speak with authority about what "we [Muslims]" need to understand. The following sequence occurred during discussion about gender roles:

Suad: I don't think though, that what we need to do is, like, begin to sort of carve out particular sort of, like, cookie cutter roles for people to fit themselves into in their relationships with people

Naimah (interjects): But what we have to realize

Suad: What works, works

Naimah: What we have to realize is that, as Muslims, we have been sent down a *furqan*, a criteria, we have been sent down social roles . . .

Suad (interjects): and it doesn't say, it doesn't say, stay home.

Naimah: No it doesn't say that, but Allah *subhana wa ta'ala* says clear in his *mushaf*, in the *Qur'an*, that the men are not like the women.

As the conversation moved through this discussion of gender roles to the discussion of polygyny, in which Naimah and I continued to argue opposing views, the other women were primarily listeners to our debate. As illustrated in the preceding dialogue, it also appeared that Naimah's use of Arabic words and phrases increased as our debate ensued. I felt the increase to such an extent that I felt compelled to respond in kind. I do not generally use Arabic to support my arguments, and so to reconcile this "weakness," I paraphrased Qur'anic verses in English in order to confer my opinions the same religious legitimacy I (and the other participants) had given Naimah's views *because* of her use of Arabic.

In the preceding speech events, Naimah adheres to one of the main language ideologies found among African American Muslims, which I call the '*Arabiya ideology*. The 'Arabiya ideology represents the beliefs of African American Muslims who posit that the use of Arabic (along with the adoption of specific modes of behavior: dress, hygiene, social etiquette) should be incorporated in everyday life in order to be "authentically" Muslim. They argue that the effort to follow

God's law as understood through interpretations of the Qur'an and the practices of Prophet Muhammad leads to the total reevaluation of former ways of
sociality, a reevaluation that has clear cultural ramifications.[30] For these African
American Muslims, all practices that are associated with the Prophet Muhammad—from prayer to wearing a long robe to speaking Arabic—are Islamic activities, and therefore their adoption is encouraged, if not deemed necessary.[31] Thus
by deploying Arabic, Naimah lends what she believes to be the force of religious
authority to her particular interpretations.

This is a technique that another dominant language ideology, the *black Americanist*, would describe as a misguided. The *black Americanist ideology* is the standing critique that challenges the discourse and practices of the 'Arabiya ideology.
The black Americanist ideology takes a position toward the Arabic language that
is grounded in a definition of African American Muslim identity that insists on
the primacy of the *Americaness* of African American Muslims. The black Americanist sees the 'Arabiya ideology as an attempt to distance one's self from her
African American cultures. Where the 'Arabiya ideology intentionally adopts the
Arabic language, this language ideology consciously resists the incorporation of
Arabic outside of a strictly religious context. Rather than proving to be religiously
authentic, they would argue that the 'Arabiya practitioner is parroting the *immigrant* Muslim in a rejection of their *indigenous* self.

In her examination of Spanish language use among white Americans in the
southwestern United States, Hill finds that "Anglos use Spanish, but in limited
and specialized ways that support a broader project of social and economic domination of Spanish speakers in the region."[32] Hill also notes that distortions in the
pronunciation of Spanish loan words by Anglo speakers are the enactment of
social distancing between the Anglo and Spanish speakers. In the case of African
American Muslims, the opposite rings true. For most African American Muslims,
incorrect pronunciation results from limited instruction in the Arabic language,
since this speech community very consciously orients itself toward the *perfection* of pronunciation. This goal of perfect pronunciation is shared by both black
Americanist and 'Arabiya adherents. However, their desire for linguistic perfection is imbued with a different meaning and a different underlying motivation.
For the black Americanist practitioner, the attainment of this goal is most often
situated within a context of ensuring scholastic access to Islamic texts. For those
who adhere to the 'Arabiya ideology, they, like Naimah, work toward a goal of the
perfection of pronunciation in order to *decrease* the distance between themselves
and Arabic speakers. Doing so, they believe, will grant them greater religious
authenticity and power.

Hill's insight directs us to consider how the use of a borrowed language
indexes the value placed on those from whom the language is borrowed. The
question posed by some African American Muslims is, does the value given to
the Arabic language extend to Arabs and Arab culture? In the framework of
the black Americanist ideology, as illustrated in the poetic verse below, using
the Arabic language outside of religious activities is a practice that emerges

from a preference for Arab culture—a preference that arises from the devaluation of African American culture.

> its like
> your leader looks nothing like u then you follow the rules
> but if he does you give him hell cause he's not taught by their schools
> so u want to shape and mold me cause you find me obscene
> you can have it cause Allah created me *fi ahsani taqweem.*

The preceding verses are from a spoken-word poem by California-based African American Muslim artist Kamilah Shuaibe. Shuaibe, raised among black Americanist African American Muslims, admonishes the African American Muslim who values the opinions and leadership of non-African Americans, precisely because they are *not* African American. She objects to their efforts to declare her immoral and reminds her interlocutors that her racial and cultural identity is divinely ordained—she, like all humans, was made *fi ahsani taqweem* (in the best mold).[33] Shuaibe's words are exemplary of the black Americanist ideology challenging practices such as Naimah's and questioning the 'Arabiya adherent's authenticity as an African American. The black Americanist interprets the practices of the 'Arabiya ideology as contemporary manifestations of the self-hate that has afflicted, and continues to affect, African Americans living in environments structured by and in the perpetuation of white supremacy. When Shuaibe declares she was "created *fi ahsani taqweem*," she joins African American Muslims who attempt to confront double consciousness with the possibility of blending their triple selves: black, American, and Muslim. This is an articulation of an African American Muslim identity that seeks to remain grounded in the history and contemporary reality of African Americans as the descendents of former slaves, as Americans, and as Muslims.

It is important to recognize that while these two ideologies have been depicted in stark contrast, there are consistent points of overlap or inconsistency within these ideological frameworks. For example, there are African American Muslims for whom the appropriation of the Arabic language becomes a conduit to connect not to an Arab Islamic tradition but to a different "foreign" tradition that is seen as familial: the African Islamic tradition. In the tradition of Afrocentricity, for these African American Muslims a diaspora framework is invoked through the adoption of the Arabic language, which stands a symbol of the grandeur of African Islamic centers such as Timbuktu, Futa Jalon, and the Sokoto Caliphate. This movement beyond national borders is seen as "an effective and expedient means of breaking free of the master without imitating him and constructing a Blackamerican identity that does not contradict their sense of authentic self."[34] Therefore, in this diasporic context, the tension is not with blackness but Americaness; the relationship to the American self is ambiguous, ambivalent, or antagonistic.

Black Americanist resistance to Arabic is responding to the idea that an "authentic" Muslim identity that can only be created through the adoption of foreign cultural forms and practices. Yet even adherents of this ideology use

Arabic outside of a strictly religious context. They routinely give their children Arabic names and may use Arabic in greetings and as terms of endearment as well. In a conversation between Hannan, an African American convert of over thirty years, and her second-generation Muslim daughter, Hannan expressed a wish that her granddaughter is taught to call her own mother *umi* rather than mommy. Hannan's argument was based on the idea that "we are Muslims," and Arabic was useful in this instance because it would mark her family members living in a non-Muslim majority country as Muslims. However, despite this particular wish, Hannan fiercely rejects the frequent inclusion of Arabic typically associated the 'Arabiya ideology. Frequency of Arabic language use may be the distinction that most clearly marks the difference between the two ideologies, and, for some, it is the flip side of another distinction—class.

Two second-generation African American Muslim women, Sabra and Latifah, attributed higher incidences of Arabic language use to class. They made a distinction between their middle-class families and other Muslims who were more likely to use Arabic frequently—African Americans Muslims of lower socioeconomic class. For them, these African American Muslims use Arabic to overcompensate—to make up for what they lack in financial and educational stature and, most importantly, in cultural self-confidence. Sabra described their own use of Arabic as either solely related to religious practice or "words that are affectionate or complementary because we [African Americans] are an affectionate and complementary people." In their interpretation, when they *did* use Arabic, it was conscious and used only in ways that would reflect self-love—any other use of the Arabic language was a laughing matter. Latifah is well known for calling friends and family while donning an obviously exaggerated Arabic accent to poke fun at the practice. Although their analysis unfairly generalizes African American Muslims of low socioeconomic status, their observation of a correlation between class and the use of the Arabic language may not be unfounded.

As noted by scholars on African American Islam, the Islamic impulse in these communities is shaped by ideological trends within the broader African American context.[35] Class and the Arabic language appear to dovetail when segments of this community respond to American racism through a separatism that is not (black) nationalist but accepts the framework of Immigrant Islam. In this instance, one's religious piety is in opposition to one's Americaness: to be a good Muslim means to reject America, including the institutions that are the standard pathways to economic success in the United States.[36] This rejection of America is then coupled with the adoption of the Arabic language, in limited ways, and other customs typical of the *immigrant* Muslim.

Scholars also note the irony of the fact that these African American Muslims eschew the mechanisms of success that their immigrant counterparts eagerly participate in, implying that these African Americans have become victims of a false consciousness.[37] Yet rather than simply be victims of false consciousness, these African American Muslims could be seen as making a rational choice as their class positions do not give them the freedom to choose *how* they are going to be American. The only America they have ever known perpetuates poverty and injustice in their everyday lives, in ways middle-class African American Muslims may no longer know as intimately. Of course, their rejection is severely limited—they may not vote or attend college but they use metropolitan transit and accept U.S. currency for their incense and oils.[38] Yet while their desire to reject America is not unproblematic, the choice might also be seen as creative: they imagine themselves as different, escaping injustice by traveling, through language and other practices, to those places their empty pockets, prison records, and poor education prevent them from reaching.

"AWW MAN, THOSE BINTS ARE CRAZY!"

Throughout this chapter, we have encountered examples of the way Arabic functions in African American Muslim speech events in gender-segregated environments, yet these techniques are used by both women and men. In fact, African American Muslim men may feel the powers of authority wielded by Arabic with more intensity than their female counterparts. In a patriarchal social context where men are primarily expected to be leaders for the African American Muslim community and in its interactions with the *immigrant* Muslim community, African American Muslim men are called upon to show a facility with Arabic more often. Moreover, in a U.S. context where an empowered notion of black masculinity seems elusive, Arabic language use may also be entangled in notions of femininity and masculinity among African American Muslims and how they structure relations of power between the genders.

In a conversation with Shaheed, an African American Muslim male who converted to Islam thirty years ago, he explained that "back in the days of ignorance we [African American Muslim men] used to use to say—'Aww, man those *bints* is crazy!'" Notably, he remarked that bint was one of the first words he learned upon his release from prison. Its meaning was explained as a relationship between the Arabic word *bint* (daughter or girl) and a woman as a *bent rib*. The idea of women as bent ribs is derived from a saying attributed to the Prophet Muhammad in which women are said to be created from the rib of men. This description is followed by advising men that a woman is like a bent rib and if a man tries to straighten the rib, he will break it; therefore, men should leave women as they are.

Shaheed acknowledged that the two terms have no technical linguistic relationship and he explained that the concept derived from a belief among African American men that women are emotional and this could not be changed. He also laughingly noted that the term was often used by the brothers who were considered most "knowledgeable" in the *deen* (Islam) at the time, which meant they had been Muslim for three to five years and had two to three wives on public assistance. Therefore, *bint* and *bent* were seen as linguistically related because of the similarity of their sounds and the persons whom they signify. He also insisted I include another interpretation of that saying by the twelfth-century Islamic scholar Imam al-Ghazali. He informed me that al-Ghazali wrote that woman was not created from the feet of man to be below him, nor from the chest to be ahead of him, nor from the head to be on top of him, but from his heart to be loved by him and be his companion.

In a final Ramadan scene, an *iftar* in Brooklyn, New York, consisting primarily of young African American Muslim professionals, the topic of how the Arabic language is used was discussed. A group of men informed the women present about the ways *bint* is used to refer to them. One claimed that men would use that word in phrases such as "those *bints* are crazy," and "you better get your *bint* in check." The men interpreted this usage in two ways: (1) that *bint* became a substitute for the word "bitch" and (2) that *bint* was used as a power play—that through language the women in discussion and all women in particular are placed in a subordinate position in relation to men. Based on their interpretations, the use of *bint* is reflective of patriarchal behavior that can be attributed to particular interpretations of Islamic texts, particular beliefs within African American cultural contexts, and those found in broader American culture. As *bints*, adult women are, by nature, always girls, subordinate to men, and when "acting out," bitches.

While greater study of these practices is warranted, the use of Arabic terms seems to reflect negotiations of power that occur both on the personal and everyday level, as well as in more public and scholastic spaces[39]—spaces in which, scholars have shown, African American Muslims confront, combat, and concede to power by directly engaging religious texts.[40] Through this engagement, they are providing interpretations through language that undoubtedly inform everyday speech. In *bints*, *majnoon* individuals causing *fitna*, African American *umis* and multicolored *khimars*, time and time again, located within Arabic language use are the major discourses of the African American Muslim community in the United States.

CONCLUSION

African American Muslims use Arabic in specialized and strategic ways that express the discourses and desires of a community that is always functioning in a context that is simultaneously black, Muslim, and American. In this chapter, I have attempted to trace the life of the Arabic language for this community that merges multiple identities. Word choice—why *zawj* is used to substitute for spouse while *sayaara* is not chosen for car—may emerge from an engagement with the Arabic language through texts; yet once the words leave their textual context, they come to life in relationships reflecting an engagement with Islam that informs the everyday—its highs and lows.

The discourse around the Arabic language within the African American Muslim community reflects debates of authenticity. Each ideology—ʿArabiya and black Americanist—attempts to draw a theoretical line in the sand, so to speak, declaring who is the real African American Muslim. To use Arabic excessively is the performance of "cultural apostasy"; to use Arabic too sparingly is the rejection of the "true Islamic identity." However, in the practice of everyday life that seeks to respond to the African American Muslim here and now, those lines are consistently crossed and smudged.

Albeit in a slightly different direction, I will end by invoking the suggestion of John Jackson, Jr., that analysts consider sincerity an alternative framework to authenticity in the analysis of identity.[41] Whether they use Arabic a little or a lot, these African American Muslims are the real deal. Through language and other practices they experience and bring into life identities, like those in the communities of the Amadous of West Africa and Symas of South Asia—identities born from the everyday negotiations of culture and faith.[42]

APPENDIX

TERMS OF ENDEARMENT

Ak/Aki	brother
Ukt/Ukti	sister
Shaykh/Shaikh	elder/respected friend
Habibi/Habibti	loved one

FAMILY RELATIONSHIP TITLES

Um/Umi	mother
Abu/Abi	father
Ibn	son
Bint	daughter
Zawj	spouse
Jadda	grandmother

PLACES

Bayt	house
Hamaam	bathroom
Masjid	mosque
Suk	marketplace

CLOTHES

Khimar/hijab	female head covering
Kufi	male head covering
Jilbab	long, loose female dress
Niqab	face veil
Abaya	long, loose gown
Thobe	long, loose male dress

TITLES

Amir	leader
Wali	guardian
Imam	religious leader
Salafi	follower of Salafi movement
Bint	female/bitch

COURTESIES/CIVILITIES

Shukran	thank you
Afwan	you're welcome
La	no
Na'am	yes
Mabrook	congratulations
Jazakallah	thank you
Kaifal hal?	How are you?
Tayyib	good

EVENTS AND ACTIVITIES

Iddat	after-divorce waiting period for women
Nikkah	wedding/wedding ceremony
Walima	wedding reception
Aqiqah	newborn ceremony
Iftar	meal to break Ramadan daily fast
Suhoor	meal to begin Ramadan daily fast
Salat	ritual prayer
Du'a	supplication
Jummah	Friday prayer
Taleem	religious lecture, in many mosques typically held on Sunday

THEORETICAL CONCEPTS

Fitna	social chaos
Kafir	nonbeliever/non-Muslim
Majnoon	insane
Deen	religion/Islam
Dunya	this world
Niyah	intention
Sunnah	practices of Prophet Muhammad
Da'wah	call to Islam/prostelyzation
Biddah	innovation
Haram	forbidden
Mu'tah	temporary marriage

Hijra	migration from United States to the "Muslim World"
Adaab	etiquette
Aqeedah	creed

THINGS

Stinja	from verb *istinjaa*: water bottle used to clean private parts after using the bathroom
Faloos	money

COMMANDS

Takbeer	Say: Allahu Akbar
Ta'al, ta'al huna	come/come here
Ijlis	sit
Uskut	be quiet/shut up

NOTES

1. See the appendix for introductory listing of Arabic terminology with translation, as used among African American Muslims in the United States.

2. For some examples in the non-U.S. context, see John R. Bowen, "Does French Islam Have Borders?: Dilemmas of Domestication in a Global Religious Field," *American Anthropologist* 106, no. 1 (2004): 43–55; John R. Bowen, "Salat in Indonesia: The Social Meanings of an Islamic Ritual," *Man* 24, no. 4 (1989): 600–19; Magnus Marsden, *Living Islam: Muslim Religious Experience in Pakistan's North-West Frontier* (Cambridge: Cambridge University Press, 2005).

3. I respect and acknowledge the arguments against the perpetuation of U.S. regional hegemony through using the term "American" to solely refer to U.S. citizens and "America" to the United States. However, for the sake of brevity and grammatical ease, in this chapter, the terms U.S. and America will be used interchangeably and the related "American" will be used to refer to U.S. citizens only.

4. See John J. Gumperz, "The Speech Community," in *Linguistic Anthropology: A Reader*, ed. Alessandro Duranti (Oxford: Blackwell, 2001), 43–52; Dell Hymes, "On Communicative Competence," in *Linguistic Anthropology: A Reader*, ed. Alessandro Duranti (Oxford: Blackwell, 2001), 53–73; H. Samy Alim, *You Know My Steez: An Ethnographic and Sociolinguistic Study of Styleshifting in A Black American Speech Community* (Chapel Hill, NC: Duke University Press, 2005).

5. Although I use the term conversion in this chapter to refer to the process of converting from one belief system to another—in this case, Islam—it is important to note that among African American Muslims, the term reversion is often used as an alternative. Reversion is meant to reflect both the belief that (a) all human beings are born Muslim, that is, in submission to God, and (b) because a number of Africans enslaved in the Americas were Muslim, African American Muslims are seen as reverting back to their ancestral religion.

6. Susan Harding, "Convicted by the Holy Spirit: The Rhetoric of Fundamental Baptist Conversion," *American Ethnologist* 14, no. 1 (1987): 178.

7. Edward Sapir, *The Collected Works of Edward Sapir*, ed. R. Darnell and J. Irving (Berlin: Mouton de Gruyter, 1999), 90–91.

8. I recognize that the term lacks nuance, yet I find it a useful trope to describe the relationships between these two communities, *indigenous* and *immigrant*, as I have seen it in the field. Therefore, the terms immigrant and indigenous when used according to these definitions will be italicized in the chapter.

9. The earliest incidence of the use of this term that I have found so far is cited in Sulayman S. Nyang, *Islam in the United States of America* (Chicago: Kazi, 1999), 143–47. Nyang cites the use of the term in the news publication of the Darul Islam, *Al Jihadul Akbar*, with dates of publication in the late 1960s.

10. Pierre Bourdieu, *Language and Symbolic Power*, ed. John B. Thompson (Cambridge: Harvard University Press, 1991), 170.

11. Although not addressed in this chapter, another question that emerges from this issue is the notion of legitimate political authority and representation in the relationship between American Muslims and different spheres of U.S. society, particularly the state and the media.

12. "Immigrant Islam embodies the habit of *universalizing the particular.* It enshrines the historically informed expressions of Islam in the modern Muslim world as the standard of normativeness for Muslims everywhere . . . And in this process Immigrant Islam's interpretations are effectively placed beyond critique via the tacit denial that they are in fact interpretations." Sherman A. Jackson, *Islam and the Blackamerican: Looking Toward the Third Resurrection* (New York: Oxford University Press, 2005), 12.

13. Jamillah Karim notes, "Many young African American Muslim women felt that Muslim immigrants question African American Muslim legitimacy not because they know Islamic teachings better but because they consider invalid any expression of Muslim identity outside their own cultural notions of Islam," Karim, "Between Immigrant Islam and Black Liberation: Young Muslims inherit Global Muslim and African American Legacies," *The Muslim World* 95, no. 4 (2005): 503.

14. Nyang notes that conflict between indigenous and immigrant Muslims stems in part from the fact that "most immigrants [in the late 1960s] were . . . South Asian and Middle Easterners who generally perceive themselves as 'whites' or 'browns.'" Nyang, *Islam in the United States of America*, 144. For another analysis of this type of racial identification among South Asian Americans, see Vijay Prashad, *The Karma of Brown Folk* (Minneapolis: University of Minneapolis Press, 2000).

15. See Karim, "Between Immigrant Islam and Black Liberation"; Jackson, *Islam and the Blackamerican.*

16. Elinor Ochs and Bambi B. Shiefflein, "Language Acquisition and Socialization: Three Developmental Stories and Their Implications," in *Linguistic Anthropology: A Reader*, ed. Alessandro Duranti (Oxford: Blackwell, 2001), 264.

17. Jane H. Hill, "Hasta La Vista, Baby: Anglo Spanish in the American Southwest," *Critique of Anthropology* 13, no. 2 (1993): 153.

18. Hill, "Hasta La Vista, Baby," 145–76.

19. These constraints included the inability to pay for advanced Arabic language classes that would likely be held at academic institutions, the inability to afford overseas travel to study the Arabic language, and the inability, particularly for single mothers, to attend classes that conflicted with work and childcare demands.

20. The same is true for the terms used by African Americans Muslims referring to fathers, *abi* (my father) and *abu* (father) and those used by Arab speakers, *baba* (father).

21. Qur'an 24:31 refers to khimar in its plural form, "*wal yadribna bi khumurihina 'ala juyubihina*" (and drape their coverings over their chests).

22. Mona Siddiqui, Mona, "Veil," in *Encyclopedia of the Qur'an*, ed. Jane Dammen McAuliffe, Vol. 5 (Leiden: Koninklijke Brill, 2006), 412.

23. Debra Spitulnik, "The Social Circulation of Media Discourse and the Mediation of Communities," in *Linguistic Anthropology: A Reader*, ed. Alessandro Duranti (Oxford: Blackwell, 2001), 97–99.

24. Spitulnik, "The Social Circulation," 99.

25. Here I do not mean to imply that Islamic and African American folk beliefs are the only traditions that argue the existence of spirits or other supernatural beings, but rather that these are the sources of concepts of the supernatural that are particularly important to African American Muslims.

26. Alan Dundes, *Mother Wit from the Laughing Barrel: Readings in the Interpretation of Afro-American Folklore* (Jackson: University Press of Mississippi, 1990), 378.

27. Erving Goffman, *Forms of Talk* (Oxford: Blackwell, 1981), 145.

28. In the Arabic language, the word for Arabic is *'arabiya*; in the field, I have noted the use of the Arabic word *'arabiya* as opposed to the English translation, "Arabic."

29. Cornel West, *The Cornel West Reader* (New York: Basic Civitas Books, 1999).

30. Nyang describes this type of African American Muslim as an assimilationist who takes on a consciousness that negates American cultural norms and adopts foreign Muslim cultural practices as "an alternative, and sometimes superior, identity to his original ethnic identity." Nyang, *Islam in the United States of America*, 73.

31. It should be noted that the question of what is specifically required by the mandate to follow the *Sunnah* (tradition) of Prophet Muhammad remains contested among Muslims across the globe.

32. Hill, "Hasta La Vista, Baby," 146.

33. This phrase comes from the Qur'an 95:4.

34. Jackson, *Islam and the Blackamerican*, 153.

35. See C. Eric. Lincoln, *The Black Muslims in America* (Grand Rapids: Wm. B. Eerdmans, 1994); Aminah B. McCloud, *African American Islam* (New York: Routledge, 1995); Yusuf Nuruddin, "African-American Muslims and the Question of Identity: Between Traditional Islam, African Heritage, and the American Way," in *Muslims on the Americanization Path?* ed. Yvonne Y. Haddad and John L. Esposito (Atlanta: Scholars, 1998), 267–330; Edward E. Curtis, *Islam in Black America: Identity, Liberation, and Difference in African-American Islamic Thought* (Albany: State University of New York Press, 2002); Richard B. Turner, *Islam in the African American Experience* (Bloomington: Indiana University Press, 2003).

36. Jackson, *Islam and the Blackamerican*.

37. Ibid.

38. As I have observed, and as other African American Muslim interlocutors living in major U.S. cities have described to me, peddling incense and scented oils is a common alternative occupation of African American Muslim men.

39. In the case of African American Muslim women, one term that might reflect relations of power between the genders is *aki* (my brother). Many African American Muslim men call each other *aki* to reaffirm the importance of fraternity between African American Muslim men. I have noted instances where women, outside the company of men (though potentially within their earshot), mimic this form of address to poke fun at the practice. In their enunciation of the word and their gestures, they act out a caricature of a caricature. They reproduce an exaggerated masculinity performed by men whom they see as pretending to be the ideal patriarch. Perhaps, as their wives, mothers, and daughters—as the women who are the primary targets and potential recipients of patriarchy—they know intimately where their men meet and fall short of this ideal. The women not only invoke the term playfully but also appear to offer a critique of the performance of masculinity embodied in the use of this Arabic term.

40. See, for example, Carolyn M. Rouse, *Engaged Surrender: African American Women and Islam* (Berkeley: University of California Press, 2004).

41. John L. Jackson, Jr., *Real Black: Adventures in Racial Sincerity* (Chicago: University of Chicago Press, 2005).

42. Amadou is the version of the Arabic name Ahmad common in West Africa; Syma is the version of the Arabic, Saa'ima, common in South Asia.

CHAPTER 11

LIGHTS, CAMERA, SUSPENSION

FREEZING THE FRAME ON THE MAHMOUD ABDUL-RAUF-ANTHEM CONTROVERSY

ZAREENA GREWAL

INTRODUCTION

IN MARCH 1996, MAHMOUD ABDUL-RAUF OF THE DENVER NUGGETS CAUSED A national stir when the National Basketball Association (NBA) suspended him for refusing to stand during the national anthem before games. Across the country, Americans responded to Abdul-Rauf in a variety of ways, ranging from patriotic outrage to sympathy and admiration for his principled stance. Abdul-Rauf explained his dissent was an act of his "Muslim conscience."[1] As the story of the NBA suspension broke in the national media, Muslim Americans were put on the defensive: did their belief in Islam compromise their ability to be loyal American citizens? The reaction of Muslim American spokespeople interviewed by the media (most of whom were immigrant Muslims) was striking; they approached the controversy as a public relations disaster and responded to the requests to explain Abdul-Rauf's position with an embarrassed disavowal.[2]

Muslim American spokespeople's desire to distance themselves from Abdul-Rauf's stance reflects the climate of political fear in American mosques, the routine scrutinization of these populations, as well as their internal structures of ethnic and racial hierarchy. It is difficult to know how many Muslims live in the United States since the U.S. Census Bureau collects no information on religion, but there may be as many as five to eight million, making Islam the second religion in the

U.S. after Christianity.[3] Today, most Muslim Americans are first- and second-generation immigrants from over sixty nations in South and Central Asia, the Middle East, Africa, the Far East, and the Balkans, who came in the largest numbers after the 1965 appeal of the Immigration Act, although some immigrant communities date back to the turn of the century. For most of the twentieth century, most Muslims in the United States were black, and blacks still make up the largest subset of Muslim Americans, with estimates ranging from 30 to 40 percent.[4] Muslim American communities are the most diverse in the world and mosques attract more heterogeneous congregations than any other houses of worship in the United States. The fact that Muslim Americans from different walks of life regularly pray next to one another, however, does not mean that they necessarily understand one another. Typically led by highly educated, upwardly mobile immigrants from the Middle East and South Asia, Muslim American mosque communities and national religious organizations are riddled with racial and class tensions and cultural misunderstandings.

One particularly strong culture gap that crystallized in the Abdul-Rauf controversy is the markedly different historical trajectories of African American Muslims and immigrant Muslims and their often radically different orientations to the state. The immigrants, often political and economic refugees, are torn between their loyalty to the United States for privileges and freedoms (including religious freedom) often denied in their home countries, the intense pressures to assimilate to the dominant American culture, their social marginalization due to rampant prejudices, and their sense of alienation from the U.S. government's often brutal foreign policies in Muslim-majority countries, particularly in the Middle East. For many African American Muslims, Islam serves as a tablet on which to inscribe their painful history in the United States as well as their radical political desires for reshaping the structure of American society such that it is more egalitarian and just. Abdul-Rauf's stance, although consistent with the political orientation of many African American Muslims (few of whom were interviewed in the mainstream media), perplexed many immigrant Muslims consumed by a different set of political issues, an international rather than domestic focus, and who, in large part, do not possess the cultural reference to understand Abdul-Rauf's decision in the context of a long history of American dissent. Sherman Jackson notes that the arrival of the post-1965 immigrants transformed religious discourse in American mosques and that "the West" became the new "countercategory" of American Islam. Black Muslims often mistook immigrants' resentment and disdain for the West as a common opposition to white supremacy. The pursuit of the (racial) American dream that brought the immigrants to the United States often led them to distinguish whiteness (a status many immigrants coveted) from Westernness. And thus, the anti-Western sentiments expressed in impassioned Friday sermons rarely connected to domestic social and political problems or to explicitly racial politics. Issues such as police brutality, unemployment, and drugs were overlooked because, in the eyes of immigrants, they paled as social justice concerns compared to oppressive Western policies in the immigrants' countries of origin.[5]

In this chapter, I reexamine the 1996 Abdul-Rauf anthem controversy and the accompanying representations of Islam, Muslim Americans, and African American political consciousness. The media became the stage where the competition for religious authority between blacks and immigrants was enacted, revealing the fractures within Muslim American communities as well as the particular secular and racial logics that structure mainstream media representations of Islam, constructions of race, and national debates about patriotism, freedom of speech, and freedom of religion. By placing this moment in a broader historical context of national discourses, we see the ways Islam is both raced and erased in the business of making black Muslim athletes into national heroes or villains.

BY THE "MISSISSIPPI SON'S" LIFE:
AMERICAN DREAM OR AMERICAN NIGHTMARE?

In the fall of 2000, as a graduate student in anthropology and history at the University of Michigan, I decided to make a short documentary film about Muslims in the United States in lieu of a seminar paper. Although my memories of the Abdul-Rauf-anthem controversy were fuzzy, I came across a recently published article that quickly brought them into focus. Daniel Pipes, a neoconservative columnist, author, and frequent media terrorism analyst often accused of being an "Islamaphobe," had written yet another one of his alarmist essays on the threat of Islam to Americans, arguing that Islam undermined the culture, customs, laws, and polices of the United States in more fundamental ways than fascist or Marxist-Leninist ideologies. In this particular piece, he focused on the special threat to the United States posed by African American converts to Islam because their "protest" temperaments and special susceptibility to the contempt for America imported by Muslim immigrants made them dangerous.[6] As evidence of the ways Islam turned American converts against their own country, Pipes cited the example of the 1996 suspension of Abdul-Rauf by the NBA for not standing for the national anthem. He also applauded the opposition to Abdul-Rauf voiced by the Muslim spokespeople as a rare expression of moderate Islam.[7]

Pipe's representation of Abdul-Rauf as the prototypical "un-American" Muslim became one point of departure for my film, *By the Dawn's Early Light: Chris Jackson's Journey to Islam*, which quickly snowballed from a class project into a feature-length educational documentary broadcast on the Documentar Channel.[8] By taking Abdul-Rauf's life and career as a case study and the anthem controversy as the climax, the film provides an ethnographic window into Muslim American communities, particularly the cultural misunderstandings and rifts between immigrants and blacks. As a documentary in the true sense, the film documents multiple and competing perspectives by drawing on a wide range of archival footage sources, personal interviews, and expert analyses by Muslim American public intellectuals and scholars. Rather than relying on a one-dimensional narration through a voice-over, the film's montage of voices provides a rich, complex picture that weaves together Abdul-Rauf's biography, analyses of the different histories and fractures in Muslim American communities, and various debates about the nature of patriotism, freedom of speech, and the complexities of racial politics in the United States at the turn of the century. Yet the film remains a simple story of one man's spiritual journey. The argument I develop here offers a more sustained analysis of the coverage of the controversy than I was able to explore in the film. Here I draw on the narrative arc of Abdul-Rauf's life in order to highlight how he went from being an icon of the American dream to the embodiment of Pipe's American nightmare.

Long before the anthem controversy made him a national news story, Abdul-Rauf was an established favorite as a human-interest story for sports journalists with a penchant for "American dream" angles. Born Chris Jackson to a single mother juggling two jobs to keep food on the table for her three boys, Jackson's humble beginnings in the crime-ridden, drug-infested slums of Gulfport, Mississippi are almost an "up from your bootstraps" cliché. Journalists eagerly reproduced Jackson's gripping descriptions of his loneliness as a child and his hungry desperation to change his family's condition that drove him to spend countless, mean hours perfecting his skills on the basketball court, playing against an "invisible man" who was always one step quicker, perhaps the father he had never known and who could never be mentioned or a kind of phantom manifestation of his own unforgiving perfectionism. His perfectionism and obsessive-compulsive practicing was indirectly tied to an undiagnosed neurological condition he suffered from, Tourette's syndrome, which also caused involuntary grunting and body spasms that could become so intense he would often collapse in speechless exhaustion. Jackson's condition went neglected and untreated for most of his childhood (in fact, he was even wrongly placed in the special education track for a period) until his high school coach referred him to a physician who finally diagnosed him properly and started him on a drug therapy. Despite suffering a disease that sometimes made it difficult to carry on a conversation or read a book, let alone thread a bounce pass through a crowd of looming opponents, Jackson's remarkable skill made him a local legend and Mississippi's Player of the Year for two years (earning him the nickname "The Mississippi Son") as well as a recruiting magnet for college talent scouts from around the country.[9]

At Louisiana State University (LSU), "Action Jackson" set three NCAA records for freshman, averaged 30.2 points per game, and made nearly every post-season all-America first team, landing him on the cover of *Sports Illustrated* as a freshman. The next year he averaged 27.8 points per game, led the southeastern conference in scoring, and once again earned Player of the Year honors in the conference despite the looming presence of his friend and teammate Shaquille O'Neal. He was the third pick in the 1990 NBA draft, and in his rookie season with the Denver Nuggets he averaged an impressive 14.1 points in 22.1 minutes coming off of the bench. It is as if by sheer will alone he rose from his humble conditions in the face of his debilitating disease, the crushing poverty, and the entrenched racism of Gulfport, Mississippi, to awe crowds as an NBA star, and journalist after journalist remarked on how he was "a perfect example of the American dream."

His soft-spoken and unassuming manner and his lean six foot one inch frame almost conveyed frailty in the world of NBA giants, yet time and time again, Jackson proved himself on the court with his explosive first step past defenders and his incredibly accurate shooting record. By the 1992/1993 season, he was the beloved, fan-favorite starting point guard of the Denver Nuggets and scored a career-high 19.2 points per game.

In addition to his remarkable talent, Jackson was also a spiritual seeker. Although Jackson was first exposed to Islam through *The Autobiography of Malcolm X* as a student at LSU, the stresses of the transition to the NBA ignited his desire to learn more about Islam and to read the Qu'ran. In 1991, he quietly converted but his new faith only became a news story once he legally changed his name from Chris Jackson to Mahmoud Abdul-Rauf in 1993. Despite the warnings that his new name would hurt his career since it was Chris Jackson that was the established household name and that the Arabic might be difficult for Americans to pronounce, fans and sports commentators adjusted to the change relatively easily, chanting "Rauf, Rauf, Rauf is on fire!" Although generally invisible to spectators, Islam radically transformed Abdul-Rauf. He began to see his life, his disease, and his passion for basketball in a new light. Bruce Schoenfeld, in *The Sporting News* on February 14, 1994, quotes Abdul-Rauf: "God has given me, through His blessing, Tourrette's, and he has given me basketball and Islam to cope with Tourrette's. Through basketball, I get a little peace. And through Islam I get total peace." He also began reading voraciously, everything from religious literature to history and global politics. His decision in 1995 to remain in the locker room during the national anthem developed out of his growing social consciousness, political awareness, and sense of global responsibility, all products of his new religious sensibility.[10]

For over sixty games, Abdul-Rauf remained in the locker room or hallway to the stadium while the national anthem was being played and then walked silently onto the court for the starting lineup. Although his practice was recognized without objection by the Denver Nuggets and the Players Association for the entire season, in early March 1996, fans called a local radio station and complained on air. This led to a newspaper story and a courtside television news interview the following Tuesday (March 12) during a morning shootaround in Denver that produced the sound bite that spread like wildfire in the national media. Abdul-Rauf explained to a reporter that the United States flag is "also a symbol of oppression, of tyranny, so it depends on how you look at it. I think this country has a long history of that. If you look at history, I don't think you can argue the facts." Later the same day, the NBA suspended Abdul-Rauf indefinitely without pay (a loss of $31,707 per game) for being in violation of a clause in the league's rule book that states, "Players, coaches and trainers are to stand and line up in a dignified posture along the sidelines or the foul line during the playing of the national anthem." The Players Association disputed the suspension on the grounds that the rule "was not one agreed to in collective bargaining, but was imposed by the league unilaterally in an operations manual, without any input from the players." The league countered by arguing that there is a standard clause in all player contracts stipulating league rules must be abided by. In a Denver Post poll, 72 percent of Denver-area adults took issue with Abdul-Rauf, many arguing his $2.6 million salary made it incumbent upon him to be appreciative and grateful to the United States. Abdul-Rauf remained firm in his conviction in face of the suspension and his critics: "My beliefs are more important than anything. If I have to give up basketball, I will."

THE POLITICS OF CONFUSION

Despite the widespread media attention garnered by the Abdul-Rauf-anthem controversy, the coverage of the story was marked by inaccuracies, contradictory reports, and a sense of confusion. I argue that the cloud of confusion around the controversy reflects particular secular assumptions and racial logics regularly employed in the mainstream media's representations of Islam and Muslims. First, there are the problematic false equivalences between Islam and the Middle East and Muslims and Arabs, which, in this case, led to a great deal of confusion about Abdul-Rauf's racial, national, and religious identity. Second, much of the coverage was structured by a false opposition between Abdul-Rauf's religious and political beliefs. Since Abdul-Rauf's statements did not fit neatly into a secular division of religion and politics, his reasons for not standing sometimes seemed incoherent or even contradictory in the coverage. Islam was represented as a static body of beliefs and practices and an "eastern" religion, and, as a result, Abdul-Rauf was often rendered an inauthentic or misguided religious novice.

The coverage of the Abdul-Rauf-anthem controversy reproduced a number of common misconceptions about Islam and Muslims. Americans often wrongly use the ethnic category of Arab and the religious category of Muslim interchangeably (in fact, Arabs only make up about 17 percent of the 1.2 billion Muslims in the world). This commonly held, but incorrect, view of Islam undermines the universality of the tradition and implies that Islam is not and cannot be an indigenous, American religion. Fixing Islam to the Middle East as an "eastern religion" in this way elides important facts: African Americans comprise the largest single Muslim American population, Americans are converting to Islam at a faster rate than any other religion, and Islam has a long history in the United States, beginning with the arrival of the African slaves, many of whom were Muslim.

In fact, the assumption that Muslims are a foreign population from somewhere else led many to assume that Abdul-Rauf was an immigrant. He received two garbage bags full of hate mail in the fallout from the controversy and many of the letters instructed him to "Go home!" or "Go back to Africa!"[11] In fact, even fans that came to his defense sometimes assumed he was from another country; one African American woman interviewed by a local Denver TV news reporter suggested, "Maybe that's what they do where he comes from."[12] Part of the public's failure to recognize Abdul-Rauf as African American may also be attributed to the general and striking silence on the part of African American Christian leaders during the controversy, which, like the embarrassed disavowal of Muslim immigrant spokespeople, reflects the fractures within the African American community along religious lines.[13] At one point, Jesse Jackson told a journalist he was willing to mediate the dispute; however, Abdul-Rauf's camp dismissed this gesture as a media stunt since Jackson never contacted Abdul-Rauf or his representatives.[14] There was such confusion about Abdul-Rauf's identity that he felt the need to identify as an African American as a premise to his official statement to the press: "I am an African-American, a citizen of this country, and one who respects freedom of speech and freedom of expression."

In other instances, the representation of Abdul-Rauf as a Middle Eastern immigrant was purposeful and political. Cartoonist Drew Litton of the *Rocky Mountain News* satirized Abdul-Rauf by "Arabizing" his features, giving him a Semitic nose, heavy-lidded, sunken eyes, and by lengthening his beard. A satire of Abdul-Rauf as a Sambo figure certainly would have offended readers and immediately drawn criticism, however, racial caricatures of Arabs, Middle Easterners, and Muslims as villains in the mainstream media are so common they rarely draw attention or objections. Interestingly, Litton replicated the patterns of coverage of the Middle East that so often represents political conflicts in the region as spontaneous, inexplicable, and without history.[15] By depicting Abdul-Rauf as casually tossing the flag in a dirty towel bin, Litton suggests that Abdul-Rauf's decision was spontaneous rather than considered, that his "Muslim features" are explanation enough for his "un-American" disdain. We are perhaps most familiar with this narrative formula from the coverage of the terrorist attacks of September 11. For example, the incredulous and ubiquitous question and headline "Why do they hate us?" represents the attacks in the American media and the official discourse of the state as a spontaneous, ahistorical, and incomprehensible act of evil ("Because they hate freedom."), rather than a calculated political action (or reaction) with a long (and modern) history tied to the shifting and complex political and economic relationship between the United States and the Middle East.

In addition to the confusion about Abdul-Rauf's identity, his reasons for not standing for the anthem were also often presented in an unclear way in the media because they did not fit neatly into a clean secular division of religion and politics. In order to conform to secular categories more neatly, Abdul-Rauf's explanation of his stance was distilled into two separate sound bites. The first was coded as religious: "I'm a Muslim first and a Muslim last. My duty is to my creator, not to nationalistic ideology." This "theological" explanation of Abdul-Rauf's dissent referenced the strong emphasis on God's oneness and exclusive right to be worshipped, which distinguishes expressions of monotheism in Islam from the other Abrahamic faiths. The second and much more ubiquitous sound bite was coded as political rather than religious: "[The flag is] also a symbol of oppression, of tyranny, so it depends on how you look at it. I think this country has a long history of that. If you look at history, I don't think you can argue the facts." The sentence immediately following this quote was usually left out the clip: "You can't be for God and for oppression. It's clear in the Qu'ran." The excluded sentence makes it quite clear that for Abdul-Rauf, his political and religious beliefs cannot be so neatly bifurcated.

The interpenetration of religion and politics in Abdul-Rauf's explanation characterizes not only African American Islam but also what a number of scholars have referred to as the specifically American tradition of Black Religion.[16] Black Religion should not be confused with the Black Church nor mistaken as a synonym for the supercategory or mother set of "African-American religion."[17] A more instructive definition of Black Religion is as an aggregate of black American religious experience whose most enduring feature is its radical, holy protest against the material and psychological effects of racism.[18] The secular binary opposition between political and religious beliefs obscures more than it illuminates in the coverage of Abdul-Rauf's case, just as in the early scholarly analyses of African American Islam the same secular logic was invoked to dismiss communities like the Nation of Islam as thinly guised ideological, sociopolitical movements beyond the pale of genuine religious experience. Interestingly, several articles mistakenly identified Abdul-Rauf as a member of the Nation of Islam. The representation of Abdul-Rauf as an inauthentic Muslim was propelled in part by contrasting him with Hakeem Olajuwon, a Nigerian Muslim who played for the Houston Rockets. He explained, "In general, Islamic teachings require every Muslim to obey and respect the law of the countries they live in. You know, that is—that is Islamic teachings. You know, to be a good Muslim is to be a good citizen." As an immigrant with a heavy accent, Olajuwon was assumed to be the authentic Muslim with authentic religious authority. In addition, in the *New York Times*, Olajuwon even reproduced the clean binary opposition between religion and politics. "The difference must be distinguished between worship and respect," he said. "Islam orders you to obey and respect, as long as you are not worshiping anything other than God. The Koran teaches respect for all people. That's why it's so important that people understand that there is a difference between respect and worship. People that worship the flag should also understand that there is a difference. Islam is a religion of peace. You don't attack. You explain." Olajuwon's two sound bites corresponded to Abdul-Rauf's as correctives. In one, Abdul-Rauf's "theological" logic is represented as a confusion of respect and worship. In the other, Abdul-Rauf's politics are represented as a radical departure from a religion that requires the faithful to be good law-abiding citizens. Despite the frequent references to Olajuwon in the national press, his acknowledgement of Abdul-Rauf's ability to have a valid religious interpretation that diverged from his own was not included in their coverage. On March 14, only Ed Fowler of the *Houston Chronicle* reported that Olajuwon said that "if Abdul-Rauf is certain his interpretation is the only acceptable one, he should be applauded for taking that stand at the cost of a magnificent livelihood."

The same insistence on the clean division between political and religious beliefs came out in the equally confused legal analyses of the NBA's suspension. Experts were called on to determine whether Abdul-Rauf was legally bound by his NBA contract to stand for the anthem or whether the NBA had violated

Abdul-Rauf's Title VII right not to be discriminated against as an employee due to his religion. In Jason Diamos's March 14 *New York Times* article, legal expert Martin Garbus cast doubt not only on Abdul-Rauf's potential legal case but also on his religious integrity. He explained, "There is a difference between religious and political beliefs. If it's religious, he has an absolute right to do it. He's saying religious, but a lot of the language is political. And to the extent that the religious and political become intertwined, you may find yourself in a situation where there is a clear and present danger." Garbus's alarm about the "clear and present danger" posed by Abdul-Rauf echoes Pipes. There is also another incorrect assumption at work here: that Islam is a monolithic and static body of beliefs and practices and that there is only one possible "correct" Islamic position on any given issue. Garbus's analysis seems particularly odd given that the U.S. legal system recognizes the diversity of opinions within any community of believers and, therefore, the legal system only requires that litigants establish that they hold "a sincere religious belief" and that they prove that they communicated their beliefs to their employer.[19]

When determining if one's religious beliefs are protected under the Free Exercise Clause of the First Amendment as opposed to Title VII, courts conclude that "interfaith differences . . . are not uncommon among followers of a particular creed, and the judicial process is singularly ill equipped to resolve such differences in relation to the Religion Clauses . . . [The] guarantee of free exercise is not limited to beliefs which are shared by all of the members of a religious sect . . . Courts are not arbiters of scriptural interpretation."[20]

Yet again and again, in legal analyses of the potential case, Muslim immigrants' claims that standing for the flag did not contravene Islam was presented as evidence that Abdul-Rauf was either trying to hide his political beliefs in the guise of religion in order to acquire some kind of legal protection (a kind of perversion of Islam) or that he, as a black convert, failed to understand Islam the "right" way, as authentic Muslims from the east could.

The same assumption about a singular "correct" or "authentic" understanding of Islam emerged in the coverage of the resolution of the suspension. Abdul-Rauf was blasted in the media first for his refusal to stand and then for finding a way to reconcile his religious beliefs and honor his contract. The suspension was lifted after one game because Abdul-Rauf agreed to be on the court during the anthem ceremony and to stand and the NBA agreed to let him pray silently rather than sing the anthem. He told a reporter, "This is what I believe, and, and, I'm not wrong for the stance that I took, and in no way am I compromising, but I'm saying I understand and recognize that there is a better approach and in Islam . . . after making a decision, if you see that which is better, you do that." In another interview, he emphasized that his religious feeling and opposition to injustice transcended racial, national, and religious lines: "I'll stand but I'll offer a prayer, my own prayer, for those who are suffering. Muslim. Caucasian. African-American. Asian. Whoever is in that position, whoever is experiencing difficulties."

Abdul-Rauf's willingness to adjust his practice did not satisfy his critics. In his first game back on the court following the suspension, the Chicago crowd showered him with boos as he scored nineteen points. According to Roscoe Nance's March 14 article in *USA Today*, the Denver Nuggets had already received more than two hundred phone calls from fans threatening to boycott games if Abdul-Rauf remained with the team or threatening to cancel their season tickets. The league decision to suspend him was influenced by the fear that the economic consequences of dissatisfied fans might translate into a reduction in support from advertisers like Nike and television networks like NBC. Michael Hiestand, in the October 22, 1996, *USA Today* revealed that Nike was also considering buying Ascent Entertainment, which owned the Denver Nuggets in the same period. Given these circumstances, it is not particularly surprising that Abdul-Rauf was traded to the Sacramento Kings the following season, despite being the Nuggets' playmaker and lead scorer and one of the better free-throw shooters in NBA history. His basketball career never fully recovered from the damage of the controversy and he eventually began playing in Europe.

RACING AND ERASING ISLAM: RELIGION IN THE WORLD OF SPORTS

Of course, Abdul-Rauf is not the first professional athlete to bring religion into the sacred space of the stadium. As depicted in the Academy Award winning movie *Chariots of Fire* in 1981, Eric Liddell, a member of Great Britain's track team at the 1924 Paris Olympics, opted out of the 100-meter race once he learned he would have to run a heat on the Sabbath. His stance shocked the sports world and he eventually won a gold medal in the 400-meter race. In 1965, Sandy Koufax, a pitcher for the Dodgers, refused to play on opening day of the World Series because it fell on Yom Kippur. The league accommodated Koufax by shifting the series by a day. Steven Maranz of *The Sporting News* detailed on November 28, 1994, how Shawn Bradley, a Morman basketball player, delayed his potential earnings from a lucrative professional basketball career with the Philadelphia 76ers in order to fulfill his obligation to his faith to carry out a two-year mission. While all of these athletes should be admired for the strength of their convictions in the face of enormous pressure, what makes them different from Abdul-Rauf is that their religious beliefs did not interrupt the secular division between religious and political beliefs the way Abdul-Rauf's did. Arguably, their ability to fit easily within secular discursive frames facilitated both the processes of accommodating them as well as representing them as principled heroes. There is another important iconic athlete who challenged the division between religion and politics in much more profound and radical ways than Abdul-Rauf: Muhammad Ali. In the name of Islam, Ali had refused to serve America in the time of war and, as a result, was nearly imprisoned, lost his title, was banned from competing, and was condemned by the national media. In 1967, Ali refused to step forward to be inducted into Vietnam and within the hour, before being been charged with any crime, the New York State Athletic Commission suspended his boxing license and stripped him of his title as heavyweight champion, with all other jurisdictions in the United States quickly following suit.[21] Yet, today, Ali is a national icon, the quintessential American. Louisville, the city that once legally renounced him, now boasts a highway in his name.

I am not invoking Muhammad Ali only because he is the most famous example of a dissenting athlete or because, like Abdul-Rauf, he is black and Muslim. Rather, the comparison to Ali brings Abdul-Rauf's case into relief because Ali's image was reinvented in 1996, overlapping with the Abdul-Rauf-anthem controversy. In the opening ceremony of the 1996 Olympics, Muhammad Ali raised a trembling arm to light the fuse to the Olympic cauldron, followed by a recitation of an excerpt from Martin Luther King Jr.'s March on Washington speech. The widely broadcast image suggested a kind of fulfillment of King's dream in the national embrace of Ali as an icon of harmony and goodwill. Outside the stadium, Hosea Williams, a former colleague of King's, led a small protest against the Georgia state flag's Confederate stars and bars fluttering over games that celebrated human equality but few reporters covered her demonstration. At halftime during the Olympic basketball final, Ali was presented with a replacement gold medal for the one he had flung into the Ohio River thirty-six years earlier precisely because of the gaps between the Olympic ideals and American social realities. *USA Today* credited Ali's Olympic cameo with sparking "a renaissance for the Greatest" and he took *Sports Illustrated's* cover for a record-breaking thirty-fourth time.

For all the sudden interest in Ali's legacy in 1996, the fact that, in his heyday, he had always been more popular abroad than at home was seemingly forgotten. After all, his anti-American defiance is what made him an international hero in much of Africa, Asia, Latin America, and Europe in that period. Equally significant, Ali's Islam seemed to be erased from collective memory as well, or at least irrelevant in the new coverage that rarely focused on his Muslim identity except as a biographical detail. Analysts of the Black Power movement of the 1960s frequently divide the radicalism of the movement between the elements that were co-opted and sanitized and those that were destroyed. Muhammad Ali embodies both of those processes; he has been co-opted and depoliticized as a national icon decades after being the most reviled figure in American sports and, in some sense, physically destroyed. The stripping of his title and his resultant lengthy absence from the ring meant that when Ali finally returned to the ring, he was slower and had to rely on his ability to absorb punches, which is undoubtedly related to his current medical condition.[22]

The Ali offered up for veneration in the 1990s is not the Ali of the 1960s, and the image of the 1960s that is celebrated or damned in the 1990s is a mere caricature of the original. In both cases, a complex and contradictory reality has been homogenized and repackaged for sale in an ever-burgeoning marketplace for cultural commodities. In the nineties, we were told that the causes and complaints of the sixties were redundant, that the conflicts that once surrounded Ali have been resolved. Somehow the rights and wrongs of the hard choices he made have been declared peripheral to his legacy—as if racism and warfare, Islam and the West, personal identity, black leadership and the use of U.S. military might in the poorer, darker countries were yesterday's issues, no longer pertinent, no longer divisive.[23]

Ali's fame was renewed and reinterpreted in such a way that his Islam was either erased or represented as benign, apolitical in order for him to be converted into a symbol of national identity. Since being reborn as a benign, national icon, Ali's image has been used to sell everything from Sprite to Microsoft to the war on terror in the Muslim world by the State Department. In this sense, the story of the evolution of Abdul-Rauf's image is the inverse of Ali's. Ali, once demonized for his Islam, has been transformed into a national icon through the muting of his religious identity. Abdul-Rauf, on the other hand, once heralded as an icon of the American dream as the "Mississippi Son," has come to represent all that is wrong, threatening, and dangerous about Islam by muting his identity as an African American.

Before the anthem controversy, Abdul-Rauf was involved in another minor controversy that made sports page news briefly when he went head-to-head with Nike, who had initially given him a shoe contract. As recounted by John Mossman of *The Baton Rouge Advocate* on March 15, 1996, after a scuffle with Nike that led to the loss of the contract, Abdul-Rauf began covering the logos on his shoes with masking tape and refused to identify the brand when asked by journalists. In this regard, Abdul-Rauf is the inverse of that other American icon so often compared to Muhammad Ali: Michael Jordan. Jordan, however, is no Muhammad Ali. Where Ali refused "to carry a sign," Jordan only carries the signs he gets paid for and his patriotic feelings seem to be inseparable from his capitalist acumen. During the 1992 Olympics in Barcelona, Spain, Michael Jordan and fellow face of Nike Charles Barkley used the American flag to cover a Reebok logo on the Olympic medal stand as a symbol to their commitment to life, liberty, and the pursuit of endorsements, and when Nike was blasted in the media over its Indonesian sweatshops controversy, it was the patriotic "CEO Jordan" ad campaign that salvaged the company's image.[24] Where Ali was the idealistic hero carrying the banner of humanitarian internationalism in the global struggles against inequality, poverty, and injustice, Jordan is the American emblem of globalization. Jordan is a symbol of "corporate America and its winner-takes-all ethic. His blackness has been deliberately submerged within his Americanness, which is reduced, in the end, to his individual wealth and success."[25] In contrast to Jordan, Abdul-Rauf no longer has the option to submerge his blackness with his Americanness because, in a sense, his Islam negates both his blackness and his Americanness. If the transformations of Ali's image are any indication, it seems Black Muslim athletes can either be represented as (black) American heroes or as Muslim villains.

Conclusion

On August 1, 2001, two days after my camera crew had packed up and left Gulfport, Abdul-Rauf's home, which was under construction, was burned to the ground. Abdul-Rauf reported finding KKK graffiti on his property, and although the arson investigation is ongoing, the FBI's prime suspect in the case remains the KKK. The documentary ends on the image of Abdul-Rauf's burning house, anticipating another burning building and another terrorist attack that would take place just over a month later. Since September 11, the culture of hyperpatriotism, the suppression of dissent, and the scrutinization of Muslim Americans has been dramatically increased, although all of these elements were present in the Abdul-Rauf-anthem controversy as well.

Perhaps most disturbing, before September 11, racial profiling had largely been dismissed as an "inefficient, ineffective, and unfair" policy and, in fact, it was explicitly condemned by President Bush and John Ashcroft, but in the months after September 11, polls indicated that the American public consensus was that racial profiling was not only a good thing but essential for the nation's survival. A Gallup poll found that one-third of New Yorkers supported the internment of Arab Americans. Another Gallup poll surveying African Americans in the weeks after September 11 found the majority (71 percent) supported racial profiling of Middle Easterners and Arabs. As other minorities endorsed the racial profiling of Middle Easterners, Arabs, and Muslims, their endorsement Americanizes them and consolidates the new configuration of a singular, multicultural nation. "[Although] racial policing continues apace in all communities of color, and we can anticipate that this new multiculturalist national identity is a momentary phenomenon, whites, African Americans, East Asian Americans, and Latinas and Latinos are in a certain sense now deemed safe and not required to prove their allegiance. In contrast, those who inhabit the vulnerable category of looking 'Middle Eastern, Arab, or Muslim,' and who are thus subject to potential profiling, have had to, as a matter of personal safety, drape their dwellings, workplaces, and bodies with flags in an often futile attempt at demonstrating their loyalty."[26] The exaggerated appearance of American flags in Arab and Muslim American communities is predictable because flags are so often the means of normalizing and disciplining "others." Like marginal communities before them, Muslims have a range of motivations in displaying them, from defiant assertions of patriotism to talismanic shields protecting them from hostility.[27]

It becomes obvious very quickly that pressures are not being applied evenly across Muslim American communities. Veiled women, as the most visible and, therefore, vulnerable segment of the community, have suffered the brunt of hate crimes and harassment. Arab and Middle Eastern Muslims are profiled much more frequently than African American Muslims. Blackness often acts as a layer of insulation for African American Muslims and guarantees their status as social citizens in ways not shared by their immigrant coreligionists. We might imagine how the media, the Justice Department, and the Immigration and Naturalization Service might have reacted had the 2003 "D.C. Sniper" been a Middle Easterner rather than an African American. (The fact that John Muhammad is a Muslim received little comment in the press.) Certainly, the American public would have responded with outrage had the attorney general rounded up hundreds of African

American Muslims and held them for months without charges the way Arab and Middle Eastern Muslims have been. Since the association of Islam with Arabs and the Middle East in the American imagination is so blindingly strong, blacks generally remain an invisible part of the Muslim American community and are often immune from social stigma and exclusions.[28] Abdul-Rauf's case demonstrates that the layer of social insulation to African American Muslims provided by their blackness could easily slip. In this sense, Abdul-Rauf's case is both an important historical moment that takes on greater relevance in a post-September 11 America as well as a cautionary tale about the fluidity and the brutality of race in the United States.

NOTES

1. Mahmoud Abdul-Rauf, interview with author, July 20, 2001.
2. National Muslim organizations made public statements insisting that standing for the national anthem did not contravene Islam; only the Islamic Society of North America's (ISNA) statement was neutral. Jason Diamos reported in *The New York Times* on March 15 that ISNA's secretary general Sayyed M. Syeed told the Associated Press that the decision to stand for the anthem or the flag is a subjective one, and believers are responsible to their own consciences.
3. Karen Isaksen Leonard, *Muslims in the United States: The State of Research* (New York: Russell Sage Foundation, 2003).
4. Fareed H. Nu'man, *The Muslim Population in the United States* (Washington, DC: American Muslim Council, 1992).
5. Sherman A. Jackson, *Islam and the Blackamerican: Looking Towards the Third Resurrection* (New York: Oxford University Press, 2005), 72.
6. Daniel Pipes, "In Muslim America: A Presence and a Challenge" *National Review*, February 21, 2000.
7. Pipes, "In Muslim America."
8. Zareena Grewal. *By the Dawn's Early Light: Chris Jackson's Journey to Islam* (Cinema Guild, 2004).
9. Mahmoud Abdul-Rauf, interview with author, July 20, 2001.
10. Abdul-Rauf, interview.
11. Ibid.
12. Zareena Grewal, *By the Dawn's Early Light.*
13. Sherman A. Jackson, *Islam and the Blackamerican.*
14. Shareef Nasir, interview with author, July 20, 2001.
15. Melani McAlister, *Epic Encounters: Culture, Media, and U.S. Interests in the Middle East 1945–2000* (Berkeley: University of California Press, 2001).
16. Gayraud Wilmore, *Black Religion and Black Radicalism: An Interpretation of the Religious History of Afro-American People* (Maryknoll, NY: Orbis, 1983).
17. Sherman A. Jackson, *Islam and the Blackamerican*, 25, 29. Black Religion is one among many black religious traditions in the United States, although its hegemonic (but not universal) influence has been unparalleled in the United States over the past two centuries.
18. Sherman A. Jackson, *Islam and the Blackamerican*, 25, 29.
19. Kelly B. Koenig, "Mahmoud Abdul-Rauf's Suspension for Refusing to Stand for the National Anthem: A 'Free Throw' for the NBA and Denver Nuggets, or a 'Slam Dunk' Violation of Abdul-Rauf's Title VII Rights?" *Washington University Law Quarterly* 76, no. 1 (1998).

20. Thomas v. Review Bd. of the Ind. Employment Sec. Div., quoted in Koenig.

21. Thomas Hauser, *Muhammad Ali: His Life and Times* (New York: Simon and Schuster, 1991), 172.

22. Dave Zirin, *What's My Name Fool: Sports and Resistance in the United States* (Chicago: Haymarket Books, 2005).

23. Mike Marqusee, *Redemption Song: Muhammad Ali and the Spirit of the Sixtie* (New York: Verso, 1999) 5.

24. Jes Cortes, "Star Spangled Business: Labor, Patriotism, and Michael Jordan in a Nike Advertisement." http://to-the-quick.binghamton.edu/issue%201/nike.html (accessed June 16, 2006).

25. Mike Marqusee, *Redemption Song*, 295.

26. Leti Volpp, "The Citizen and the Terrorist," *September 11 in History: A Watershed Moment?* (Durham, NC: Duke University Press, 2003), 151.

27. Andrew Shryock, "New Images of Arab Detroit: Seeing Otherness and Identity through the Lens of September 11," *American Anthropologist* 104 (2002): 919.

28. Sherman A. Jackson, *Islam and the Blackamerican*, 132–33.

CHAPTER 12

PROTECT YA NECK (REMIX)

MUSLIMS AND THE CARCERAL IMAGINATION IN THE AGE OF GUANTÁNAMO

SOHAIL DAULATZAI

The small man builds cages for everyone he knows.
While the sage, who has to duck his head when the moon is low,
Keeps dropping keys all night long for the beautiful rowdy prisoners.
—Hafiz, "Dropping Keys"

THE SCREENING OF THE GILLO PONTECORVO'S 1965 FILM *BATTLE OF ALGIERS* AT the Pentagon in 2003 revealed a great deal about American imperial ambition and the measures that they were willing to take in order to fulfill the neocon prophesy of "full spectrum dominance." As a result of the Pentagon screening, the film has garnered a great deal of attention and quite predictably has also been completely reappropriated by the Beltway belligerati not as the embodiment of the struggles for national liberation against racist colonial violence but instead as a blueprint and training manual for "terrorism" in a post-9/11 world. In framing the film in this way, American officials engaged in an act of selective memory and collective amnesia as they attempted to erase the history of colonialism and violence that is endemic to the European and American encounter with the Third World. Not only that, this attempt at historical revisionism also sought to erase America's own complicity with, and extension of, European colonialism as the Cold War unfolded (Iran, Vietnam, the Congo, etc.) while also attempting to reaffirm their current occupation of Iraq and Afghanistan as another example of American imperial benevolence.

What else was the United States hoping to gain from viewing the film? How to deal with an insurgency? How to win the battle but not lose the "hearts and minds" of the colonized? The film anticipates many of the forms of control and violence exhibited within the current American imperial project such as the military-media alliance, the multiple forms of surveillance of the Muslim body (check points, cameras, searches), the use of counterrevolutionary methods (bombings of civilians, psychological warfare, disinformation), and the empty rhetoric of "democracy" when done at gunpoint. But it is the central role of torture and the prison as technologies of colonial violence within the film that eerily anticipate the revelations at Abu Ghraib, Guantánamo, Bagram and other U.S. detention centers throughout the world.

How are we to understand the place of torture and incarceration as forms of control in the current global order of imperial violence? While torture and colonial imprisonment used by Europeans and the United States as forms of counterinsurgency against Black and Third World peoples is not new, the sands seem to be shifting. Because in this new age of empire where the formal elements of colonization no longer persist, but their neoliberal residue does, we are now seemingly in a situation where the nation-as-prison has become the operative model for continued Euro-American dominance. Where the nation was once the site for liberation, it is now the space for violent containment—as economic control vis-à-vis International Monetary Fund and World Trade Organization policy, strict control over migration flows, increased militarization, and Western-backed repression within the global south have severely controlled the lives of the majority of the worlds peoples. But just as violent is the possibility that the nation-as-prison is no longer only expressed through the ambient force of postcolonial power and the bloody borders drawn by the West. Instead, the erection of prison walls, electric fences, minefields, concrete barriers, and barbed wire now emerging around the West Bank, the Afghan-Pakistan border, and the U.S.–Mexico border have become the de facto practice that now tragically separates peoples, histories, and even the possibilities for change. The metaphoric is now the literal as the border watchers, checkpoint battalions, and anti-immigrant hysteria running rampant in the West still presumes that we from the global south are guilty, yet on furlough, and allowed to work, survive, or even cling to life—but only if deemed "useful" in a highly racialized neoliberal global order.

These contemporary configurations of power have, according to Achille Mbembe, historic roots within a European juridical order based upon a racialized hierarchy of nation-states. For Mbembe, the colony—that space existing at the frontiers of "civilization"—has its roots within the slave plantation. Through the idea of "necropolitics," Mbembe views modern war as an extension of previous "topographies of cruelty" emanating from Europe and the United States—such as the *colony* and the *slave plantation*—where race becomes the arbiter that not only determines but also sanctions who can be killed and who will be allowed to live as imperial power unleashes itself.[1] Where Mbembe links the plantation and the colony through violence and race, Loic Waquant links the *plantation* to the *prison* as an extension of racial violence within the United States.[2] I am interested in the cartographies of American power that link the *colony* to the *prison*—where the domestic politics of race and captivity that sit at the heart of the empire become the means by which those in the colony become legible under brutal American power. The linking of American imperial reach today, through the space of the prison, becomes a way of exploring and exposing the lie that is "American exceptionalism"—a brutal charade about a mythic American universalism that not only fractures America's historic complicity with European empire building, but also masks America's own violent history as an expansionist power through slavery, Native genocide, and its imperial ambition throughout the globe up to today.

In exploring the prison as an emerging form of transnational power and empire state building of *Pax Americana*, my interest lies in the increasingly blurred lines between the foreign and the domestic, nations and their fragments, and nationalisms and their "others." In doing so, I borrow from the work of Amy Kaplan, who writes that "domestic and foreign spaces are closer than we think, and that the dynamics of imperial expansion cast them into jarring proximity."[3] For Kaplan, U.S. imperialism demands a coherent and stable national identity but U.S. imperial culture has continually created anxiety about this coherence as it expanded its power abroad. She states, "the idea of the nation as home is inextricable from the political, economic, and cultural movements of empire, movements that both erect and unsettle the ever-shifting boundaries between the domestic and the foreign, between 'at home' and 'abroad.'"[4] I want to examine these "movements of empire" through the space of the prison in order to explore the domestic and international regimes of imprisonment that have forged a kind of transnational "carceral imagination" around the most prominent figure of racial otherness today—the Muslim. In attempting to locate this "carceral imagination," this piece will explore the links between the global mapping of imprisonment and control of Arabs and South Asians as evinced by the emergence of Abu Ghraib, Guantánamo, secret detention centers, and the Orwellian euphemisms "extraordinary rendition" and "ghost detainees" with the domestic assault on African American Muslims in U.S. prisons who have been very recently targeted as a potential "fifth column" of recruitment and radicalization against the United States in the "War on Terror."

The racial anxieties about the threats posed by Muslim "terrorists" abroad are projected upon its domestic racial analog—African American Muslims—an anxiety that is more acute due to the ambivalent place of Blackness within the history of U.S. expansionism and imperial culture. Made legible through the historic gaze of American imperial power, the presence of African American Muslims within the United States threatens a coherent national identity that upsets racial hierarchies and domestic unity. As Kaplan says, "underlying the dream of imperial expansion is the nightmare of its own success, a nightmare in which the movement outward into the world threatens to incorporate the foreign and dismantle the domestic sphere of the nation."[5]

The racialized figure of the Muslim haunts the geographic and imaginative spaces of American empire. As the site of the phantasmatic, the excessive and the perverse, the tortured figure of the Muslim—be it the immigrant and foreign (Arab, Asian, Persian or African) or indigenous and domestic (African American)—fluidly and flexibly mobilizes various constituencies, nationally and internationally, in the service of American imperial designs. Whether it's "freeing Muslim women" abroad, creating fractures within and between African Americans and immigrants in the United States, mobilizing Christian evangelicals, liberal multiculturalists, right-wing extremists, New Right culture warriors, and anti-immigrant xenophobes, the figure of the Muslim has been used to cultivate tremendous ideological ground by containing and limiting the scope of dissent, forging an imperial citizenry, masking structural inequalities and massive economic instability, and reinforcing the philosophical basis of "colorblindness" by promoting an American triumphalism in which a panracial enemy (the Muslim) threatens a multicultural America.

In a global order structured along the lines of race, this is not the first time in American history where race has been the thread that negotiates the interplay and fluidity between the foreign and domestic realms of American power. With the end of World War II and the dawning of the Cold War, the United States used the threat of "Communism" as a proxy for race as it sought to replace Europe as a global power in the era of decolonization and national liberation movements—using "anticommunism" (via the Truman Doctrine) to fulfill American imperial desire through interventions within the Third World to destabilize the burgeoning anticolonial and anti-imperialist movements taking place in Asia and Africa. The "Communist" threat also achieved its domestic goals on race by fracturing very powerful alliances being forged by African Americans who sought to tie their claims toward racial justice to those within the Third World. Instead, the rhetoric of "anticommunism" birthed a heightened American nationalism as influential African American liberals who sought to gain concessions on civil rights domestically supported violent American foreign policy throughout Africa and Asia. This fractured an emergent Black internationalism by containing antiracism within a national framework and created new kinds of imperial citizen-subjects during a new period of empire building. But it also isolated and increased the domestic repression of African Americans (such as Du Bois and Robeson) who linked the legacies of slavery with colonialism in the Third World as a means toward systemic change rather than reform against Jim Crow segregation, disenfranchisement, and civil death.[6]

Similarly, in a post–Cold War era, the rhetoric of "terrorism" has also become a proxy for race, generating tremendous political and ideological capital. The sine qua non for "terrorism" has been the Muslim—a highly racialized figure that has been mobilized to reinforce American hegemony abroad, while also containing antiracist and economic justice movements domestically. This threat of "terrorism" to American interests abroad has justified a violent reassertion of American power and militarism to extend Cold War alliances, further American geopolitical dominance, and refashion the United States as the sole power in a unipolar world through "preemptive war," covert intervention, aggressive militarism, and unilateralism. Domestically, the threat of "terror" from the *immigrant* Muslim has justified a highly racialized and vicious crackdown on immigrants in the United States, as the Immigration and Naturalization Services (INS), now under the Department of Homeland Security, has normalized deportations, detentions, and disappearance. In addition, the increasing prominence of the figure of the *indigenous* Muslim—the African American Muslim—within the security apparatuses and mainstream political discourse carries tremendous symbolic and historic significance, as African American Muslims have historically been constructed as fundamental threats to American democracy and racial harmony dating back to slavery. The November 2007 PBS *America At A Crossroads* series episode "Homegrown: Islam in Prison" on African American Muslims revealed that these fears are alive and well, as the documentary's alarmist tone echoes within the U.S. security establishment around Black Muslim communities. This notion of an "enemy within" plays on highly racialized fears deeply embedded within the white imagination, while also heightening tensions within African American communities along class, religious, and political lines, as it polices the boundaries of race through its attempts at forging an imperial Blackness.

The ideological elasticity of the Muslim and the rhetoric of "terrorism" has morphed and begun to outline the contours of the threat posed by the figure of the African American Muslim. The racialized discourse of empire is now producing a subject where the foreign and the domestic collapse upon themselves, as fears of "terror" are conflated with "Black criminality," gangs, prison culture and urban violence. The carceral logic and captive power that has historically been forged around Blackness not only makes legible this new emerging "threat"—as it was the "logic" used by the Los Angeles Police Department in its recent testimony in Congress to "map" Muslim communities—but it also becomes the template for the exporting of this prison regime to the colony in the "War on Terror." This fear of "homegrown terrorism" that collapses the domestic and the foreign around the figure of the Muslim can be clearly seen in the recent, near unanimous passage of House Resolution 1955 ("The Violent Radicalization and Homegrown Terrorism Prevention Act of 2007"). This bill seeks to establish grant programs, assigning a university-based "Center of Excellence for the Study of Radicalization and Homegrown Terrorism in the United States," and conduct research into how other countries—ostensibly in Europe—work to prevent "homegrown terrorism and radicalization." In addition, Michael Downing, the LAPD's commanding officer for "Counter-Terrorism" testified before the Committee on Homeland Security and Governmental Affairs in the U.S. Senate—one week after the passage of H.R. 1955—on the need to "map" Muslim communities in Los Angeles as a way to prevent—coincidentally–"homegrown terrorism" and "violent radicalization." In his testimony, Downing highlighted the need for this kind of surveillance, mentioning 9/11, the current wars, attacks in Europe and their policing of Muslims in their midst, and, not surprisingly, he cites gang culture in Los Angeles and the ways in which they have been policed as the blueprint and analog for mapping Muslim communities—"their history, demographics, language, culture, ethnic breakdown, socio-economic, status, and social interactions."[7]

Through the figure of the Muslim, then, I will trace the intimacies between the colony and the prison within the unfolding narrative of American imperial hubris by exploring the relationships between U.S. prisons and the emergence of imperial imprisonment in Iraq, Afghanistan, and Guantánamo and the connections being forged between the military and local police in the United States. I will also examine recently published reports by the Departments of Justice and Homeland Security, and Congressional committee testimonies to historicize the state's racialization of African American Muslim inmates. I will then close with a discussion of *Rap DNA*, the yet to be published prison writings of the most prominent incarcerated African American Muslim in the United States today, Imam Jamil Al-Amin (formerly H. "Rap" Brown), who was transferred in August 2007 from Georgia to the federal government's highest security prison—the Federal Supermax in Florence, Colorado. As a counternarrative to the State's captive power, his writings embody a stunning act of Michael Eric Dyson calls "Afrecriture."

In exploring these diverse narratives, it is important to view the broader architecture of power that is in place: that to "map" Muslims, police thought, invade, occupy, bomb, incarcerate, torture, render, and murder—these are not reflective of just a will *to* power, but is raw naked power in full effect, and it is this structure of power that is central to the whole notion of what I am calling the "carceral imagination." The American legal, political, and security apparatuses are animating this "carceral imagination" by constructing a domestic threat of "homegrown terrorism" around the figure of the African American Muslim and his potential links to "gang Islam," "prison Islam," and global "terrorist" networks. While the foreign Muslim has been incorporated within the prison regime through the exporting of American carceral power vis-à-vis Abu Ghraib, Guantánamo, and secret detention centers, the specter of the African American Muslim—as the "enemy within"—haunts the domestic homeland by shattering the divisions between the national and the global as the spaces of imperial power continue to converge.

FROM ABNER LOUIMA TO ABU GHRAIB

In *Discipline & Punish*, Michel Foucault argues that prior to the prison, violence inflicted upon the prisoner, or the executed, brought pleasure to the crowd as a parade of body parts became central to popular violence and the imagination that fueled it, and so the prison transformed punishment and its relationship to power. With the "birth of the prison" in the nineteenth century, "the great spectacle of physical punishment disappeared; the tortured body was avoided; the theatrical representation of pain was excluded from punishment. The age of sobriety in punishment had begun."[8] I am intrigued by Foucault's discussion about the birth of the prison, the public display of the body, and the "imagination" that it fuels in the Age of Abu Ghraib and the Birth of Guantánamo. For Foucault and others, the prison ushered in an "age of sobriety" around punishment when it seems that the post-9/11 imperial state has revealed a new "age of excess" around torture, display, and penal power around the Muslim body.

For me, this "age of excess" is central to the globalization of imprisonment and the carceral logic of America's "War on Terror." In exploring the transnational logic of incarceration and the institutional links between the prison regimes around African American Muslims in the United States and the emergence of military prisons in Iraq, Afghanistan, and Guantánamo, history may have come full circle. According to Alan Gomez, a 1961 gathering called "The Power to Change Behavior: A Symposium Presented by the U.S. Bureau of Prisons" was a key moment in "the politicization of institutionalized brainwashing, behavior modification and torture within the prison regime" that resulted in a national directive to "experiment with these techniques, originally used against American POW's in Korea, on the black Muslim prison population." With these experiments, the federal government utilized physical and psychological torture to suppress political activity and dissent and to redefine accepted notions of "cruel, inhuman and torturous treatment."[9] With the emergence of the Nation of Islam and Malcolm X (as well as increased Muslim immigration and the anticolonialism of the Third World), the FBI and law enforcement agencies monitored Black Muslims as a potential threat to the emerging orthodoxy of the Civil Rights Movement, so that the use of Black Muslims in prison in the early 1960s for use in state-sanctioned torture and experimentation raises profound questions, not the least of which are its possible connections and historical precedents to the more recent tortures of Muslims in Abu Ghraib, Guantánamo, Bagram, and elsewhere.

The collapsing of the ideological and territorial boundaries between the foreign and the domestic are embodied in the increasing consolidation of the two major security institutions within American imperial culture—the military and the prison—revealing the emergence of a new kind of formation of power and domination used to implement a violent global order that has challenged and usurped both national and international juridical foundations through torture, the abrogation of habeas corpus, the Birth of Guantánamo, military tribunals, the Patriot Act, preemptive war, and the legal declaration of Muslims as not "falling within the definition of 'person.'"[10] This confluence between the military and the prison regime runs deep in the current "War on Terror." According to Avery Gordon, Mark S. Inch, the corrections and internment branch chief at the Office of the Provost Marshal General, stated that "the synergy between the reservist's civilian employment in the corrections field and his or her duty to confine enemy combatants in Afghanistan, Cuba [Guantánamo] and Iraq . . . could not be more evident and essential to mission success."[11] According to Gordon, the 300th Military Police Brigade, the 327th Military Police Battalion, and the 800th Military Police Brigade, designed Camp Delta at Guantánamo Bay, have run various detention centers throughout Afghanistan and have reorganized the prison systems in Iraq for enemy combatants and prisoners of war, respectively. In addition, prison guards and administrators at state and federal penitentiaries throughout the United States continue to contribute a great deal to America's global archipelago of imprisonment—from John Vannatta, the superintendent of the Miami Correctional Facility in Indiana who works in Guantánamo, to many Chicago policemen and prison guards throughout Afghanistan, and to Captain Michael Mcintyre and Master Sergeant Don Bowen who work at the federal penitentiary at Terre Haute, Indiana, who helped to build the Iraqi prison system.[12]

According to Leah Caldwell, in May of 2003, just two months after the invasion of Iraq, Attorney General John Ashcroft gathered a group of American prison officials under the auspices of the International Criminal Investigative Training Program (ICITAP) and sent them to Iraq to prepare existing Iraqi prisons to house additional prisoners. ICITAP, which is based within the Department of Justice and funded by the State Department, has been in existence since 1986 and has been sent throughout the world to "rebuild" criminal justice systems and to support police and prison regimes in American-backed client states, including Haiti, Indonesia, and the former Soviet Union. According to Caldwell, ICITAP is a successor to a training program run by the Agency for International Development, which was stopped in the mid-1970s after it was revealed that this program was used to train police forces and prison officials abroad in murder and torture against mostly leftist insurgencies.[13]

What is especially revealing is that many of the prison officials sent by Ashcroft to Iraq were not only employed by private prison firms around the United States, but were also heads of various state departments of corrections throughout the country, including Terry Stewart (Arizona), Gary Deland (Utah), John Armstrong (Connecticut), and Lane McCotter (Texas, Mexico, and Utah). In fact, all have been involved in a range of human rights abuses and legal cases from inmates in the United States, including denial of medical treatment, harsh conditions, sexual harassment, torture, and even death. McCotter, who was forced to resign as head of Utah's State Board of Corrections due to the death of an inmate shackled naked to a chair, was, according to Gordon, chosen by then Attorney General John Ashcroft to head the reopening of Iraqi jails under American rule and also to train Iraqi prison guards just one month after the Justice Department released a report—following the death of a prisoner—about the lack of medical and mental health treatment at one Management and Training Corporation's jails, a private firm where McCotter was an executive.[14] To add insult to irony, McCotter and Deland were at the ribbon-cutting ceremony at Abu Ghraib when it first opened, as McCotter said of Abu Ghraib, it was "the only place that we agreed as a team was truly closest to an American prison."[15]

While the administrative and institutional framework for United States colonial imprisonment in Iraq, Afghanistan, and Guantánamo has deep ties to the domestic prison establishment, state corrections, and even private prison corporations, it is also the case that some of the guards involved in the Abu Ghraib tortures had careers within U.S. prisons. According to Anne-Marie Cusac, Ivan L. (Chip) Frederick II and Charles Graner worked in the United States as corrections officers. Frederick was a guard in Virginia while Graner, who was described as one of the "most feared and loathed of American guards"[16] at Abu Ghraib and who is infamously known for his sinister glasses and wicked "thumbs up" signs in the released photos of torture, was an officer at the Greene County Prison in Pennsylvania.

According to Cusac, the Taguba Report investigating torture at Abu Ghraib found that U.S. military personnel placed detainees in sexually explicit positions, engaged in forced sodomy on Iraqis with chemical lights and possibly even a broom handle, beat them, and wrote vulgar epithets on their bodies. Cusac goes on to reveal that in 1998, Greene County guards, where Graner worked, were charged with sodomy with a nightstick, unlawful nude searches, and using prisoners' blood to write "KKK" on the floor of the prison.[17] In fact, according to Gordon, Graner was repeatedly "implicated in violence against prisoners at the Pennsylvania super-maximum security State Correctional Institute at Greene, where he was employed . . . and given supervisory roles at Abu Ghraib because of his guard experience."[18]

Ironically, part of Bush II's reconstruction plans for Iraq include the building of a "Supermax" prison where Abu Ghraib now stands as a way of erasing the stain and stigma of the tortures that went on there. The "Supermax" prison has proliferated throughout the United States, becoming the penal blueprint for imprisonment, and is widely seen as the most repressive kind of carceral power available. According to Gordon, "the Abu Ghraib photographs did not expose a few bad apples or an exceptional instance of brutality or perversity. They exposed the modus operandi of the lawful, modern, state-of-the-art prison. Nowhere is this clearer than in the growth over the past 25 years of what is called super-maximum imprisonment, the cutting edge in technology and the prototype for re-tooling the military prison for the war on terror."[19] In fact, Supermax prisons emphasize the state's desire for ultimate control, as the "language of security has authorized Supermax imprisonment by treating it not as punishment but as a set of administrative procedures for managing high-security populations."[20] More specifically, the policies and procedures used in Supermax facilities in the United States, "now legally sanctioned as ordinary and acceptable norms of prison life, were once considered violations of the U.S. Constitution's Eighth Amendment prohibiting cruel and unusual punishment. The Supreme Court's Eighth Amendment cases are the legal and linguistic basis for the "detainee interrogation" memos prepared for the War on Terror."[21]

As the American prison establishment exports to the outposts of empire new technologies of carceral power, the lines between the domestic and the foreign blur. And with the establishment of Abu Ghraib, Guantánamo, and other structures of imperial imprisonment in the "War on Terror," the legibility of "terrorists" as detainees is made possible through the domestic politics of race and incarceration. But the incorporation of the foreign Muslim within the imperial prison regime raises the haunting specter of a domestic analog: the African American Muslim.

HOME IS WHERE THE HATRED IS

In addition to the links between Abu Ghraib, Guantánamo, and other sites abroad with prison structures here in the United States, there is also an emerging connection being forged between local policing in the United States with the military. LAPD Chief William Bratton has been advocating and lobbying that Los Angles become the national nerve center for an official agency collaboration between Homeland Security and local law enforcement throughout the country. And he has made tremendous institutional strides. Known for his part in the draconian implementation of the Manhattan Institute's "Broken Windows" policy in New York in the early 1990s under Rudolph Giuliani, which led to police abuse cases such as Abner Louima and Amadou Diallo, Bratton has declared a "War on Gangs" in Los Angeles, referring to gang activity as "homeland terrorism" and lamenting that the federal government "needs to get preoccupied with the internal war on terrorism as well."[22] This linking of local policing with global war, and the highly racialized language of "terrorism" linking Black black and Brown brown communities in the United States with their neocolonial counterparts in the Global South, forges an ideological link between domestic control and imperial power that generates tremendous political capital in a post-9/11 environment.

But Bratton takes the ideological and material links between domestic policing and colonial war much further. According to Bratton, there are about seventy LAPD officers in Iraq training with the United States military at any one time, while Los Angeles police officers have also trained Marines how to gather evidence at a bomb scene and also to give their guidance on urban policing for the United States military abroad.[23] In addition, the military is helping the LAPD prepare for the "eventuality" of suicide bombers and improvised explosive devices (IEDs) in Los Angeles. In fact, according to Bratton, a team of soldiers from the U.S. military in Baghdad visited Los Angeles to share with the LAPD their knowledge about "IED's and suicide bombers and the tactics employed by death squads and insurgents in Iraq."[24] Bratton continues, "We're always wondering why we don't have suicide bombers and IED's here. We're trying to learn from each other. It's only a matter of time before we are experiencing the issue here. We have no idea why it hasn't occurred. What they're dealing with is what we may face." For Bratton, there are "many similarities between what is going on in Baghdad and here. The similarities to gang warfare are strong."[25]

While the racial and ideological calculus behind such comparisons helps generate the necessary fear and political will needed for increased domestic repression, Bratton is suggesting something more—a rhetorical strategy that has historically resonated within the United States and has resoundingly reappeared in a post-9/11 climate that links Muslims, Blackness, prisons, and gang culture.[26] Bratton mentions the "growing influence of converted Muslim radicals in the U.S. prison system" who, upon release, will spread their new ideas to their companions such that gangs will conspire with "terrorist" organizations to carry out attacks on U.S. cities. As Bratton says, "There is a potential of some gangs who are disaffected to latch on to the Islam movement. We've seen movement in that direction."[27]

This kind of alarmist rhetoric echoes a larger concerted chorus of voices who are constructing an "indigenous" component to the global "threat" posed by Muslims. The projection of racial anxiety from the foreign Muslim to the domestic complement—the African American Muslim—mobilizes national fears, as African American Muslims in prisons are perceived as a "fifth column" of potential recruitment into "terrorist" activities within the United States. Not surprisingly, high level officials, numerous think tanks, policy institutes, nongovernmental organizations, and official state agencies have published reports and presented to several congressional committees and other law enforcement agencies their fears about African American Muslims within United States prisons, and their potential for "radicalization." This kind of institutional power bearing down upon what is being constructed as an emerging frontline in America's "War on Terror" is an extension of the policing and surveillance of Muslims in the United States over the last forty years or so and also a continuation of the pernicious forms of racial profiling that have impacted Black communities dating all the way back to slavery—what Loic Waquant argues is the historical continuity between the slave plantation and mass incarceration of African Americans within the United States.[28]

A 2004 Department of Justice report issued a series of warnings about the possibilities of African American Muslim "radicalization" within prisons, the emergence of "Prison Islam," and the lack of monitoring of and by Muslim chaplains that could lead to "extremist" ideologies circulating within prisons.[29] In 2005, FBI director Robert Mueller told the Senate Intelligence Committee: "Prisons continue to be fertile ground for extremists who exploit both a prisoner's conversion to Islam while still in prison, as well as their socio-economic status and placement in the community upon their release,"[30] and Republican Senator Susan Collins from Maine has said that radical Islam within U.S. prisons was "an emerging threat to our national security."[31] In April 2006, Attorney General Alberto Gonzalez also weighed in on the topic of Muslims in prison and the "challenges to detection" that this poses to authorities.[32] A 2006 study by the Homeland Security Policy Institute at George Washington Institute in conjunction with the University of Virginia was released at a Senate Homeland Security and Governmental Affairs Committee hearing on "homegrown" terrorists. It argues that, due to limited funds, there is little supervision of Muslims within prisons so that "radical Islam is spreading and raising a new generation of potential terrorists," and warns that "Jailhouse Islam," which is "based upon cut-and-paste versions of the Qu'ran and incorporates violent prison culture into religious practice,"[33] is a threat to prison security. In addition, Charles Colson, who was special counsel to Richard Nixon during Watergate in 1974, and who now runs the Prison Fellowship Ministries, says that al-Qaeda training manuals "specifically identify America's prisoners as candidates for conversion because they may be "disenchanted with their country's policies."[34] Colson asserts that "terrorism experts fear these angry young recruits will become the next wave of terrorists. As U.S. citizens, they will combine a desire for 'payback' with an ability to blend easily into American culture."[35] In addition, Roy Innis, the national chair and executive director of the Council of Racial Equality (CORE), has also met with federal officials and has been vocal about the threat he perceives coming from African Americans Muslims who are not properly monitored both within and outside prisons.

As a "U.S. citizen" who can "blend easily into American culture" and who, according to Attorney General Gonzalez, poses "challenges to detection," African American Muslims become the site upon which the projection of racial anxiety from the foreign Muslim—who is deemed more recognizable by his "foreign-ness"—is branded. But because of the historic relationship of Blackness within the United States, and the familiarity that this contempt bred, African American Muslims are potentially more threatening and not racially legible in the same ways that the foreign Muslim is. The carceral regimes in Iraq, Afghanistan, and Guantánamo are then, in Kaplan's formation, the "success" which has birthed a "nightmare" that now threatens the domestic front of American imperial culture and which is woven through the very rhetoric about African American Muslims in prisons, as they are threats not only to the "security" of the prison regime (as prison officials have declared) but also to the very organizing logic of the state's architecture of power.

This specter of threat is deeply rooted within fears of African American conversion to Islam, particularly as prisons have become the site for transformations in selfhood of all types, where a whole new aesthetics of being erupts. For African American Muslims like Malcolm Little—who got his "X" inside—or H. Rap Brown–who, in the 1970s, became Jamil Abdullah Al-Amin—the prison becomes the crucible for a rejection of that imposed master narrative of nonbeing and the subjection of Blackness upon which—through chattel slavery, disenfranchisement, and social death—America ultimately rests. For if prison is about disappearance and erasure, silence, and violence, then epiphany, conversion, and politicization are a kind of ontological resurrection against social and civic death—redefining one's existence and challenging the panoptic power of the state.

Conversion then, is what Richard Brent Turner calls "signification,"[36] and is a stunning effacement of the state's attempt to contain blackness within the ideological boundaries of the United States and its history of enslavement. As Chuck D argues, "if people with African roots are connected to Islam, then you got a problem of taking your slaves away, 'we lost our slaves—they're international now!'"[37] As Chuck D's poignant argument suggests, this *loss* demands that African American Muslims be monitored, policed, and disciplined as fundamental threats to the American racial project and its imperial ambition. The fear of their incorporation stems from their forging of an alternative Black identity and community of belonging rooted in an expressed solidarity with Third World liberation struggles, their fundamental and radical challenge to America's racial hierarchies rooted in white supremacy, and their remapping of a Black diasporic identity that Amiri Baraka referred to as "post-American."[38] In occupying a liminal space within the United States, the presence of African American Muslims, especially in the post-9/11 moment, shatters the coherence of American nationhood and haunts the tattered edges and imaginative spaces of the empire.

THE LAST REAL NIGGA ALIVE

Rap became a version of Malcolm and Martin.

—Nas, "Last Real Nigga Alive"

Who frame Rap Jamil Al-Amin?

—Amiri Baraka, "Somebody Blew Up America"

Imprisoned in 2000 and sentenced to life in Georgia State Prison at Reidsville, Imam Jamil Al-Amin (formerly H. "Rap" Brown) was transferred to the Federal Supermax prison in Florence, Colorado in August 2007—a prison where the federal government houses its most "dangerous inmates" and is openly referred to as the "domestic Guantanamo." The October 2007 *60 Minutes* piece on the Colorado Supermax, titled "A Cleaner Version of Hell," focused on the security and secrecy of the prison and talked almost completely about the immigrant Muslims who were inside—"The Shoe Bomber" Richard Reid, the supposed twentieth 9/11 hijacker, Zacarias Moussaoui, the convicted "mastermind" of the 1993 World Trade Center bombing, Ramzi Youssef, and others. While Imam Jamil was not mentioned, it is important to understand that this transfer (done while an appeal is underway in the State of Georgia) reflects so much of the geographies of power around the Muslim in post-9/11 America.

While Al-Amin's transfer to this prison is no doubt endemic of the state's desires to violently continue the narrative of Black captivity that sits at the heart of the empire, it is also difficult to underestimate the post-9/11 climate and the ways in which, as an African American Muslim, Imam Jamil is viewed as the "homegrown terrorist." Even in the State of Georgia prison system, Imam Jamil was deemed a "high security" captive who was subjected to around-the-clock surveillance and solitary confinement, while his trial and conviction have been called into question by several prominent legal experts and international human rights organizations. Now in the Supermax and under a "behavior modification program," Imam Jamil's history within the Civil Rights Movement (as a member of the Student Non-Violent Coordinating Committee [SNCC]), his Black Power activism and his work as a Muslim leader to thirty-four different urban communities in the United States over several decades embodies the multiple histories and identities that have consistently challenged American militarism, economic injustice, and racism nationally and internationally. During his trial, he was constructed by the state as the embodiment of the "homegrown terrorist," but his is not the case of the captive who becomes "radicalized" in prison but the revolutionary who is imprisoned by the state. His captivity can, in many ways, be seen as a metaphor for the continued attempts by the racial state to rewrite the history of Black radicalism and Third World internationalism over the last fifty years, the sustained attacks on Black community-based leadership, the suppression of political dissent in the United

States, and the increased surveillance on the immigrant, but especially African American Muslim, as the haunting figure within American imperial culture.

While in prison, Al-Amin has completed a manuscript titled *Rap DNA* that extends the legacy of powerful Black political prison writings that narrate a resistance to captivity and transcend the boundary between the free and the unfree—that vicious structuring logic of America that is centered around Black captivity and containment. Like his first book—the vastly overlooked Black Power manifesto *Die Nigger Die!* (1969), which is a brilliant political autobiography and narrative of rebellion—*Rap DNA* is also a stunning act of what Michael Eric Dyson has called "Afrecriture,"[39] a poignant act of the writing of Black presence into history. Written in rhyme form and the embodiment of the talking book, *Rap DNA* is a powerful glimpse into a revolutionary aesthetic that invokes and challenges a range of Black thought within the violent crucible of American slavery, as the opening lines state: "I am the seed of the survivors of the Middle Passage, the harvest from those who could not be broken, would not be broken, never a slave." With this as the opening salvo, *Rap DNA* also explores Fanon, Langston Hughes, Rakim, the murder of Amadou Diallo, critical race theory, his experiences in exile, the Patriot Act, American militarism, the "War on Terror," and an incredible range of ideas through lyricism and irony, insight and wit.

But it is his reclamation of the moniker as "Rap" that is most compelling, as this becomes, in many ways, the voice through which he narrates. This is striking because it suggests an embrace of his multiple identities and histories that so often get overlooked within narratives of conversion that are meant to inaugurate a new ontology of the self and a rejection of one's "slave name." Though not an embrace of Hubert Brown (his former name), the reclaiming of "Rap" works on multiple levels: as an embrace of his persona as the fiery Black Power orator of the past and also an embrace of the *act* of speaking. To "Rap," then, is a radical process that is a scream against the silence imposed by the captive power of the prison. And as indicated by the title of the writings, *Rap DNA* is not only a gesture suggesting that this radical act of speech as "Rap" continues beyond conversion and is imprinted upon his very being, but it also becomes a suggestion that these writings and his history are the ideological blueprint and life line for rap music as well. In fact, throughout the writings he brilliantly challenges contemporary hip-hop artists in rhyme style, battling them on page to use hip-hop as forum for transcendence, redemption, and rebellion, while also laying claim as the father of hip-hop, as he says, "if Christian music is alive / It's the only music named after someone besides Rap that survives"[40] and then later where he politicizes the gangster figure when he says, "No gangster rap music/Rap was the gangster of his time."[41]

In a piece called "Conflict Or Conciliation," Al-Amin invokes his history with the American racial state to inform contemporary generations about Black history, referencing how the U.S. Congress ("passed a law named after me, trying to stop the flow of the R-A-P") and the FBI ("for the first time in history of their ten most wanted list . . . changed it to eleven, and before the manhunt could even start, they moved Rap to the top of the chart") sought not only to sanction and criminalize him but also to incarcerate him. In doing so, Al-Amin also critiques the hip-hop generation, writing "Rap became public enemy number one, and I ain't talking about Billboard son,"[42] arguing that material excess and record sales are not tied to Black uplift and transformation but to Black repression and degradation. In Al-Amin's view, rap music should aspire to be at the top of the FBI's charts as a barometer of its Fanonian possibilities, not the music industry's attempts at promoting a commodified rebellion in the form of hip-pop.

Rap DNA also gives eloquent testimony to his life of struggle and his persistent attempts to connect Black radical praxis not within the confines of a collective bourgeois racial identity or a narrow nationalism that edifies white supremacist power but in relationship to a global struggle against the racial and economic legacies of slavery and colonialism. A battle cry from behind enemy lines, titled "Seconds," highlights the painfully enduring question about the role of Blackness within the American imperial project, as he writes, "always talking that we, us, our, my country, our war our team our dream, USA USA how many kids you kill today? Negroes greatest dream? Take one for the team." Al-Amin continues critiquing the complicity of certain forms of Blackness with empire when he writes, "more than willing to kill in a foreign land, women, children another man, Buffalo soldiers like their fathers of old, put the sin of empire on their soul." He powerfully interrogates the historic role of African Americans as imperial citizens who have supported the expansionist project of the United States not only in the distant past, but more recently during the Cold War and in the "War on Terror," when he writes, "give their sons and daughters to spread tyranny and slaughter, in the name of empire's new world order."[43]

As simultaneously Black and Muslim, endemic threat and exogenous insurgent, Jamil Al-Amin's imprisonment links the domestic politics of race with the global terrain of the American warfare state, highlighting the central place that race continues to occupy within the security apparatus around who is to be incarcerated, surveilled, declared war upon, tortured, and killed. And in this time when the euphoria around Barack Obama's election is seen by many as a stamp of legitimacy for the American imperial project, giving the United States both the allure of democracy and the seduction of benevolence, Al-Amin's writings ask a profound question: in this moment where Blackness and American-ness are closer than they ever were, how are Black communities and constituencies, in all their plurality and diversity, going to align themselves in relationship to American empire—as promoters of an imperial Blackness or as critical resisters against it? As Malcolm would say, time will tell.

NOTES

1. Achille Mbembe, "Necropolitics," *Public Culture* 15, no. 1 (Winter 2003): 11–40.
2. Loic Waquant, "From Slavery to Mass Incarceration," *New Left Review* 13 (January–February 2002).
3. Amy Kaplan, *The Anarchy of Empire In the Making of U.S. Culture* (Cambridge: Harvard University Press, 2002), 1.

4. Ibid.

5. Ibid., 12.

6. See Penny M. Von Eschen, *Race Against Empire: Black Americans and Anticolonialism, 1937–1957* (Ithaca, NY: Cornell University Press, 1997).

7. Testimony by Michael P. Downing, Commanding Officer for Counter Terrorism/Criminal Intelligence Bureau, Los Angeles Police Department before the Committee on Homeland Security and Governmental Affairs of United States Senate, October 30, 2007.

8. Michael Foucault, *Discipline and Punish: The Birth of the Prison* (New York: Vintage Books), 14.

9. Alan Gomez, "Resisting Living Death at Marion Federal Penitentiary, 1972," *Radical History Review*, no. 96 (Fall 2006): 60.

10. See the Center for Constitutional Rights at http://www.ccrjustice.org/newsroom/press-releases/judges-dismiss-suit-seeking-damages-guantanamo-torture.

11. Avery Gordon, "Supermax Lockdown," http://www.middle-east-online.com/english/?id=18365.

12. Ibid.

13. Leah Caldwell, "From Supermax to Abu Ghraib," http://www.counterpunch.org/caldwell10152004.html.

14. Gordon, "Supermax Lockdown."

15. Caldwell, "From Supermax to Abu Ghraib."

16. Quoted in Anne-Marie Cusac, "Abu Ghraib, USA," *The Progressive*, August 6, 2004, http://www.alternet.org/story/19479/.

17. Cusac, "Abu Ghraib, USA."

18. Gordon, "Supermax Lockdown."

19. Ibid.

20. Ibid.

21. Ibid.

22. Megan Garvey and Richard Winton, "City Declares War on Gangs," *Los Angeles Times*, December 4, 2002, A1.

23. Pamela Hess, "Analysis: Police Take Military Counsel," found at http://lapd.com/article.aspx?a=4150.

24. Ibid.

25. Ibid.

26. These alarmist rhetorical strategies have a history that date back to the rise of the Nation of Islam, particularly in the 1950s and 1960s as well as during the uprisings in Los Angeles in 1992 when the gang truce was brokered with the help of the Nation of Islam. In addition, the case of Jeff Fort and the El Rukns in the 1980s, an offshoot of the Blackstone Rangers in Chicago, is another prominent case in point.

27. Hess, "Analysis: Police Take Military Counsel."

28. Waquant, "From Slavery to Mass Incarceration."

29. "A Review of the Federal Bureau of Prisons' Selection of Muslim Religious Services Providers," from the Office of the Inspector General, Department of Justice, April 2004.

30. Hess, "Analysis: Police Take Military Counsel."

31. Quoted in "US: Prisons Are Breeding Grounds for Muslim Terrorists" found at www.westernresistance.com/blog/archives/002992.html.

32. Testimony before the World Affair Council of Pittsburgh, "Stopping Terrorists before They Strike: The Justice Departments Power of Prevention," April 16, 2006.

33. "Out of the Shadows: Getting Ahead of Prisoner Radicalization," A Special Report by the George Washington University Homeland Security Policy Institute, 2006, iv.

34. Bill Berkowitz, "American Muslims: A Clear and Present Danger?" http://zmagsite.zmag
    .org/Jun2003/berkowitz0603.html.

35. Berkowitz, "American Muslims."

36. See Richard Brent Turner, *Islam in the African-American Experience* (Bloomington: Indi-
    ana University Press, 1997) for an excellent history of African American Islam.

37. Personal interview, August 2006.

38. Marvin X and Faruk X, "Islam and Black Art: An Interview With Leroi Jones," *Dictionary
    of Literary Biography: Black Arts Movement*, ed. Jeff Decker (Detroit: Gale, 1984), 128.

39. Michael Eric Dyson, *Open Mike: Reflections on Philosophy, Race, Sex, Culture and Religion*
    (New York: Basic Civitas Books, 2003), 27.

40. Imam Jamil Al-Amin, Rap DNA, unpublished.

41. Ibid.

42. Ibid.

43. Ibid.

PART III

URBAN ENCOUNTERS

CHAPTER 13

OVERLAPPING DIASPORAS, MULTIRACIAL LIVES

SOUTH ASIAN MUSLIMS IN U.S. COMMUNITIES OF COLOR, 1880–1970

VIVEK BALD

LIVES IN MOTION

IN THE SPAN OF A FEW YEARS IN THE 1920S, TWO BROTHERS, BAHADOUR AND Rostom, and their elder sister, Roheamon, all got married in Harlem. The three siblings were recent arrivals to New York City from the U.S. South. Roheamon and Rostom had been born in New Orleans and Bahadour in Waveland, Mississippi, all in the first years of the century. Sometime in 1910, their mother Ella brought them north, along with three other siblings, and, by 1920, the mother and six children were living with a family relation in the Bronx.[1] At the moment they had reached their teens and early twenties, Roheamon, Rostom,

I wish to thank Alaudin Ullah for his generosity in sharing his family's history in discussions over the course of the last six years and for his continuing insights regarding this work, and Noor Choudry and Habib Ullah, Jr., for their openness in discussing their lives and recollections in recent interviews. I also wish to thank Brent Hayes Edwards, Adam Green, Vijay Prashad, Kym Ragusa, Suresht Renjen Bald, Andrew Ross, and Gayatri Spivak; my colleagues Rich Blint, Miabi Chatterji, Dawn Peterson, Sujani Reddy, Emily Thuma, and Manu Vimalassery; and the members of NYU's Faculty Working Group on Diasporas—Aisha Khan, Lok Siu, Gayatri Gopinath, Jacqueline Nassy Brown, Juan Flores, Vanessa Agard Jones, Michael Birenbaum Quintero, and Anantha Sudhakar—for reading and commenting on earlier drafts of this chapter, and Gary Okihiro, Sohail Daulatzai, Junaid Rana, Jigna Desai, Sunaina Maira, Rajini Srikanth, and Gita Rajan for feedback on portions of the article which were delivered at the 2006 Association for Asian American Studies Conference in Atlanta, Georgia.

and Bahadour were thus poised at the edge of Harlem, and it was not long before all three moved the short distance across the river into a neighborhood beginning to crackle with possibility for people of color from all over the country and all over the world. Bahadour moved right into the center of Harlem's growing black community to an apartment on West 133rd Street between Seventh and Lenox avenues, and married a young woman who had also come from the South, twenty-four-year-old Margaret Carree from Savannah, Georgia. On their marriage certificate, Bahadour gave his age as twenty-three and his occupation as "actor" and both Margaret and Bahadour were officially classified as "colored."[2] Rostom moved just two blocks away to an apartment on 135th Street and married seventeen-year-old New York–born Pearl Pierce. Their 1924 marriage certificate also records husband and wife as "colored" and shows that Rostom had found a far less glamorous occupation than his younger brother; Rostom was working as a "fireman"—a term that, in this era, for workers of color, most likely referred to the job of shoveling coal and stoking furnaces in the basement boiler rooms of hotels and other large buildings.[3] Around this same time, sister Roheamon married Arthur Straker, a former boarder in their family's Bronx apartment. In 1930, according to census records, the two were living on St. Nicholas Avenue between 151st and 152nd streets. Their occupations were recorded as "housewife" and "laborer," respectively, and both were racially classified as "negro." Younger brother Bahadour was now living with them, along with his second wife, Thelma, who was listed as a twenty-six-year-old "negro" actress from Texas.[4]

In many respects, the lives of these three siblings, as captured in the fragments of archival records, suggest stories typical of the early years of the Great Migration: a family that moves first from one part of the South to another, then thousands of miles away to one of the great expanding cities of the North; young black women and men from Louisiana, Mississippi, Georgia, and Texas, converging, colliding, and coming of age in 1920s Harlem. What sets Roheamon, Rostom, and Bahadour apart from most of their contemporaries, however, was the fact that their father, Moksad Ali, was an Indian Muslim, most likely from East Bengal, the region that would eventually become Bangladesh. According to census records, Moksad Ali came to New Orleans in the year 1888, worked as a street peddler, and, in 1894, married a local African American woman, Ella Blackman. The two had eight children in all: four in New Orleans and four more after moving to Mississippi.[5] Moksad drops out of the historical record at this point, but the archival traces of the couple's movements—Moksad's from India to Louisiana to Mississippi and Ella's from Louisiana to Mississippi to New York—provide a glimpse of two lives in motion and two historical trajectories brought together in a single family. Roheamon, Rostom, Bahadour, and the other children of Moksad and Ella Ali indeed embody the overlap and the convergence of two separate diasporas—African and South Asian—on U.S. soil.

What makes this particular family even more significant is the fact that their connections to both New Orleans and Harlem tie together two little-known histories of South Asian Muslim migration to the United States and two sites of contact, interaction, and intermarriage. While much pathbreaking work has been done in the emerging scholarly field of South Asian American studies over the past three decades, there remains a gap in the historical narrative of South Asian immigration to the United States—a gap in which the stories of Moksad Ali, Ella Blackman, and their children lie largely hidden, along with those of hundreds of other families and individuals. In one sense, this "gap" can be measured quantitatively as the roughly fifty-year period between the consolidation of U.S. anti-Asian immigration laws in 1917 and the "reopening" to Asian immigration brought about by the Hart-Celler Act of 1965. It has been largely assumed that during this period, South Asian immigration to the United States was reduced to a mere trickle of students and professionals. However, evidence from a range of archival sources shows that this was not the case—that, in fact, large numbers of working class Indian Muslim men were moving in and out of, and often settling in, U.S. port cities from the late nineteenth century through the 1910s and 1920s and that this continued and quite possibly increased during the high years of Asian exclusion from the 1920s through the 1940s. These migrants remain largely unaccounted for in the official statistics that we have used to determine the size, nature, and significance of South Asian immigration to the United States in the first half of the twentieth century.

But coming to terms with the gap in the historical narrative, I believe, goes beyond correcting dates and statistics. In a number of significant ways, this particular migration was qualitatively different from others that have previously defined the story of South Asians in the United States. First, the largest group of men among those who were settling in and around U.S. port cities during this period, like Moksad Ali, were Muslims from Bengal. Second, most of these men appear to have been part of a larger population of British maritime workers who were circulating between multiple metropolitan sites in Britain, the United States, the Middle East, and Asia, participating in both the global economic circuits of the British empire and the more localized urban and industrial economies in and around different port cities, moving on and off ships in different locations as opportunities shifted, legal regimes changed and onshore networks became more established. Third, in cities like New Orleans and New York, a significant number of Indian Muslim men ended up living, working, marrying, and starting families within local working-class communities of color—among African Americans in New Orleans and among Puerto Ricans and African Americans in Harlem.[6] Finally, the children who emerged from this early history of South Asian migration and settlement in the United States—who constituted the first sizable "South Asian second generation" east of California—were not only beginning to appear over a century ago, but were, by and large, working-class, Muslim-descended, and racially mixed, and were growing up primarily in urban African American and Latino communities.[7]

This chapter, then, is part of a larger project of historical recovery. In the pages that follow, I will first draw on archival sources and then upon interviews with the descendants of early South Asian immigrants to New York City. My work here is an effort to open up for reconsideration our understanding of the South Asian presence in the twentieth-century United States as well as our understanding of the relations between South Asians—and in particular, South Asian Muslims—and other U.S. communities of color. Such a project must start from an understanding of cities like New Orleans and New York as unique sites of intersection and interaction, where the members of multiple migrations come to share the same spaces—buildings, blocks, neighborhoods, workplaces—and through the encounter with and negotiation of difference, generate new senses of individual and collective identity. This insight is in itself not new, nor is it new in considerations of the relationship between Muslim immigrants and U.S. communities of color. In his groundbreaking *Islam in the African American Experience*, for example, Richard Brent Turner focuses on the processes of "urbanization, migration, and immigration," which he argues were key to the rise of the various African American Islamic movements of the twentieth century. Turner asserts that groups like the Moorish Scientists and the Nation of Islam embodied the creative, syncretic, and heterodox responses of African Americans to the experience of uprooting during the Great Migration and to the need for new self-definitions in urban contexts—New York, Chicago, Detroit, Pittsburgh—which were marked not only by continued oppression but by exposure to a multiplicity of new peoples and ideas.[8] In his treatment of the influence of South Asian and Arab Muslims in these processes, Turner looks primarily at individual figures whose impact was made in the arena of religious and political movements—from Marcus Garvey's Sudanese-Egyptian adviser Duse Mohammed Ali to the Indian emissary Mufti Muhammad Sadiq, who introduced Ahmadiyya Islam into black communities throughout the cities of the North and Midwest.

What I hope to add here, even in preliminary form, is a consideration of another arena of encounter between Muslim immigrants and U.S. communities of color—one which was also taking shape over the first half of the twentieth century but which stemmed from the settlement of groups of working-class Indian men in African American and Latino neighborhoods. This entails a shift in focus to a more local and personal register—to the daily exchanges across difference that occur in urban spaces shared by members of multiple migrant groups. Here we find experiences of encounter, interchange, and transformation that did not directly manifest themselves in political or religious movements. In the case of the Indian Muslims who were settling in places like New Orleans and New York, these experiences were, in fact, often accompanied by the muting of any strict or explicit Muslim religious identity. I believe a consideration of these histories and processes is equally important to an understanding of the presence of Islam, and of Muslims, in African American and Latino/a communities in the United States.

Following on the work of a range of scholars—Gilroy, Edwards, Prashad, Gopinath, Lewis[9]—this project also seeks to contribute to a broadening in the conceptualization of "diaspora" as an analytic category. In presenting the histories that follow, I wish to shift (1) away from a linear, binary conception of diasporic movement (from "homeland" to "diaspora") toward a conception diaspora as a set of movements and relations between multiple locations; (2) away from an overdetermining focus on "homeland" as either a place of origin or a place of real or imagined return toward an emphasis on the complex histories and dynamics of settlement in the places to which migrants have come; and (3) away from a singular emphasis on particular racial, ethnic, religious, or national groups "in diaspora" toward an understanding of diaspora itself as a process of encounter, intermixture, and the negotiation of difference across all these lines.

"HINDOOSTANIS" IN NEW ORLEANS

New Orleans is not typically the first place that comes to mind in relation to the history of South Asian immigration to the United States—but it is here where we find traces of one of the first settlements of South Asians in the country. From the 1890s through the 1930s, census records show a growing and expanding community of Indian men, primarily Bengali Muslims, living in what it records as "black" sections of New Orleans' fourth and fifth wards.[10] When Moksad Ali settled in New Orleans in the closing decades of the nineteenth century, he was not alone. Alef Ally, for example, had arrived six years earlier than Moksad, in 1882. Eshrack Ali had arrived in 1893, Sofur and Majan Ali in 1895, and Solomon Mondul in 1896. The 1900 U.S. Census recorded about a dozen Indian Muslim men living in New Orleans; all of them were listed as "peddlers" and the majority were living together at two adjacent addresses, 1420 and 1428 St. Louis Street, on a predominantly African American block just west of Congo Square and a short distance north of the French Market.[11] The earliest arrival of these men, Alef Ali, married a local black Creole woman, Minnie Lecompte, some time in the 1890s, and by the time of the 1900 census, was widowed with two young daughters, Viola and Mary Sadie Ali.[12] Ten years later, when the 1910 census was conducted, Alef Ali was fifty years old and had been a resident of New Orleans for more than half his life, almost thirty years. Now, in addition to continuing work as a "merchant," he was running 1428 St. Louis Street as a boarding house for more recent arrivals from India. Besides himself, his younger daughter Viola, and a sixty-five-year-old African American woman, Costina Gardner,[13] the census shows twenty-one other Indian men residing at Alef Ali's address on St. Louis Street. These men ranged in age from twenty-two to sixty years old and most were recorded as having been in the United States for between one and five years.[14] Spread across several other New Orleans addresses, there were now, in 1910, around fifty other Indian men, most with clearly Muslim names and a number with names that suggested they were both Muslim and specifically Bengali. All of these men were engaged in the peddling of "dry goods," "fancy goods," "oriental goods," or simply "retail merchandise."[15]

The 1920 and 1930 censuses record a continuing expansion of the population of Indian Muslim men living in the fourth and fifth wards. By 1930, there were at least fifteen addresses, in a rough semi-circle to the west and north of

Congo Square, where either groups of Indian men or mixed Indian-black families resided. While in previous censuses men from this population had been listed with the birthplaces "Hindoostan" or "India" and with a mother tongue of "Hindu," "Hindoostani," or, occasionally, "Arabian," by the 1920 and 1930 censuses, these same men and others who had arrived more recently were now more often listed with "India" or "East India" as their place of birth and with a mother tongue of "Bengal," "Bengali," "Bangola," or "Bengalese."[16] Street peddling continued to be the most common form of work listed for these men, though by 1920 and 1930, a growing number were moving into more stable and settled work as merchants or shop owners.

By 1930, an increasing number of these men had also married and started families with local African American and Creole women. In some cases, the mixed-race sons of these families had followed their fathers into work as street peddlers. And quite significantly, a number of the men who were living in New Orleans in 1910 had now moved out of the Crescent City. In addition to Moksad Ali's family, for example, a number of other Bengali men and their families took part in the Great Migration, ending up in places like Chicago and even moving westward from Louisiana to Los Angeles.[17] Equally significantly, by 1930, several of the men who were living and working as street peddlers in New Orleans in the first decade of the century had now moved their business out to other parts of the American South. So, for example, the 1930 census shows groups of Bengali men working as peddlers in Jacksonville, Florida, in Atlanta, Georgia, and in Chattanooga and Memphis, Tennessee.[18]

The settlement of working-class Indian men in and around New York City also gained momentum during this period, picking up during and immediately after World War I and continuing steadily over subsequent decades. By the 1920s and 1930s, a wide range of sources—census records, city directories, marriage records, newspaper articles, and interviews with family members of early immigrants—indicate that groups of Indian men, again predominantly Muslims from Bengal, were now living and working in the Lower East Side of Manhattan and in Central and East Harlem as well as in waterfront areas like Hell's Kitchen and the Syrian Quarter and across the water in New Jersey. While the migration to New Orleans may have begun earlier, the migration to New York was larger and more extensive. In New York, therefore, certain aspects of this migration become clearer. New York was also a particular kind of destination in the 1920s and 1930s. The working-class men who came to New York from Bengal and other parts of the subcontinent in these years were entering a city that had become a unique space of overlap between multiple diasporas. In neighborhoods like Harlem, these men lived alongside the increasing numbers of African Americans who were arriving from the U.S. South and black and Latino/a (im)migrants who were arriving from Puerto Rico, Jamaica, Cuba, and other parts of the Caribbean. I will draw out some of the contours of this intersection below. In New York, it also becomes increasingly evident that these men, and likely others settling in New Orleans and other U.S. port cities at this time, were part of a larger history of Indian labor on, and desertion from, British merchant marine vessels. At the same moment that Roheamon, Rostom, and Bahadour Ali were moving into Harlem as young people caught up in the Great Migration, there was a steady flow of Indian Muslim men moving into Harlem after deserting British merchant ships. The migration of *these* men was likely more closely linked to that of their father Moksad's in New Orleans and, indeed, to the migration of hundreds of other Indian men who were settling at this time in port cities like London, Glasgow, Southampton, and Hull. Theirs was a migration of Indian laborers escaping from the circuits of British imperial trade.

JUMPING SHIP

During the first two decades of the century, there had already been a steady flow of Indian merchant sailors moving in and out of New York port areas. These men typically came in on ships that carried cargo like jute, hemp, and cotton from Calcutta and Bombay or which plied the shorter and more heavily traveled passenger and trade routes between New York and various English and Scottish ports. Indian sailors had been used on British merchant vessels since the eighteenth century, but as Tabili, Visram, Balachandran, and others have noted,[19] a major change occurred in the latter half of the nineteenth century with the advent of steam navigation and the opening of the Suez Canal. The move to the use of steamships initiated the industrialization of maritime work, creating, as Visram puts it, "totally new categories of labour in the engine room, those of [the] firemen and trimmers needed to stoke the furnaces."[20] British shipping lines made use of the large pools of colonial labor available to them in India, Egypt, Yemen, Somalia, and elsewhere to do these jobs, which involved working below deck in intense heat for hours and days at a time. In the racist logic of the British shipping companies and indeed the British government, these men were considered naturally suited to such work because they came from a "tropical" climate. Of course, there were also clear economic advantages to using a colonized work force and, ultimately, the British used these sailors both below and above deck—not just as firemen and trimmers but as cooks, stewards, and servants.[21] The largest group of colonial maritime workers on British ships were Indians, who, by the 1920s comprised roughly 25 percent of the British maritime work force.[22] The vast majority of these Indian sailors were Muslim and the largest regional group was Bengali, usually hailing from rural inland regions such as Sylhet and Noakhali. These men would sign on from the port city of Calcutta and then circulate globally almost everywhere the British traded goods.

Significant numbers of Indian merchant sailors deserted their ships in British ports during the first decades of the twentieth century—a phenomenon that became the basis for some of the earliest communities of Indians in Britain.[23] Most of the existing scholarly work on Indian merchant marine desertion has focused on this British context. But what is striking about the British scholarship is how closely it describes the contours of the working-class Indian Muslim communities that were beginning to emerge in places like New Orleans and New York. Peddling, for example, was one of the first and most widespread onshore occupations for Indian maritime workers who had jumped ship in Britain. While this began in the nineteenth century, by the 1920s and 1930s, there were Indian ex-sailors working as peddlers of clothes, silks, perfumes, and household goods in street markets and door-to-door all over England, Scotland, Ireland, and Wales.[24] From World War I onward, it was also common for escaped Indian seamen to find employment in factories onshore, where the work was sometimes similar to what they did on steamships—firing and stoking furnaces—but in a less captive environment, with marginally better and more regular wages.[25]

In virtually every major British port, boardinghouses were also established by some of the earliest Indian sailors to arrive and desert. The proprietors of these boardinghouses played a key role in helping sailors find work, navigate the process of establishing legal residence, and adjust to life onshore.[26] Some of the first Indian restaurants and cafes in Britain were established by Sylheti sailors who had jumped ship and these became spaces, according to Visram, "where news of jobs, friends, and family were exchanged" and where networks began to take shape.[27] Many of the Indian men who chose to stay in England after jumping ship also married English women from the working-class neighborhoods where they settled. These women engaged in a range of work—from domestic and reproductive labor to their work as cultural and literal translators—which facilitated the assimilation of Indian men into local economies and communities. For those men who were proprietors of the boardinghouses and restaurants that became early spaces of community formation, the labor of their wives was absolutely crucial. "As partners," Visram writes, these women, "helped to run [their] businesses and keep accounts" as well as navigating legal paperwork and, when necessary, dealing with police and municipal officials.[28]

The evidence in both New Orleans and New York indicates that these patterns of settlement were not in fact limited to Britain but were occurring simultaneously across the Atlantic in U.S. ports. New Orleans census records, for example, show most of the same dynamics at work: peddling as a key means of economic support; marriages between Indian men and local working-class women; a boardinghouse run by one of the earliest Bengali settlers, Alef Ali, and the presence of an older African American woman living with Alef Ali, who quite likely was involved in the labor of running his boardinghouse and looking after his daughter. In New York, the picture we get is even fuller and the ties to the maritime trade more explicit. A series of news reports, deportation cases, and labor disputes from the early 1920s point to a growing presence of Indian merchant marine deserters who were either seeking out or being recruited into factory work in New York, New Jersey, and Pennsylvania. Census records from 1920 and 1930 show groups of Indian Muslim men working in silk-dying shops in Paterson, New Jersey,[29] and in the steel works in Lackawanna, New York, where they worked the furnaces as "firemen."[30] On the shop floor, these men worked alongside other immigrant laborers—Arab, Greek, Italian, and Eastern European workers, among others—and were also thrust into the racial politics of American labor, being used as strikebreakers alongside African American and Chinese workers.[31]

We also get a clearer picture of the lives that these men were building for themselves in New York, and, here again, boardinghouses and restaurants played a crucial role. Some New York City boardinghouses were conduits channeling Indian men into factory work, while others connected them to day labor and restaurant work. The 1930 census, for example, shows almost fifty Indian men living together in two adjacent buildings on Eighth Avenue around Forty-seventh Street; about half have given their occupation as simply "Laborer—Odd Jobs" and a handful of others list their occupation as either a "Seaman," "Deck Steward," "Waiter," or "Cook" on steamships. Interspersed with these men, about a dozen others list occupations connected to restaurants—there is one restaurant owner, one owner of a "tea garden," six waiters, two busboys, two cooks, a counterman, and a dishwasher.[32] This census record is suggestive both because it implies another continuity between maritime and onshore occupations—in this case, ship's cooks and stewards becoming cooks and waiters onshore—and because the Eighth Avenue boarding house was itself located just blocks away from the Ceylon Restaurant and the Ceylon India Inn, two of the first Indian restaurants in the city, which were important gathering places for the small but growing Indian community in New York in the 1910s through 1930s and hubs of Indian nationalist activity.[33]

A NEW YORK COMMUNITY

Ibrahim Choudry, one of the key figures among the group of Bengali Muslim men who settled in Harlem, appears to have made a transition from an involvement in Indian nationalism to a much deeper engagement with local community building. Through his story, we can track the movement of Indian Muslim men from waterfront areas into working-class residential neighborhoods in the city. Choudry had been a student activist in East Bengal in the early 1920s, where he had been involved in the pan-Islamic *Khilafat* movement at the time of its alignment with the movement for Indian independence. Because of these activities, he had to flee the subcontinent, and after working his way across to the United States on a merchant marine vessel, jumped ship in New York.[34] Some time in the mid-1930s, Choudry moved to East Harlem, where he married a Puerto Rican woman from the neighborhood and had two children. By the 1940s, he was at the center of a growing Bengali Muslim community in Harlem, and while he did not run a boardinghouse per se, he played many of the roles that boardinghouse keepers had taken on in other locations. Choudry's son Noor remembers his father always rushing off to someone in the community who needed help dealing with immigration officials, securing housing, writing letters, filling out forms. "That particular group," says Noor Chowdry, "were very, very tight. If they called and they needed something, my dad was boom, gone, now. He would say, son, there's a can of soup in [the cupboard] . . . I gotta go."[35]

Such a role was formalized in 1943, when Ibrahim Choudry became the first manager of New York's British Merchant Sailor's Club for Indian Seamen. The club, which occupied the two upper floors of a four-story building at 100 West 38th Street, included a small prayer room, a recreation room, and a mess hall where visiting seamen could get fresh Indian meals, served three times a day by a staff cook, Secunder Meah.[36] The records of the Indian Seamen's Club give us a sense of both the sheer number of Indian sailors going in and out of New York by the 1940s and the range of work Ibrahim Choudry was doing among this population. By the end of its first year, the club had 66,221 visits by Indian seamen and served 198,200 meals. Choudry himself regularly went out to the docks to meet with incoming Indian crews and bring them in groups by city bus to the midtown club. He oversaw weekly prayer meetings, "special parties . . . on Mohammedan festivals . . . general recitations and songs, with drum accompaniment on Saturdays [with] local residents participating . . . [and f]ilm shows three nights a week" as well as arranging "excursions to the Bronx Zoo and Coney Island," tickets to the theater, and "bi-weekly English lessons." Choudry also took dictation in his office from nonliterate seamen who wanted to send letters back home, and he arranged visits for incoming sailors "to [the] homes of local Indian residents."[37]

Perhaps what is most striking in the records of the British Merchant Sailor's Club for Indian Seamen is the acknowledgment of both a large Indian seafaring population and a smaller "resident" Indian population in New York, of which men like Ibrahim Choudry, Secunder Meah, and others listed as employees of the club—Sekunder Yassim, Salim Darwood, Wahed Ali—were all themselves a part. This "resident" community was taking shape over the course of the 1920s and 1930s in two main locations—the Lower East Side and Harlem. Census records from 1930 show clusters of Indian Muslim men in both locations. Downtown, fifteen men ranging in age from nineteen to thirty-one years old were living at three addresses within a few blocks of each other in the Lower East Side—127 Eldridge Street, 99 Suffolk Street, and 42 Rivington Street—working in restaurants and living among Jewish, Italian, Greek, and Turkish immigrants.[38] Uptown at the same time, census records show more than sixty-five Indian men living at roughly forty-five different addresses throughout Harlem—in small groups, alone as boarders, or with their African American or Puerto Rican wives. Again, most of these men had recognizably Muslim names, were in their twenties, and were working as "laborers" or as hotel and restaurant workers. Here they were living primarily among immigrants from Puerto Rico, the British West Indies, and Italy, and among African Americans from the U.S. South.[39] There is some indication that the Indian men who were initially living on the Lower East Side were moving over time up to Harlem. Habib Ullah, a friend of Ibrahim Choudry, for example, had lived on the Lower East Side when he first arrived in New York in the 1920s after deserting from a ship in Boston. Some time in the 1930s, Ullah moved straight up the East Side subway line to 102nd Street and Lexington Avenue, where he met and married his first wife, Victoria Echevarria, a young woman who had recently migrated to Spanish Harlem from Puerto Rico.

Marriages between Indian men and Latina and African American women were occurring in both the Lower East Side and Harlem, but appear more frequent uptown, where a more settled community was coalescing in the 1930s. New York City marriage records from the late 1920s through the 1930s show a whole range of these unions: in 1926, Nawab Ali, a twenty-four-year-old laundry worker from Calcutta living at 222 East 100th Street married Frances Santos, an eighteen-year-old from Puerto Rico; in 1930, twenty-three-year-old Wohad Ali, another Indian laundry worker living on East 100th Street, married Maria Louisa Rivera, a nineteen-year-old from Vega Baja, Puerto Rico; in 1934, Nabob Ali, a twenty-eight-year-old cook from Calcutta, India, living on East 110th Street, married twenty-two-year-old Maria Cleophe Jusino from Sabana Grande, Puerto Rico; in 1937, Eleman Miah, a twenty-seven-year-old "cook's helper" from Calcutta living on West 112th Street married twenty-two-year-old Emma Douglass, a "colored" woman from Indianapolis.[40] Noor Chowdry and Habib Ullah, Jr., describe their fathers' marriages as part of a larger process in which Harlem was becoming home to them—these were the women, in other words, with whom Indian men came in contact on a daily basis after they settled uptown, the women they got to know in their buildings, on their blocks, and in their neighborhoods. This is echoed in the marriage records from this time, in which bride and groom often appear to have been living within a two to three block radius of each other. There were also spaces beyond their immediate blocks where Indian men were interacting socially with local Latino/a and African American men and women. Ibrahim Choudry's younger brother Mawsood describes going with his Bengali friends to parties in Puerto Rican neighborhoods in the Bronx,[41] and Alaudin Ullah, Habib Ullah's youngest son, recounts family stories about the parties his father and friends threw in rented halls in Spanish Harlem, where Indian food mixed with Puerto Rican and Cuban music.[42] These are spaces where longer-term relationships may have begun. At the same time, Alaudin and Habib Ullah, Jr., both suspect that naturalization played a role in some of these marriages—that their father's circle wanted to marry women whose status would allow them to stay in the United States.[43]

BENGALI HARLEM

It is difficult to know exactly how many Indian Muslim men lived with Latina and African American women in Harlem during this period, or, for that matter, how many chose to live with other men in a variety of domestic settings, from sailors' boardinghouses to shared apartments and homes. There were presumably a number of ex-seamen who lived with and even started families with local women but who were never officially married, and, indeed, given their legal status after the 1917 Immigration Act, most of the ex-seafaring population would have felt the imperative to remain out of view of local officials, and thus, ultimately, out of the archival records. Yet interviews with the children of early twentieth-century Indian, Latina, and African American migrants to New York City—a "second generation" who are now in their sixties and seventies—suggest that by the 1940s and 1950s, a community had quietly taken shape in the New York area, with Harlem as its apparent center of gravity. This group included roughly 100 to 150 Muslim men, predominantly from those areas of East Bengal, which, after the partition of the subcontinent in 1947, became East Pakistan, as well as almost an equal number of Latina and African American women and a growing number of U.S.-born mixed-race children. While the men of this community were connected to each other through preexisting ties of kinship, language, labor, region, and religion, their daily lives and those of their families, neighbors, and the others with whom they came in regular contact were such that all were forced to move beyond the realm of what had previously been familiar to them and to create new familiarities and new forms of kinship specific to the places in which they settled. The stories of the second generation give us a glimpse of these processes by providing a sense of the different spaces that their parents created and adopted in 1940s through 1960s New York and of the lives which they forged in and through these spaces.

As in Britain, restaurants became some of the primary locations in which Indian Muslim ex-seamen both reconstituted their ties with one another and interacted on a daily basis with members of other communities. By the 1940s, an increasing number of Bengali men were opening Indian restaurants either in Harlem, close to where they lived, or, if they could afford it, in the midtown theater district. For their proprietors, restaurants were a means of escaping and moving on from the types of jobs they had labored at since jumping ship—as doormen, laundrymen, line cooks, street vendors, and factory workers. This was the case, for example, for Habib Ullah, who, after years of work as a dishwasher in the kitchens of hotels, opened the Bengal Garden Restaurant at 144 West 48th Street with the financial backing of Ibrahim Choudry and another Bengali friend who had saved up $20,000 selling hotdogs from a pushcart on 110th Street.[44] The Bengal Garden was jointly operated by Victoria and Habib Ullah—Victoria worked the register and did the books while Habib cooked in the kitchen—and was a space marked by its heterogeneity. On a day-to-day basis, it catered to a non-South Asian crowd of midtown office workers and evening theatergoers. Often, the first person these customers encountered upon arrival at the Bengal Garden was a Puerto Rican woman—Victoria Ullah. On key occasions, the restaurant transformed into a meeting place, filling up with Muslim ex-seamen from the subcontinent, brought together under the banner of an organization cofounded by Ibrahim Choudry in 1947: the Pakistan League of America. Beginning in the late 1940s, many of the Pakistan League's banquets and special events were held at the restaurant, including communal celebrations of Eid and public discussions of U.S. immigration policy and events on the subcontinent with invited guests that included U.S. Senator William Langer and the writer Hans Stefan Santesson.[45]

The restaurants which Bengali ex-sailors opened up in Harlem during in the 1940s through 1960s were, in some ways, even more complex social spaces because of their groundedness in local communities—because, in other words, they developed into places where the multiple populations who shared the larger neighborhood of Harlem spent time and came in contact with one another on a daily basis. These restaurants included Said Ali's on East 109th Street, Ameer's on West 127th Street, Eshad Ali's, also known as the Bombay India Restaurant, on West 125th Street between Morningside and Amsterdam avenues, and another unnamed restaurant in the basement of a building in the East 100s. As was the case with the Bengal Garden, the proprietors often involved family members in the daily operation of these establishments, so their African American and Latina wives and mixed-race children were often the ones greeting and serving customers. Said Ali, for example, ran his East Harlem restaurant with the help of his African American wife and two sons, while Eshad Ali's wife Mamie operated the Bombay India Restaurant on 125th Street for many years after her husband's death.

These restaurants were, of course, important spaces for those men from the subcontinent who had settled in Harlem. They could come to places like Said Ali's to eat familiar food, converse in Bengali, and discuss events that were unfolding "back home." At the same time, it is apparent from a number of sources that these Indian restaurants became integrated into the lives of Harlem's African American residents. Miriam Christian, a community activist who came to Harlem in the 1950s and ultimately wrote for the New York Amsterdam News, recalled fondly in a 1996 interview that in the mid-1960s, she and her circle—which included other black journalists, artists, organizers, and local politicians—used to have special occasions catered by one of the neighborhood's Indian restaurants.[46] In an interview shortly before his death, the poet and East Harlem resident Sekou Sundiata similarly described frequent visits to "Ali's place" on 109th Street in the late 1960s and early 1970s, where he remembered offering one of Said Ali's sons a sympathetic ear as he struggled with the prospect of taking on his father's business.[47] Miles Davis is said to have regularly patronized one of the neighborhood's Indian establishments, most likely the Bombay India Restaurant on 125th Street, where percussionist James Mtume claims Davis soaked in the sounds of the music the proprietor played.[48] But perhaps most significantly, since these were some of the earliest *halal* restaurants in Harlem, they became spaces where Bengali and African American Muslims frequently met and interacted. Ameer's on 127th Street, for example, became a regular spot for both groups; here, according to stories passed down to Habib Ullah's son Alaudin, immigrant Muslims engaged in discussion and debate with members of Malcolm X's Organization of African American Unity about their differing practices and interpretations of Islam.[49]

Beyond these restaurants, the neighborhood of Harlem itself—or more accurately the neighborhoods that comprised Harlem—were spaces of ongoing interaction and exchange between Bengali Muslim immigrants and their African American and Latino/a neighbors, friends, and family members. Half-brothers Alaudin and Habib Ullah, Jr., provide a vivid picture of their father Habib, Sr.'s relationship to his adopted neighborhood of Spanish Harlem. "It was bi-cultural," says Habib, Jr., who was born in 1942 and lived with his father on East 102nd Street after his mother Victoria died ten years later, "[My father's] friends were basically Indian *and* Spanish—and he got along well equally with both groups."[50]

Alaudin, whose childhood memories date from the early 1970s, describes accompanying his father on his rounds through the neighborhood, where he would engage in "the art of conversation"—either with his Bengali friends who congregated at an Indian-owned jewelry store on East 103rd Street, with his Puerto Rican neighbors, or with the proprietors of the various shops he visited: "It seemed like he used to venture [out] a lot by foot because I remember when I was really young, maybe four or five years old, I would hold his hand and we would go from store to store. Like we'd go to Paul's place, the jewelry store, we'd go to Said Ali's restaurant on 109th, and he would just [talk to people]—it was like the art of conversation for him is what he thrived on, what he loved."[51] Habib, Jr., describes his own experience, from the early 1950s, of going with his father to buy food and supplies on Saturdays at Spanish Harlem's huge covered market, *La Marqueta*:

> In those days, it went from 110th Street up to 116 Street. We would walk through the whole thing and Pop would have his certain stalls that he would go to. And there was a *botanica* there, and Pop would go there and . . . the lady behind the counter knew him. And this is how he made his own curry, because he would go there and he would say "give me a pound of that, give me a pound of this, give me a pound of that" . . . he knew the spices. And she would put it in, mix it up for him, and that was his own curry. That's why Pop's curry was always distinct.[52]

Habib Ullah, Sr.'s walks through East Harlem—which comprised a routine that he repeated day after day in one variation or another over the course of thirty years—are significant not merely because they were part of the process by which he made a home out of the physical space of his adopted neighborhood. His circuits through Spanish Harlem were also the means by which he forged and maintained a new set of human relationships—often across differences of race, ethnicity, language, religion, and so on—in this new home.

Most of the second generation grew up navigating this kind of multiplicity within the very context of their immediate and extended families. Ibrahim Choudry's son Noor, whose mother was born in Cuba to Puerto Rican parents who later settled in Spanish Harlem, spent his childhood moving between a series of multiracial spaces. For the first few years of his childhood, he lived with his mother and maternal grandmother in East Harlem, speaking only Spanish until the age of six. Later he lived with his Bengali uncle, African American aunt, and their college-age son Hassan in Belleville, New Jersey. Here, Noor says, he fell in love with jazz music after spending hours listening to his cousin play saxophone: "I very seldom even saw him. I heard him and when the door was closed we were told just leave—don't bother him. By the night time he got out of there, you know, doing his thing. I remember he played at clubs at night—jazz clubs in the city. And he just got me to the point where I said 'I want to play the sax like him.'"[53] Noor would also spend time with his father's friend Abdul, who had married and settled in another African American community in New Jersey and had a son Noor's age. Abdul ran a small business making perfumes and lotions which he sold door to door—an echo of the earliest Bengali settlers in New Orleans—and

Noor and Abdul's son often accompanied the father on his own routine circuits through working-class black neighborhoods, where, moving from one home to the next, he would meet with his regular customers and make his sales.[54]

Noor Chowdry's descriptions of the extended family on his father's side give us a sense of the larger multiracial community that had taken shape by the 1950s. His father and his friend Habib had married Puerto Rican women from Spanish Harlem, two of his father's cousins had married African American women in New Jersey, and his father's younger brother had married an English woman during a two-year stay in London. When these men and women gathered for Thanksgiving each year at Noor's uncle Idris's home in Montclair, New Jersey, they brought together a group of more than a dozen American-born cousins who were Bengali and Puerto Rican, Bengali and African American, Bengali and white, who grew up, among other things, hearing multiple languages and eating multiple kinds of home-food, and who were exposed through different family members both to Islam and various strains of Christianity. This was a microcosm of a community that, by the 1950s and 1960s, Noor Chowdry and Habib Ullah, Jr., estimate numbered somewhere between four and five hundred men, women, and children.[55] By this time, Ibrahim Choudry and others had started the yearly tradition of a group boat trip from Battery Park up the East River to Rye Beach Playland. "I remember standing on line," says Ullah, "there were literally hundreds of people out there waiting to get on that boat . . . Bengali, Indian, Pakistani" men, their Latina and African American wives, and all their children. "There was a big picnic and there was a band on the boat and they would dance . . . [T]his was how they renewed their bonds as a community."[56]

The relationship of this large mixed community to Islam appears to have been a varied and complicated one. On the one hand, according to Habib Ullah, Jr., a lot of his father's friends relaxed their religious observances after coming to New York. For these men, Islam consisted largely of the maintenance of dietary strictures, the observance of the yearly Ramadan fast, and the celebration of Eid at the end of fasting. Some of the Latina and African American women who were part of this community converted to Islam. Others did not convert but participated in prayers on occasions like Eid. Still others appear to have been vocally opposed to aspects of Islamic practice they believed subordinated women. When it came to the young people of this community, their relationship to Islam largely depended on their fathers. Habib Ullah, Sr., for example, de-emphasized both Islam and the Bengali language in bringing up his children because he wanted them to "fit in and become American." Significantly, "fitting in" was, in this case, less of an assimilation into white American society than one into the immediate neighborhood of Spanish Harlem.[57] But there were other Bengali men, such as Ibrahim Choudry, who had a more active and explicit religious life. According to Noor Chowdry, his father often played the role of an imam in the earlier years of the community, leading prayers and officiating at religious events. Later, he was involved in the formation and activities of a series of religious and cultural organizations; in addition to cofounding the Pakistan League of America in the late

1940s, Ibrahim Choudry was also secretary of the Islamic Council of New York and an active member of a citywide interfaith organization in the 1950s, and in the early 1970s, helped found one of New York City's first mosques, on Second Avenue and Tenth Street.[58] Not surprisingly then, Choudry tried to instill in his son the importance of daily observance. Noor took in what his father taught him but recalls his ambivalence about some of the experiences that this observance entailed—such as praying on the side of the highway with his father and other members of the community during long road trips: "He'd ask me to do the call to prayer, and I'd do it. You know, you see all these cars slowing down and looking at you while everybody's praying. But then you just get back on the bus and keep going."[59]

Noor Chowdry and other members of the second generation are more enthusiastic in describing the Pakistan League's yearly Eid celebrations, which brought out a larger, more diverse group and were as much social as religious events. While the Pakistan League played multiple roles—social, political, and religious—these yearly Eid celebrations were perhaps its most important contribution to the life and coherence of a larger community. These events were initially held at the league's offices in the Lower East Side and then, as they got larger, in a series of banquet halls in the midtown hotels where its members were employed as kitchen workers and doormen. By the1950s and 1960s, the Pakistan League's Eid celebrations brought out hundreds of people—as with the boat rides, primarily Bengali men, their Latina and African American wives, and their children. "In the early days, when I was younger," remembers Chowdry,

> we'd go to the Eids and there'd have to be three or four hundred people maybe even more. The men would get there earlier and they'd have to do a lot of praying and it seemed like there were at least a hundred, hundred-and-fifty men only. And after that it came time to have the big feast and then the women would come, so then it was wall to wall people . . . And I can remember at one time they had twelve rows of tables, and these guys would come marching out of the kitchen—had to be eight or nine of them with everything, putting it on the table, family style—the rice, the bowls of chicken, and stuff like that, everything. If you went home hungry it was your own fault. Eid was always a great celebration . . . it was a time when everybody was joyful.[60]

To be sure, the life of this community was not always so joyful, nor was it without significant conflicts, strains, and fissures. These included failed marriages, men who abdicated care of their children, business partnerships that went sour, and members of the community—Bengali, Puerto Rican, African American, and mixed—who left Harlem as soon as they were in a financial position to do so, gradually weakening what was once a collective center of gravity.

The Pakistan League of America's Eid celebrations are, however, a window into a different kind of Muslim presence within twentieth-century U.S. communities of color—different, for example, from the histories of both Ahmadiyya communities and the Nation of Islam. The Pakistan League itself is perhaps most interesting for all the complexity, multiplicity, and localized dynamics that existed just beneath the surface of its deceptively narrow invocation of the nation of Pakistan. The organization was founded by men who, when they came to the United States in the 1920s and 1930s had been Indians—albeit subjects of a colonized, British-controlled India—and then had stayed in the United States. Because they came from Muslim majority regions of British India, these men became "Pakistani" in 1947—but in absentia, when their home regions were included in the newly formed Pakistan.[61] Since the majority of the Pakistan League's members were Bengali—and thus specifically "East Pakistani"—they would again change their national identity in absentia when East Pakistan gained its independence and became Bangladesh in 1971. For these men, the yearly Eid celebrations were a means of maintaining a connection to the lives they had once lived in a place that had gone through vast transformations in their absence and a way of carrying forward past cultural and religious practices into a new social and national context. At the same time, these Eid celebrations brought out a community that included more African American and Puerto Rican women and U.S.-born mixed-race children than Muslim men with roots in the new nation of Pakistan. For these women and children, Eid presented one of the few formalized settings in which they could gain some access to their husbands' and fathers' past worlds. But this was not ultimately a form of ethnic consolidation, a closing in around a particular identity or set of practices passed on by the men of the community. For the women and children, these events were part of lives that accommodated, navigated, and opened up to multiplicity. Through its Eid celebrations, in other words, the Pakistan League played the role of creating and sustaining a multi-racial, -ethnic, -religious, -linguistic, and -generational community—one that arguably had more to do with New York City than with the nation which gave the organization its name.

However, this unique mixed community—which was at once Bengali, Puerto Rican, and African American—slowly dissipated after the generation of Muslim immigrant men who gave rise to it began to pass away. Ironically, this generation's preexisting bonds and shared experiences of migration, and their efforts to maintain a connection both with each other and with elements of their shared pasts, may have been the glue which held together something that went far beyond themselves—and, at the same time, the specificity of these connections may have been the reason that the larger community did not outlast the men around whom it formed. The Hart-Celler Immigration Act of 1965, which favored professionals and allowed for the immigration and reunification of entire families from the subcontinent, meant that a larger, wealthier suburban South Asian community quickly overshadowed those who were already here. This was a community that was more inclined, and in more of a position, to become insular, and one that largely played the role of upholding, rather than transgressing, the existing racial divisions of the United States. This change in the community is openly lamented by some of those who lived through the earlier and lesser known history of South Asians in the United States. "There was a different atmosphere, a different attitude, a different way of life in those days," says Habib Ullah, Jr. "And I think its sad. To a certain extent I understand keeping your culture, and I think my father's generation did to a great extent. But now [South Asians] . . . keep their culture and put a wall around what is around them, and I think that's a disservice to the coming generations."[62]

A PHOTOGRAPH

Alaudin Ullah, Habib Ullah, Sr.'s younger son and a lifelong resident of Spanish Harlem, remembers a photograph that he saw as a child, in the early 1970s, perhaps in the apartment of Ibrahim Choudry. The image is of Choudry and Malcolm X standing together amidst a group of other African American and South Asian Muslims.[63] Ullah has been trying, for years, to locate this photograph, which may now simply be a fading memory of a snapshot that no longer exists. The histories that I have begun to lay out here, which exist in stories, memories, and scattered archives, have also been fading and disappearing. But the significance of these histories is multiple. On the one hand, they raise a series of productive questions about diaspora: What does it mean that the first South Asian "second generation" in the United States was everywhere racially mixed: Punjabi and Mexican in California,[64] Bengali and African American in New Orleans, and Bengali, African American, and Puerto Rican in New York? What would it mean to consider the Latina and African American women who married Indian Muslim immigrants in New York and New Orleans as part of the South Asian diaspora? What would it mean to consider these escaped Indian maritime workers as part of the black diaspora? How do we speak of the mixed second generation in a way that sees them not as anomalous, exceptional, or inauthentic but rather as complex embodiments of equally complex intersections of history? These and other scholarly questions, however, capture only part of what is important here. Alaudin Ullah's quest to find the lost photograph of Ibrahim Choudry and Malcolm X is driven by more than the image's ability to link his father's community to an iconic figure of African American Islam and liberatory political change. For Ullah, this image represents a past of shared everyday lives of which his father was a part—and a sense of possibility that is still connected to this past. As important as it is to Alaudin Ullah that he find the lost photograph, it seems crucial for us to understand this possibility of interracial community and solidarity that the image represents to him, and which was realized, however momentarily, in the lives of a now disappearing generation.

NOTES

The term "overlapping diasporas" is drawn here from Earl Lewis's article "To Turn as on a Pivot: Writing African Americans into a History of Overlapping Diasporas," *American Historical Review* 100 (June 1995): 765–87. Lewis employed the term to describe the meeting of culturally and linguistically different African populations in the context of plantation slavery in the U.S. South and to explore the processes by which these multiple

African groups and later their descendents in Northern cities came to forge a more singular African *American* community/identity/culture. I use the term to explore a different context: that of the neighborhoods, buildings and other spaces where migrant populations from *multiple* continents—Asia, Africa, the Americas, and Europe—met and intermingled in the early twentieth-century United States. As Juan Flores, Rich Blint, Michael Quintero, and others have pointed out in their readings of earlier versions of this chapter, such a shift away from Lewis's original use of the term must account for the ways different (im)migrant groups who share the same spaces are differently racialized and thus come to have different relationships and access to power. With this in mind, it is important to specify here that in the press and in government documents of this era, the racialization of working-class Indian immigrants shifted between an identification of these men with African Americans (evidenced by their classification as "black," "negro," or "colored" in a variety of official and popular sources) and an identification with other legally excluded Asian immigrant populations (evidenced by their classification and description at other times as "Asiatics," "Hindoos/Hindus," or "aliens"). In both cases, their access to power and social mobility was severely limited by their racialization and criminalization (as "illegal" immigrants). I will be exploring these processes in greater detail in future work; here, I use the term "overlapping diasporas" as part of an effort to think through the experiences of encounter and transformation that occur among and between different migrant groups in diaspora. I also follow on Brent Hayes Edwards's insistence, in considering the movements and encounters of black artists, intellectuals, and political figures in 1920s Paris, that "[i]f *diaspora* is an appropriate term to describe these circuits, it is partly because it . . . forces us to understand a context like Paris as multiple, as heterogeneous, in a manner that makes it impossible to consider any single history of migration and exile without considering 'overlapping diasporas'—simultaneous, transnational patterns that influence one another." Brent Hayes Edwards, "The Shadow of Shadows," *positions: east asia cultures critique* 11, no. 1 (2003): 13.

1. United States Department of Commerce, Bureau of the Census (hereafter, USDC/BC), *U.S. Census, 1900*: Population Schedule, New Orleans, Louisiana, Orleans Parish, 9th Precinct, Ward 11, Supervisor's District 1, Enumeration District No. 115, Sheet No. 18. USDC/BC, *U.S. Census, 1920*: Population Schedule, Bronx, New York, Supervisor's District 6, Enumeration District No. 271, Sheet No. 7.

2. New York City Municipal Archives (hereafter, NYCMA), *State of New York Certificate and Record of Marriage*, August 20, 1923, Certificate No. 26075.

3. NYCMA, *Certificate and Record of Marriage*, October 29, 1924, Certificate No. 33040.

4. USDC/BC, *U.S. Census, 1930*: Population Schedule, New York, Ward AD21, Block L, Enumeration District No. 31-1019, Supervisor's District 24, Sheet No. 6B.

5. USDC/BC, *U.S. Census, 1900*: Population Schedule, New Orleans, Louisiana, Orleans Parish, 9th Precinct, Ward 11, Supervisor's District 1, Enumeration District No. 115, Sheet No. 18. USDC/BC, *U.S. Census, 1920*: Population Schedule, Bronx, New York, Supervisor's District 6, Enumeration District No. 271, Sheet No. 7.

6. There is also evidence of Indian men living and working in and around Galveston, Texas; Atlanta, Georgia; Jacksonville, Florida; Baltimore, Maryland; Philadelphia, Pennsylvania; and Detroit, Michigan, and evidence of intermarriage within the black community in Detroit.

7. This history complements the stories of Punjabi-Mexican intermarriage in California in the 1910s through 1940s, as described in Karen Leonard's *Making Ethnic Choices*. See Karen Isaksen Leonard, *Making Ethnic Choices: California's Punjabi Mexican Americans* (Philadelphia: Temple University Press, 1992).

8. Richard Brent Turner, *Islam in the African American Experience* (Bloomington: Indiana University Press, 1997, 2003), 74.

9. Earl Lewis, "To Turn as on a Pivot: Writing African Americans Into a History of Overlapping Diasporas," *American Historical Review* 100 (June 1995): 765–87; Paul Gilroy, *The Black Atlantic: Modernity and Double Consciousness* (Cambridge: Harvard University Press, 1993); Paul Gilroy, *Postcolonial Melancholia* (New York: Columbia University Press, 2004); Brent Hayes Edwards, *The Practice of Diaspora: Literature, Translation, and the Rise of Black Internationalism* (Cambridge: Harvard University Press, 2003); Gayatri Gopinath, "'Bombay, U.K., Yuba City': Bhangra Music and the Engendering of Diaspora," *Diaspora* 4, no. 3 (1995): 303–21; Vijay Prashad, *Everybody Was Kung Fu Fighting: Afro-Asian Connections and the Myth of Cultural Purity* (Boston: Beacon, 2001).

10. Although from one U.S. Census to the next, racial categories changed, such that, in some years, categories like "mulatto" and "quadroon" were available to census-takers, it appears from most of the census sheets examined here that the variety, multiplicity, and mixture of New Orleans' "non-white" population, was flattened into the category "black." Indian men were themselves often recorded in these documents with the racial categorization "B" or sometimes "Mu" (mulatto), which was then in many instances scratched out and replaced with "H" or "Hin" for "Hindu" (the broad racial categorization used for anyone from India). It is unclear when and where these recategorizations from "B" to "H" took place and who made them—that is, whether it was the census-taker or a local supervisor or a Washington D.C.-based clerk.

11. USDC/BC, *U.S. Census, 1900*: Population Schedule, New Orleans, Louisiana, Orleans Parish, 3rd Precinct, Ward 4, Supervisor's District 1, Enumeration District No. 36, Sheets No. 6 & 7.

12. USDC/BC, *U.S. Census, 1900*: Population Schedule, New Orleans, Louisiana, Orleans Parish, 3rd Precinct, Ward 4, Supervisor's District 1, Enumeration District No. 36, Sheet No. 6.

13. This name, which is difficult to read in the original census record, could also be "Christina Gardner."

14. USDC/BC, *U.S. Census, 1910*: Population Schedule, New Orleans, Louisiana, Orleans Parish, 2nd Precinct, Ward 4, Supervisor's District 1, Enumeration District No. 59, Sheet No. 5.

15. See, for example, USDC/BC, *U.S. Census, 1910*: Population Schedule, New Orleans, Louisiana, Orleans Parish, 2nd Precinct, Ward 4, Supervisor's District 1, Enumeration District No. 59, Sheet No. 7 and Enumeration District No. 60, Sheets No. 8 & 12.

16. See, for example, USDC/BC, *U.S. Census, 1930*: Population Schedule, New Orleans, Louisiana, Orleans Parish, 6th Precinct, Ward 5, Block 182, Supervisor's District 11, Enumeration District No. 36-76, Sheet No. 14B.

17. See, for example, USDC/BC, *U.S. Census, 1930*: Population Schedule, Chicago, Illinois, Cook County, 6th Precinct, Ward 25, Block 393, Supervisor's District 6, Enumeration District No. 16-2638, Sheet No. 5A.

18. USDC/BC, *U.S. Census, 1930*: Population Schedule, Chattanooga, Tennessee, Hamilton County, 7th Precinct, Supervisor's District 13, Enumeration District No. 33-22, Sheet No. 13B. USDC/BC, *U.S. Census, 1930*: Population Schedule, Memphis, Tennessee, Shelby County, District 6, Ward 11, Block 417, Supervisor's District 9, Enumeration District No. 79-36, Sheet No. 3B.

19. Laura Tabili, *"We Ask for British Justice": Workers and Racial Difference in Late Imperial Britain* (Ithaca: Cornell University Press, 1994); Rozina Visram, *Asians in Britain: 400 Years of History* (London: Pluto, 2002); G. Balachandran, "South Asian Seafarers and their Worlds: c. 1870–1930s," Paper presented at Seascapes, Littoral Cultures, and

Trans-Oceanic Exchanges, Library of Congress, Washington D.C., February 12–15, 2003. http://www.historycooperative.org/ proceedings/seascapes/balachandran.html (accessed July 9, 2006); Conrad Dixon, "Lascars: The Forgotten Seamen," in *Working Men Who Got Wet*, ed. Rosemary Ommer and Gerald Panting (Newfoundland: Maritime History Group, Memorial University of Newfoundland, 1980).

20. Visram, *Asians in Britain*, 54.

21. Ibid., 54–56.; see also Dixon, "Lascars: The Forgotten Seamen," 269; Balachandran, "South Asian Seafarers and their Worlds," par. 10–11.

22. Balachandran, "South Asian Seafarers and their Worlds," par 8.

23. Visram, *Asians in Britain*, 64–69, 196–224; Balachandran, South Asian Seafarers and their Worlds," par. 18, 31–37.

24. Visram, *Asians in Britain*, 260.

25. Ibid., 196–97.

26. Ibid., 257.

27. Ibid.

28. Ibid., 258.

29. See, for example, USDC/BC, *U.S. Census, 1930*: Population Schedule, Paterson, New Jersey, Passaic County, Ward 1, Supervisor's District 2, Enumeration District No. 16-8, Sheets No. 1B & 14A; Enumeration District No. 16-9, Sheet No. 6A.

30. USDC/BC, *U.S. Census, 1920*: Population Schedule, Lackawanna, New York, Erie County, Ward 1, Supervisor's District 21, Enumeration District No. 306, Sheets No. 53A, 62B.

31. In 1922, the Erie Railroad attempted to break a railroad shop workers' strike by hiring African American workers to work in their Jersey City rail yards, along with "fifty Hindus and thirty-one Chinese . . . most of whom," according to the *New York Times*, had "deserted ships in port here." At the time, the *Times* reported, there were "about 5000 Chinese and Hindus stranded in this port" and looking for work "because of dull shipping." See "Strikers Migrate to Take Rail Jobs," *New York Times*, July 12, 1922, 1.

32. USDC/BC, *U.S. Census, 1930*: Population Schedule, New York, Borough of Manhattan, 10th Assembly District, Block E, Supervisor's District 21, Enumeration District No. 31-1198, Sheets No. 49A, 50B.

33. In the 1920s, the connections between the Indian maritime population and New York-based Indian radicals—particularly Sailendranath Ghose and his Friends of Freedom for India—were in fact quite significant. These connections are discussed in greater depth in my essay, "'Lost in the City': Spaces and Stories of South Asian New York, 1917–1965," *South Asian Popular Culture* 5, no. 1 (2007): 59 76, and are the subject of ongoing research and writing.

34. Mawsood Choudhury, interview by the author, New York, February 2004.

35. Noor Chowdry, interview by the author, Los Angeles, CA, January 2005.

36. F.O. 371/42909.

37. Ibid.

38. USDC/BC, *U.S. Census, 1930*: Population Schedule, New York, Borough of Manhattan, 2nd Assembly District: Block B, Supervisor's District 21, Enumeration District No. 21-717, Sheet No. 8A; Block F, Supervisor's District 21, Enumeration District No. 31-93, Sheet No. 15B; Block O, Supervisor's District 21, Enumeration District No. 31-106, Sheet No. 20A.

39. NYCMA, State of New York Certificate and Record of Marriage (various), 1923–37; USDC/BC, *U.S. Census, 1930*: Population Schedule, New York, Borough of Manhattan, 16th Assembly District: Block J, Supervisor's District 23, Enumeration District 754B, Sheet No. 10A; 17th Assembly District: Block C, Supervisor's District 23, Enumeration

District 31-1242, Sheet No. 16A; Block D, Supervisor's District 23, Enumeration District 31-786, Sheet No. 31A; Block D, Supervisor's District 24, Enumeration District 31-922, Sheet No. 6B; 18th Assembly District: Block 6, Supervisor's District 23, Enumeration District 31-810, Sheet No. 23B.

40. NYCMA, *State of New York Certificate and Record of Marriage (various), 1923–37*.

41. M. Choudhury, interview.

42. Alaudin Ullah, interview by the author, New York, June 2003.

43. Alaudin Ullah, interview by the author, New York, December 2006; Habib Ullah, Jr., interview by the author, New York, May 2005.

44. N. Chowdry, interview.

45. *New York Times*, "Freedom of India Celebrated Here," August 16, 1948, 8; "Events Today," February 20, 1949, 64; The New York *Amsterdam News*, "Pakistan Group to Honor Sen. Langer," February 19, 1949, 32; "Sen. Langer Is Honored By Pakistanis," March 3, 1951, 18.

46. Miriam Christian, interview by the author, New York, 1996.

47. Sekou Sundiata, interview by the author, New York, May 2007.

48. Paul Tingen's biography, *Miles Beyond*, describes a series of meetings between Davis and Mtume in 1972 at an establishment that was almost certainly the Bombay India Restaurant: "Miles and I would often go to this restaurant at 125th Street in Harlem," Mtume says, "and discuss some of the influences he was hearing. It was an Indian restaurant, and obviously they were playing Indian music, and it was at that point that he was telling me about the idea of using the electric sitar and tablas." Paul Tingen, *Miles Beyond: The Electric Explorations of Miles Davis, 1967–1991* (New York: Watson-Guptill, 2001), 130.

49. A. Ullah, interview.

50. H. Ullah, interview.

51. A. Ullah, interview.

52. H. Ullah, interview.

53. N. Chowdry, interview.

54. Ibid.

55. N. Chowdry, interview; H. Ullah, interview.

56. H. Ullah.

57. Ibid.

58. "A Call to Prayer for Peace in the Holy Land: An Open Letter to the Security Council of the United Nations," *New York Times*, April 20, 1956, 14; "Holy Land to Get 'Pillar of Peace,'" *New York Times*, July 16, 1956, 18; "Islamic Groups Meet Today," *New York Times*, July 27, 1956, 7; M. Choudhury, interview; A. Ullah, interview.

59. N. Chowdry, interview.

60. Ibid.

61. Many of these men had presumably been involved in a prior New York-based organization, the India League of America (as was the case with Ibrahim Choudry), and had broken off from this organization to form the Pakistan League in 1946–47, a move which appears to have had as much to do with the class differences between Hindu and Muslim Indian immigrant communities in New York as it did with the geopolitics of the subcontinent.

62. H. Ullah, interview.

63. Alaudin Ullah, interview by the author, New York, May 2005.

64. Leonard, *Making Ethnic Choices*.

CHAPTER 14

WEST AFRICAN "SOUL BROTHERS" IN HARLEM

IMMIGRATION, ISLAM, AND THE BLACK ENCOUNTER

ZAIN ABDULLAH

INTRODUCTION

IN 1999, A MEDIA BLITZ COVERING THE POLICE KILLING OF AMADOU DIALLO, AN innocent victim from Guinea mistaken for a Black serial rapist, revealed he was not only one of many Black immigrants living in New York, but he was also a member of an emerging West African Islamic community.[1] In 2003, an undercover officer wrongfully killed another African Muslim, Ousmane Zongo, an African arts restorer from Burkina Faso, deepening the city's engagement with these recent arrivals.[2] For the first time, New Yorkers were exposed to the Muslim practices of their West African neighbors. There was even press coverage following their slain bodies from funeral services in New York mosques to their respective countries in Africa.[3] Because of a tendency to view immigrants in terms of their labor rather than their humanity, the media spread taught us another lesson. West African life in the United States cannot be fully understood by focusing on their

---

This chapter was originally published as "African 'Soul Brothers' in the 'Hood: Immigration, Islam and the Black Encounter," *Anthropological Quarterly* 82, no. 1 (Winter 2009): 37–62.

"Soul Brother" used in the title was originally a fraternal reference for an African American man with a special kind of cache. I use it here to speak to the ways West African Muslims are forging new Black and religious identities in African American communities across the country.

work habits alone. Will Herberg's classic work *Protestant, Catholic, Jew*[4] revealed how early European immigrants used religion to aid their assimilation into middle America. As Black immigrants, however, West African Muslims are already classified at the bottom of the U.S. racial hierarchy. Yet, just as Judith Weisenfeld argues that the "African American religious experience has rendered the margin a site of power and of creativity, an activity that necessarily alters the center," African Muslim migrants also challenge their peripheral status by creating religious practices they hope will shield them from a Black underclass.[5]

A recent influx of 100,000 West Africans into New York City, for example, is creating an enclave that Harlem residents call "Little Africa" or "Africa Town."[6] For the Muslims among these newcomers, their *masjids*,[7] Islamic schools, businesses, and associations are essential for how they are integrating themselves into the landscape of this predominately Black neighborhood. There has even been a street sign hung at Harlem's major intersection on the corner of 125th Street and 7th Avenue that reads "African Square." Other regions are renamed by Africans themselves, such as "Fouta Town," a heavy Fulani Muslim settlement in Brooklyn where ethnicity and religion evenly overlap.[8] This ongoing Islamic and cultural activity creates a unique kind of Muslim space,[9] a pulsating environment driving their sense of self and collective determination. These practices, then, are a major resource in their attempt to prevent a downward spiral into poor social conditions. Most observers, however, are unaware that religion plays a crucial role in this urban transformation. Because westerners view Islam as an Arab faith, many pay no attention to the Muslim identity of these Black immigrants. Yet, few miss them wearing their wide-sleeve, *boubou* robes with tasseled hats, hawking items out of brief cases in midtown Manhattan or, perhaps, strolling Harlem streets. But this costuming represents an important way African immigrants assert their Muslim presence and impact the Black public sphere. In Harlem, this performance reworks the rhythm of the metropolis, allowing them to redefine Blackness or Black identity on their own terms.

Some researchers believe that new immigrants are being absorbed into three different and sometimes disadvantageous segments of American society. Whereas traditional assimilation, for example, taught that newcomers entered the United States and followed a "straight-line" path into the dominant Anglo-American culture, current scholars claim that the integration is not straight but segmented, which recognizes that today's ethnically diverse migrants are incorporated into either the White middle class, the downward path of a Black and Latino underclass, or the ethnic community characterized by tight group solidarity and rapid economic advancement.[10] The workings of West African Muslims in Harlem, instead, reveal that immigrants can settle in poor Black communities, form solid ethnic niches and intergroup cooperatives, and not experience downward mobility, especially if they are able to create new types of religious capital.[11] In other words, rather than merely self-segregating themselves as a strategy against failure, many forge alliances and build bridges between themselves and native-born Blacks and, particularly, African American Muslims. Paul Stoller argues that Islam has historically played a major role in structuring relations between African Muslims of different ethnicities,[12] and it continues to inform their business sense in the diaspora. While money may have "no smell," as Stoller would have it, and there-fore may trump Islamic etiquette for many West Africans in New York, he asserts that "Islam has always constructed the moral framework for West African trading transactions."[13] As such, their collaboration with African American Muslims is often facilitated through their shared religious precepts. This does not suggest, however, the absence of any conflict, and while the community as a whole has advanced, their presence in Harlem has not been problem free.

Starting in the early 1990s, the rapid rise of Muslims from West Africa quickly found its way into Harlem's informal economy, creating a vibrant African bazaar with street vendors stretched along 125th Street.[14] Organized like traditional West African markets, the open-air sidewalk businesses created a celebratory atmosphere punctuated with a timbre of African languages and dialects, haggling voices, and colorful displays of traditional African, domestic, and designer prod-ucts. Prior to its dissolution and relocation to 116th Street in 1994, the African Market had become a major tourist attraction. Double-decker buses from Apple Tours brought scores of European sightseers on holiday to take pictures (within the protected confines of a company vehicle), capturing an exotic souvenir of New York's "authentic" African culture.[15] It also provided African Americans seek-ing to reclaim their African identity with an opportunity to buy handicrafts that symbolized a piece of Africa.[16]

While some store owners benefited from the way the selling of African culture attracted new customers to the area, others felt the cultural alterity of the market hampered business. Influential business owners provoked elected officials to enact the no-vending laws, a request that was made against a barrage of protests and demonstrations. Still, African street merchants and other sidewalk vendors were ousted.[17] During the conflict, each faction invoked "culture" to justify its agenda. African sellers, on the one hand, claimed that selling "African things" on 125th Street helped to revitalize the area.[18] Angry store owners, on the other hand, claimed that Africans clutter the sidewalks and restrict movement, dirty and litter the streets, and that their informal habits create chaos and induce crime. Donald Trump and the Fifth Avenue Merchants Association made a similar argument to clear midtown Manhattan of Senegalese vendors in 1985. In this way, so-called African Third World habits, portraying chaos, informality, and uncleanliness, is pitted against "First World" sensibilities representing order, regulation, and purity. Such an objectifying discourse compares African culture on a hierarchi-cal scale with that of the West, and it is deemed incapable of occupying one and the same time and space. "Oh, East is East, and West is West," Rudyard Kipling once barked, "and never the twain shall meet."[19] While Kipling's refrain in this 1892 poem actually ridiculed this polarized view of the world, it has become clear that the two spheres have indeed met. Some, despite arguments to the contrary, believe they have even clashed.[20] In either case, with thousands of new immigrants pouring into U.S. cities and, more often than not, settling into predominately

Black communities, few can argue there has not been a genuine encounter. International forces seamlessly link the East and the West, and the global and the local (not to mention the urban and rural) have become connected in ways we could have hardly imagined.

WEST AFRICAN MUSLIM IMMIGRATION

As a result of the post-1965 immigration, which signals the unprecedented arrival of new immigrants from parts of Asia, Latin America, the Caribbean, and Africa, American neighborhoods are undergoing changes that have impacted more than their demographics. The new immigration promises to bring alternative pieties and new understandings of the sacred, altering our social and religious worlds like never before. According to some estimates, Africans are arriving in the United States at rates that surpass the highest figures during the Transatlantic Slave Trade.[21] While New York remains a major point of entry for these Muslim migrants, their numbers are rapidly increasing in Boston, Washington, D.C., Atlanta, Chicago, Houston, and Los Angeles, "luring them away from New York City—especially if they have what they call 'papers,' namely, an employment authorization permit from the U.S. Immigration and Naturalization Service (INS)."[22] Their attraction to these areas is somewhat straightforward. Most migrate to large cities where employment opportunities are more or less widely available and flexible, especially for those juggling several jobs and Muslims needing special times to pray throughout the day.

Prior to urban renewal and the subsequent rise of gentrification, affordable housing in poor and working class areas is likewise appealing. More importantly, though, many early African migrants were drawn to Harlem not merely because it was economically viable but for its reputation as a major center of Black life. That is, despite its existing reputation for crime, West Africans saw it as a Black space in the center of America, and this gave them a glimmer of hope that they might be successful. This is very different, however, for Indian migrants, who have mostly settled in Queens, New York. Madhulika S. Khandelwal points out while they resided in mix communities and "interacted with people of different racial and ethnic backgrounds, . . . many Indian immigrants continue to be uncertain about their 'race.'"[23] Under the current American racial system, West Africans apparently do not have that problem. However, moving into Black Harlem (despite its changing racial landscape) means they must learn to negotiate an entirely new sense of what it means to be Black. In his recent book, *Real Black*, anthropologist John L. Jackson, Jr., rehearses a kind of racial bricolage on what it means to navigate multiple shades of Blackness, and he untangles the significance of these claims of authenticity for a place like Harlem.[24]

As I have pointed out elsewhere,[25] West African immigration to the United States can be divided into three phases: (1) the Transatlantic Slave Trade between the sixteenth and early nineteenth centuries; (2) the period between the Prohibition of the Slave Trade Act of 1808 and 1965; and (3) from 1965 to the present.

During the first two phases, Islam had been the religion of the ruling classes, the urban elite, and foreign populations under the rule of West African Muslim empires. Following this period, Islam continued to spread but this time among the masses. From the late sixteenth to the mid-eighteenth centuries, the religious campaigns of Islamic Sufi brotherhoods and other reform movements prompted mass conversion.[26] Following the collapse of these Muslim kingdoms, the entire region (with the exception of Liberia, which was nonetheless heavily influenced by U.S. economic interests) gradually fell under European control. Western nations met at the Berlin Conference in 1884 to partition West Africa into "spheres of influence" and essentially divided the area into two territories: Anglophone (English-speaking) and Francophone (French-speaking). Only the German occupation of Togoland, along with the Portuguese control of Guinea-Bissau and the Cape Verde Islands, remained beyond their reach. Thousands of agricultural workers emigrated out of French colonies to escape the official mandate of conscription. While laborers moved from one colonial territory to another, producing a type of "interterritorial" migration,[27] the next period of decolonization and African independence ushered in an international migration that is steadily becoming transnational.[28]

By 1960, the French colonial territory in West Africa was independent, and the rise of African nations brought about new boundaries. Like the colonial partitions, these state borders cut across preexisting ethnic group settlements. Family members were forced to migrate across countries to reunite with kin or to obtain employment. Establishing the free but limited movement of goods and people within French speaking states, the CEAO (Communauté Economique d'Afrique Occidentale/the West African Economic Community) was established.[29] Allowing greater movement, but only for a ninety-day period, West African countries formed the Economic Community of West African States (ECOWAS) in 1975. This freedom of movement has fostered a "floating population,"[30] referring to a pattern of labor migration that crosses countries within Africa and, I would add, extends to western nations abroad.

During independence, West Africans from former French colonies were encouraged to migrate to France as a source of cheap labor. By the next decade, however, the desire for African workers waned and France terminated legal immigration in 1974,[31] which was further enforced under the Pasqua law of 1993. At the same time, the U.S. Immigration Act of 1965, which rescinded the old quota system that favored immigrants from northern and western Europe, enacted a preference system supporting family reunification for permanent residents (74 percent), skilled labor (20 percent), and refugees seeking political asylum (6 percent).[32] This new legislation significantly opened the door of immigration to West African Muslims and many other non-Europeans. To facilitate the move between Dakar (Senegal's capital) and New York, new direct flights were added in the mid-1980s.[33] Prompted by an economic crisis in Europe and devastatingly poor conditions in West Africa, the migration of West African Muslims to France was redirected to America. Until the last two decades or so, the initial migration phase was dominated by West African immigrants with Christian leanings from Eng-lish-speaking countries like Gambia, Ghana, and Nigeria. Still, the desire to leave poor conditions does not mean migrants will have a place to go, especially if the point of destination is not willing to receive them.[34] Thus, by the 1980s, both French restrictions on immigration and a liberal American immigration policy prompted West African Muslim migration to the United States.

According to the Census Bureau's Current Population Survey (CPS), the total population count for West Africans in the United States was approximately 167,000 in March of 2000. The figures from the 1990 census estimated that for West Africans claiming foreign-born status in New York, there were 2,287 Senegalese, 1,388 Ivoirians, and 1,032 Guineans in the country. When compared with the 2000 census, only a slight difference is discernable.[35] Because the Immi-gration and Naturalization Service (INS) only records legal immigration to the United States, their numbers are invariably much lower than the U.S. Bureau of the Census, which counts both legal and illegal residents. Some informal esti-mates claim that the Senegalese in New York City, for example, number anywhere between 10,000 and 20,000.[36] Other unofficial counts have placed Murid follow-ers, a Sufi Brotherhood from Senegal, at 4,000 to 6,000 nationally, with approxi-mately 2,500 in New York City.[37] Because many West African Muslims overstayed their visiting or student visas in the 1980s, they took advantage of the lottery system granting amnesty to undocumented migrants. Once they are permanent residents, married men generally send for their wives and other family members. These opportunities will undoubtedly increase their numbers significantly over time. Nonetheless, while official figures are generally too low, informal estimates are usually exaggerated.

WEST AFRICAN ISLAM IN BLACK AMERICA

As mentioned above, West African Muslims claimed that the image of Harlem as a "Black Mecca" (i.e., a Black borough) initially attracted them to New York City.[38] In contrast, they were also led to believe American Blacks were criminals and not to be trusted. When they encountered poor Black residents, for example, and witnessed the way some were caught in a cycle of drugs and violence, the rac-ist stereotypes they had internalized prior to their migration were confirmed upon their arrival. Moreover, because the conduct of their non-Muslim neighbors would at times contradict their Islamic ethics, they tended to view these Harlem residents with contempt. Despite their strained relations with some Black residents, the Afri-can American Muslim community at Masjid Malcolm Shabazz, on Malcolm X Boulevard and 116th Street, served as their "proximal host," and this certainly has affected the nature of their incorporation. The term "proximal host" is used here to mean the indigenous group to which immigrants are assigned or voluntarily adopt due to their racial or ethnic affinities.[39] For instance, because West African immi-grants are thought to resemble Black Americans racially, they are often treated accordingly by outsiders. By the same token, many African immigrants choose to settle in Black communities and utilize Black institutions due to the racial category they share. Still, the racial designation is often resented by immigrants, especially if the group to which they are assigned is disenfranchised.

By 1990, large numbers of West African Muslims were attending the weekly *jum'ah* (Friday) prayers at Masjid Malcolm Shabazz. Realizing their growing attendance could translate into greater economic rewards, the Masjid leadership began to display signs in French instructing Africans where to pay their weekly charity. According to Hamdi, an African American Muslim pioneer and Murid convert, because of their Islamic upbringing, West African Muslims understand that their first responsibility is to support the Masjid. "They realize," he said, "if they're successful with their religious obligations, the other aspects of their life will follow suit." He continued to say that West African Murids made donations of $3,000 or more a week. He added that the Murids had such a strong religious network and work ethic that once they were able to raise $75,000 in just two weeks, which was just in time to purchase a limousine for the arrival of their *shaykh* (marabout).

Some claim the same group remits nearly $15,000 per month to their Islamic city of Touba in Senegal. Researchers report that the Banque de l'Habitat du Senegal (BHS), which opened a New York branch in 1993, handled savings and transfers for Senegalese clients amounting to $900,000 during its first year of operation.[40] By 1994, due to the 50 percent devaluation of the CFA (French African Franc), that figure increased to $4 million. The rise in bank dealings was because they were remitting more to help family members survive the economic crisis back home. As perhaps the only African bank operating in the United States, its dealings have escalated to as much as $7.5 million a year.[41] While these figures hardly compare to those for Ghanaian immigrants, who, over ten years ago, were estimated to remit between $250 and $350 million a year,[42] it does demonstrate the economic strength of West African Muslims in the greater New York area. This also speaks volumes about the role religion plays in their economic incorporation into Harlem. Besides the kind of tensions that can arise between members belonging to the same racial category, divergent religious orientations can force people to reconfigure their racial identities and sense of place as well.

Because Masjid Malcolm Shabazz grew out of the Black Nationalist sentiment of the Nation of Islam (NOI), the Friday *khutba* (sermon) usually addresses racial themes or some aspect of American race relations. West African Sunnis (orthodox Muslims) felt they had different needs and wanted a sermon advocating survival through spiritual development. They also complained that their prayer attendance was much too regimented requiring, for instance, members to stop at the door to sign in and sometimes answer questions by Masjid security. They were likewise at odds with the Black preacher style of the *khutba*, sparking frequent rejoinders from the congregation known as "call-and-response."[43] Some claimed their presence was made even more uncomfortable when the Imam gave a sermon exhorting "all Africans to go home." In 1993, when the Egyptian-led Islamic Center of New York opened on 3rd Avenue and 96th Street, many West African Sunnis left Shabazz for a more familiar religious environment. Others began attending the Mosque of Islamic Brotherhood (MIB), an African American orthodox community on West 113th Street and Nicholas Avenue. Because members of MIB are African Americans who use Qur'anic Arabic and dress in modified Islamic and African garb, African attendees felt somewhat at ease. Unlike the fees they were charged to use Masjid Malcolm Shabazz for their weddings, funerals or meetings (regarded by some as less than brotherly), MIB allowed them complete access without cost.

On the other hand, African Murids, a prominent Senegalese Sufi order, were much more familiar with the inclusion of Black themes in the *khutba* and stayed at Shabazz.[44] In the introduction to his classic ethnography on the Murids, Donal B. Cruise O'Brien states that the "brotherhood originated in the late nineteenth century as a collective response of the Wolof . . . to changes brought about by French conquest."[45] A major reaction to French colonization was to challenge their race-based policy and the colonial mandate of an *Islam noir* or a Black Islam.[46] Nonetheless, the Murid Sufi order, affectionately referred to as, Muridiyyah, emerged out of a racialized and religious context where both Blackness and Islam are equally rehearsed. As such, many Murids in Harlem actually appreciated the inclusion of "race talk" during the Friday *jum`ah* service at Shabazz.

In short, while West African Sunni Muslims left Masjid Malcolm Shabazz for a more traditional host, West African Murids remained. More importantly, their respective selection in favor of a familiar congregational culture reveals an important divide between them. Prior to establishing their own *masjids*, both groups joined Masjid Malcolm Shabazz. Because the Sufi doctrine of Muridiyyah was not in conflict with the Black Nationalist rhetoric at Shabazz, Murid followers embraced it as an appropriate platform before relocating to their own House of Islam on 137th and Edgecombe. West African Sunnis, in contrast, left Shabazz because they desired a more conventional setting or a place more suited to their own sense of religious orthodoxy. This does not mean African Sunni Muslims reject a Black discourse at the *masjid*. In fact, they have adopted their status as Black immigrants quite well and incorporated these conversations into their Islamic practice. What they sought, however, was a racial sensibility that would be totally subservient to their Islamic teachings. African Murids, on the other hand, did not recognize a separation between the two. That is, one did not supersede the other because race and religion were viewed as one and the same. This precept had already been central to the Nation of Islam's teachings and continued in many ways among African Americans at Shabazz. For both, their premigration perception of Harlem as a "Black Mecca," their practice of appropriating Black narratives into their Islamic practice, and their selection of dissimilar hosts are important ways a Black place is negotiated at African *masjids* in Harlem. Nonetheless, it is certainly not the only way they exert their identities onto the urban terrain.

The Murid own and operate most of the restaurants, variety stores, fashion boutiques, Islamic books, and supply shops on West 116th Street. As places of

cultural production, African restaurants, for example, inscribe identity by employ-
ing religious symbols or wording like *halal* to guarantee the food is authentic
and religiously "pure."[47] The stores also bear the name of their spiritual guides
(Mbacke or Kara), their holy city in Senegal, "Touba," or a combination of both.
To the outsider, however, the religious nature of these businesses is less obvious.
As one African American young man once asked a Senegalese storeowner, "Man,
who is this guy Touba? His name is all over the place. He must be rich!" In a sense,
the young man was right to recognize something very important about the signification of these businesses. Touba is the group's spiritual center.[48] Because it was
founded by Shaykh Amadou Bamba, the holy city of Touba and the founder are
viewed as one and the same. In fact, Amadou Bamba is frequently called "Serigne
Touba" or "Sir Touba." As such, the sacredness of the city is transferable to other
marabouts. When Murid leaders visit their *taalibés* (disciples) in New York or
elsewhere, for instance, the announcement, "Touba is coming to town"[49] is made
regularly. Moreover, Touba is a frame of mind. It is always carried in the heart of
the devout Murid as a point of reference.[50] By hanging pictures of the founder or
their Great Mosque in Senegal on the walls of their shops, Touba is there. Touba,
therefore, can be reproduced and transported to create Murid identity anywhere
in the world. Accordingly, "making" this kind of cultural space is one way West
African Muslims create transnational identities that link Touba with New York.

   Other cultural performances include the daily practice of wearing African
clothes. While their dress may make them easily recognizable, publicly wearing
traditional African clothing is a religious act that not only creates Muslim space
but also inscribes their Africanness onto the geography of the city. In contrast to
the Pan-Islamic view of West African Sunnis, Murids construct their own Islamic
discourse challenging Arab hegemony and notions of Black inferiority. According
to their publication, "Shaykh Ahmadou Bamba inaugurated a new era in the history of Islam and of the black man. He is the first spiritual black guide massively
followed by people from all over the world, thus showing that all men are issued
from the same soul."[51] Clothing can operate as a symbol of identity the wearer
and the observer can read.[52] It can mark a change in political orientation,[53] or, as
Roland Barthes asserted, act as a sign interpreted to suit different contexts.[54] For
Harlem residents, however, the traditional clothing worn by the Murid clearly
identifies them as African but not necessarily Muslim. In fact, most residents
described their new neighbors as simply "African" wearing some sort of African
clothing. Nonetheless, their *boubou* clothes and tasseled hat do not merely fore-
ground their Africanity alone, but they signify a commitment to an age-old, West
African Islamic tradition.

   The Shaykh Ahmadou Bamba Day parade illustrates this point as well.[55] Established in 1988, each year on this day, thousands of West African Murids march in
a procession up 7th Avenue (Adam Clayton Powell, Jr., Blvd.) from 110th Street
and Central Park North to 125th. Men, women, and children march decked out
in their traditional garb, full of color and regalia. What sets this parade apart from
many others like the St. Patrick's Day, Puerto Rican Day, and African American

Day parades is the absence of elaborate floats, loud marching bands with girls dressed in bathing-like costumes throwing batons, celebrities, or even televised coverage interpreting each event for viewers. In fact, when interviewed, no spectator knew what was going on or why. Most speculated that it was "something African" but could not figure it out.

Why, after two decades, would they still undergo so much planning and organizing to have a police-escorted parade no spectator understood? Public performances like parades allow groups to rework their identity and infuse space with special meaning. Parade participants carry large banners in different languages and march chanting Islamic slogans. These actions represent the group's attempt to communicate internally and contest competing versions of membership. It also represents their struggle to put forth a Murid identity against the backdrop of other Muslim constituencies in Harlem. At the same time, banners urge African Americans to racial pride and their own version of Blackness. Some participants carry the Senegalese and American flag along with huge portraits of Shaykh Ahmadou Bamba, Shaykh Mouhammadou Mourtada Mbacke (a prominent Murid leader), and Shaykh Salih Mbacke, the Murid's *khalifa* (supreme leader). While the parade received little media attention in New York City, journalists from Senegal (and the British Broadcasting Corporation [BBC], on a different occasion) arrived to cover the event and, according to reports, televise it in their home country for an entire week. The national symbolism of the flags and the parade's international coverage allow them to construct their Black and Muslim identities on a much larger canvas but necessarily within the context of Black America. Still, West African Muslims do not exist in a vacuum and their lives are not restricted by their association with their Muslim counterparts alone. Their Islamic sensibilities are obviously influenced by their encounter with the external world, the vast space beyond *masjid* walls and well outside the confines of public celebrations. Consequently, their engagement with local residents opens up a whole new set of issues that challenge their presence in Black America.

THE BLACK ENCOUNTER

African Hair braiding salons are sites where Africans and African Americans interact on a regular basis. While Blacks are generally consumers, West African women either work for an American proprietor or, more recently, own their boutiques. Leslie's Hair Salon in Harlem was no exception. During my initial visit, I peeked in the window and saw more than a dozen African women dressed in traditional garb, braiding hair at each station. In fact, there were so many African workers, I was certain the shop was African-owned. I was wrong. On one occasion, I was at the salon when a Black woman came stumbling out yelling at the African worker, "Come out bitch . . . , I'll cut your motherfuckin' throat." At the same time, people held the African hair braider inside as she yelled, "You better gimme my money . . . " With her hand in her pocketbook, the Black woman stood outside the door repeating her threat. While the situation illustrates just how explosive things can get between old residents and newcomers, the underlying factor appeared to be financial rather than cultural. But Rob, a twenty-three-year-old Black security worker at the salon, disagreed.

On the day of the incident, I was struck by Rob's composure during the scuffle. He sat back in his lawn chair on the sidewalk and watched things unfold. He has lived in Harlem since the age of seven, and he appeared to have grown accustomed to witnessing these outbursts. In his mind, these conflicts were indicative of cultural differences that separated Africans from Blacks. In fact, "culture," and the way it is objectified as a "thing" representing a set of values, appears to structure relations between African Americans and other immigrants as well. Robert C. Smith's work on Mexicans in New York illustrates how *mexicanidad* (Mexican-ness) or, more precisely, *ticuanensidad* (Ticuaniness), constructs a boundary of cultural difference between them and their Black and Puerto Rican neighbors. In the view of Smith's respondent, Blacks and Puerto Ricans actually suffer because they lack "culture" entirely and not so much for having a different one.[56] What makes this curious is how culture is often blamed by residents as the reason groups cannot get along. As the fight between the two women ended, Lenny, a Black man in his early thirties and a salon employee, left the shop and joined us outside. They began talking about the fight and their view of Africans in Harlem:

> "They're not like us," Rob said, shaking his head and rocking in his chair. "They're different," he added.
> "Ahhh," Lenny interjected, "Don't get me started on them."
> He continued, "They're some nasty people!"
> "And ruuude," he stressed.
> "But," Rob blurted out, staring into space a bit, "I don't know how to explain it, I guess it's just cultural differences."
> "Yeah, I don't know," Lenny replied. "I do know they don't mess with me, 'cause they know I'll tell 'em about themselves," he griped, "but they are some nasty people."
> "What do you mean by nasty?" I asked.
> Lenny thought about it for a moment and said, "Ya know, just nasty." I waited for more.
> "They throw trash and stuff all over the floor and they have an attitude," he added.

This story illustrates the assumptions some Blacks in Harlem make about what separates them from Africans. As a conversation sparked by a conflict at the salon, Rob stated there were some "cultural differences" between Africans and Blacks he was just unable to explain. Whether he could justify his statement or not is unimportant. What is worth exploring is how he recognized differences between the two groups. He was convinced "they're not like us." As such, he assumed "culture" might be the reason. Of course, people from foreign countries possess values, worldviews, and share meanings that differ in certain respects from native residents. However, "cultural differences," in and of themselves, do not create a separation. In other words, the fact that differences exist do not automatically construct a boundary forcing people to see themselves as separate and distinct. What creates and maintains the divide is the social importance or cultural meanings people attach to these differences. While Rob used culture as a way to mark the boundary that set the two of them apart, Lenny defined the line with a characteristic that gave meaning to both sides of the cultural border.

Lenny used the word "nasty" as a term many Black residents used to describe Africans. In *Harlemworld*, John L. Jackson, Jr., quotes Paula, a thirty-eight-year-old black woman who accused Africans of acting like "they are royalty" and claimed, "They all nasty."[57] He also referred to an eighteen-year-old black Latina, Elisha, who criticized them for how they wear "their African shit and act like they still in Africa. All they want us for," she asserted, "is so they can braid our hair and give us extensions."[58] As Jackson's work suggests, "nasty" is a term applied to Africans who appear pompous and antisocial. This is partially determined by the way their African dress marks a distinctions between them and others in Harlem. While some embrace their cultural practices as a public display of their African heritage, others mock it as a barrier wedged between them and their African neighbors. These are clearly fault lines where a high level of distrust exists. It is a site where even services provided by African hair braiders are held in contempt. Lenny struggled to define what he meant by calling Africans "nasty" and later associated it with unclean work habits. Being "nasty" could actually mean all sorts of things; however, whatever it means is only important when it creates a sense of emotional attachment differentiating one group from the other.

Dean, a twenty-five-year-old African American and lifelong Harlem resident, revealed his sense of the border between Africans and Blacks. He divided Africans into two types. "There are the ones that wear traditional clothes," he assessed, "they stick to themselves." He added, "I don't know if they think there're better than us or what." As for the other, he argued that "they are the Americanized Africans, the ones you see at the club." To Dean, Africans wore their garb not merely as a marker of distinction but as a sign of superiority. Ann Miles claims in her study that migrants from Ecuador living in Queens, New York, used clothing as a sort of protection that separated them from the deleterious effects of street life. Idealizing her old country ways above her newly adopted "MTV" urban world, "Rosa," Miles asserts, "proudly wore the traditional *pollera* [wide pleated skirt] that identified her as a *chola* [rural folk]."[59] Outsiders, however, rarely understand these internal cultural musings. Instead, they are at times viewed as adverse fashion attempts and, thus, presumptuous.

West African Muslims, on the other side, believe Blacks view them as "animals." Yahya, a young man in his early twenties from Guinea, claimed, "Black Americans look down on Africa." "I don't really blame them," he continued, "it is because of what they see on TV—it's what they've been taught. All they show is famine and destruction in Africa." Despite the media distortions they believe influence the view Blacks have of them, some West African Muslims admitted they treat African Americans and particularly non-Muslims with a measure of "indifference." On other occasions, West Africans are unsure how to understand the existing tensions between them and Black Americans. In *Money Has No Smell*, Paul Stoller's West African respondents complained that they were "disappointed" when treated badly by African Americans and accused of selling their "ancestors into slavery."[60]

As a familiar way to mark distinctions between ethnic groups in Africa, some African Muslims see Islam as a way to partition Harlem into *dar al-harb* (outsider) and *dar al-Islam* (insider). Scholars of Islamic studies see this dichotomy as a Muslim innovation that deviates from the principles of the religion. I have decided to translate these Arabic terms as "outsider" and "insider," respectively, because a literal rendering would miss the way they are meant to act as well-defined frontiers in Harlem. In a literal sense, they would be translated as *dar al-harb* (abode of war) and *dar al-Islam* (abode of peace). Contrary to its religious role, Islam, in this context, is used as an organizing principle, a bifurcated way of looking at the world and those who inhabit it. Like Lenny's use of "nasty" to define the boundary dividing the two groups, West African Muslims use what they believe to be a Black misperception of Africa to justify maintaining a similar distance. What helps to erect the borderline is a condescending attitude that renders Harlem residents blameworthy.[61] My conversation with Rob and Lenny quickly takes a religious turn. As we thought about Rob's answer, Lenny continued to talk about Africans. "And they call themselves Muslims?" he mocked. "I don't know what kind of Muslims they are, because the Muslims I know aren't nasty—and they're not rude!" he asserted. "Muslims are clean people and polite," Lenny added. Then he repeated, "I don't know what kinda Muslims they are."

Lenny's criticism of their Muslim affiliation is another way of disparaging the group and, thus, contesting identity. Moreover, his rejection of their self-identification reserves Muslim identity for a different group of "people," a people who are, as he would have it, not "nasty" or "rude." So, if Muslims are "polite" and "clean," West Africans, in contrast, cannot be Muslims. Otherwise, this would contradict his wayward depiction of them. Said differently, Lenny already had a positive image of Muslims in Harlem. Since the Africans at the shop claim to be Muslims, too, the only way he can maintain his negative view of them is to reject their religious claim. In this informal setting, however, he realizes his sanction has little impact on their ability to declare their Muslim identity. In essence, his statement that he doesn't "know what kinda Muslims they are" raises two points. First, it is an assertion that brings into question and ultimately rejects their right to be Muslims. Second, even if they are acknowledged as Muslims by others, the fact that he is unaware of their "kind" places them in a strange and unusual category. Because Black Muslims have a long and reputable history in Harlem, an Islamic affiliation for these Africans is hotly contested by outsiders.

Talib Abdur-Rashid, a fifty-year-old Black resident and imam of the Mosque of Islamic Brotherhood (MIB), admitted, "Coming in contact with African immigrants has challenged our ideas of Africa and Africans." He continued to say that the presence of Muslims from West Africa has compelled African American Muslims and non-Muslims to "readjust" their sense of themselves. It also forced them to rethink their idea of Africa and their relationship to it. Maxine L. Margolis's work on New York's Little Brazil unearths a similar internal shift when disparate groups meet. In her case, however, white middle-class Brazilian immigrants in New York were forced to abate their "racist attitudes" once they encountered middle-class African Americans in the United States. "White Brazilians contrast what they perceive as the 'aggressiveness' of African-Americans," Margolis argues, "with the deferential behavior they are used to and have come to expect from African-Brazilians back home."[62] She adds, "In Brazil people from the lower echelons of society are expected to act with deference towards their purported 'social superiors.' And, because African-Brazilians are disproportionately found in the lower strata of Brazilian society, Brazilians conflate such behavior with skin color."[63] By the same token, while West African street merchants may come across fewer African American professionals, using an "underclass" slur, such as when Africans chide Black residents for having an "unwillingness" to work,[64] has less to do with a sort of "racial epiphany" that there are different kinds of Black people from various social strata. It has more to do with the type of strained relations that exist between groups and their perceptions of each other as they vie for scarce resources.

Still, it is precisely this notion of "readjustment," or racial "abatement," in the case of white Brazilian immigrants, that is most interesting because it allows us to examine the dividing lines that structure our engagement with one another. Moreover, it provides a prism through which to understand the complex nature of identity formation in America. Ahmed Shahid, a Black Muslim in his fifties and a former manager at the Harlem African Market, asserted, "We had a glorious idea of Africa in our heads, until we met 'real' Africans." He added, "That helped to change our view of things." This "re-adjustment," as Imam Talib stated, or the "change of view," mentioned by Ahmed, is an essential way Blacks and Africans contest identities, conjure new meanings, and navigate their place in Harlem society. The altercation and conversation in front of Leslie's Hair Salon speak to the informal ways these meanings are recognized and sustained. In other words, everyday interactions are places where boundaries are realized and then rehearsed. Rob was forced to articulate hidden assumptions about what divided Blacks and Africans, contentions that nonetheless played an important role in defining the Black and African presence in Harlem.

CONCLUSION

While Linda Beck perceptively writes about an intra-Muslim conflict in New York between West Africans and their Arab, South Asian and African American fellows,[65] the ethnographic context of a post-9/11 climate forces us to rethink the relevance of these internal fissures in today's world. That is to say, sectarian divisions have had a long and arduous history in Islam, as rifts emerged shortly after the religion's founding in the seventh century. In the early period, the labels obviously were not "Wahhabist" or "moderate," but they reflected similar political and theological differences with terms like "Kharijite" or "Mu'tazilite."[66] What is particularly telling, then, is not the divides that separate African Muslims from their coreligionists. Rather, beyond shifting policies that have increased surveillance and restricted movement, the events of September 11 have brought about a new cultural order, marked by a unipolar world, and a war that has implicated Islam and Muslims in unanticipated ways. As such, West African Muslims have come under much closer scrutiny prior to migration and during their sojourn in the United States. Accordingly, many have undergone feelings of angst followed by bouts of depression, when pundits publically excoriate their religion and its prophet, essentially charging them and all Muslims with adherence to a faith that promotes wrongdoing or outright terrorism. Public condemnation is weighty enough for any immigrant, but it is especially daunting for these newcomers already fearful of criminal violence, racial profiling, or police brutality, in addition to a widespread anti-Muslim backlash.

As Beck has argued, West African Muslims have begun to close ranks with other Muslims (mostly at the behest of their Arab and South Asian counterparts).[67] More importantly, however, West Africans have also actively sought to educate the community and local authorities about concepts of tolerance and nonviolence in Islam.[68] They are additionally working with federal agencies as the first line of defense against terrorism, albeit with some caution against feelings of entrapment and community suspicion.[69] These actions illustrate their ongoing efforts to defend their religion against what they feel are gross misrepresentations perpetuated by both outsiders and factions within the faith. In short, a post-9/11 world and their reaction to it has cemented the presence of West African Muslims within these communities, dispelling the misplaced "myth of return" idea that they will migrate to America, grab the Golden Fleece and return home in a few years, unscathed.[70]

It is evident that West African Muslims will continue to create practices that will help to define who they are or will become in Black America. What is less apparent is the result this will yield. While West African Murids have already developed some sort of presence for themselves, one they transported from Senegal and reshaped in the New York context, their Africanness is heavily nuanced with Islamic and Black overtones and appear to operate on a somewhat equal footing. West African Sunni Muslims, on the other hand, arrived in Harlem with a Muslim identity that was much more universal and less particular. It will be interesting to see what forms their identities take as they become more ensconced in Black communities. Besides this, a major concern is whether or not they can afford to remain in a gentrified Harlem, especially when the high cost of upscale development and rising rents threaten their daily existence. Due to religious restrictions, they have chosen not to acquire interest-bearing loans, and this has stretched their resources to the limit. In fact, the African Sunni *masjids* of Aqsa and Salaam recently comprised a strategy to stave off mounting expenses by combining their congregations. While the Murids own the building where their Touba Masjid and House of Islam are located, African Sunnis are desperately looking to purchase land or a building somewhere in Harlem.

Despite the conflict between the Black patron and West African hair braider at Leslie's Salon, it is evident that African Muslims are marrying American Muslims along with some Black and Latino Christians. The manner in which Rob and Lenny understood the rift revealed crucial fault lines between African immigrants and longstanding Black Harlem residents. Even so, this episode speaks to a whole host of formulations like class differences, a clash of worldviews and values, and contrasting histories and relations with state power. The incident can certainly be unpacked in a number of ways, which would be far beyond the scope of this chapter. At the same time, political coalitions are being built that involve African leaders and Black American officials. As more turn into permanent residents and U.S. citizens, they will become concerned with their political rights and civic duties. The political campaign of Sadique Wai is a case in point. In 2001, Wai, a Sierra Leonean Muslim and, reportedly, the first continental African to run for political office in America, was supported by a West African Muslim (Sunni) leadership group called Association des Imams Africains de New York (the Association of African Imams of New York). During a meeting, the members agreed that a *masjid*-sponsored voter registration drive for a West African Muslim candidate was not only important but crucial for their continued survival. While Wai's city council bid for the 35th district seat in Central Brooklyn was unsuccessful, his run signals a growing political presence of West Africans in the political process and, subsequently, marks their involvement in the power dynamics of the city. Moreover, other Sunnis are already organizing a national organization designed to mobilize West African Muslims under a single umbrella. In fact, the first national conference was held on Labor Day in 2001 at a hotel in Atlanta. Subsequent conferences are being planned for Philadelphia and New York.

While the Murid Islamic Community in America (MICA) is still getting settled in their newly renovated building, they already have plans to find a larger place to accommodate their growing community. Because of their deep religious commitment and strong work ethic, which, according to Scott Malcomson, "can make Protestants look like pikers,"[71] the Murids will continue to make economic strides and, by strengthening their transnational networks, secure a better place for themselves in the American economy. Besides the diligent work of street vendors and traders, Murid intellectuals and professionals have been organized in New York since the late 1980s.[72] Like most West African associations in the diaspora, their parent organizations already existed in the country of origin. They have also been making sustained efforts to make Muridiyya more appealing to the American middle class. Their attempts could further the appeal of the order and ease their assimilation into Black communities and middle America.

Much of this activity is occurring among the first generation. We cannot be certain whether or not their Islamic practices will be transferred to their children—especially when African Muslims are marrying Americans (both Muslims and non-Muslims). Even for those who bring brides to New York from their home countries, pressures to maintain a two-parent income have undermined the necessary time needed to pass on traditions. For some, polygyny (having a wife in America and another in Senegal) is the answer. Still, this does not solve the problem. Many immigrants fail to realize that when they come to America for its economic or educational opportunities, they, too, are being transformed in the process. The extent to which this change happens may be debatable. Nonetheless, while the parents are adjusting to life abroad, their children are raised by relatives back home with old-country ways, and, accordingly, major problems often occur when they are reunited.[73] Of course, the solution to these problems is not simple. However, if we are able to understand something about the way West African Muslims participate in the processes of identity construction, assimilation, and encounter, perhaps we will come to grasp a bit more about the current juncture in which we live and the new global forces shaping its destiny.

ACKNOWLEDGMENTS

The bulk of the data for this chatper was derived from my ethnographic research in Harlem and the New York metropolitan area between 2000 and 2004. Updates were made for subsequent years until December of 2007. I thank all my respondents who opened their lives to me during this time. I am likewise grateful for support from the Social Science Research Council, the Center for Folklife and Cultural Heritage, and the National Musuem of African Art at the Smithsonian Institution, the International Center for Migration, Ethnicity and Citizenship at the New School for Social Research in New York, and the Rutgers Institute on Ethnicity, Culture and the Modern Experience for a course release grant. I also want to thank Andrew Shryock for his helpful comments on an earlier draft. Paul Stoller's close read of the piece and his suggestions were invaluable. Finally, I extend my thanks to all the anonymous reviewers for their very thoughtful remarks.

NOTES

1. The term Black, including its derivatives like Blackness, will be capitalized throughout this chapter. Like other ethnic, national, or linguistic designations that refer to a people (e.g., Italian, Nigerian, Chinese, etc.), I believe the name Black also speaks to the various experiences, cultures, worldviews, and unique speech patterns people share and the term should, likewise, be capitalized.

2. Amy Waldman, "Killing Heightens the Unease Felt by Africans in New York," *New York Times*, February 14, 1999; Shaila K. Dewan, "Police Shooting of Immigrant Happened in Modern-Day Maze," *New York Times*, May 26, 2003, Metropolitan Desk, Late Edition–Final. Standing in the vestibule of his apartment building, Amadou Diallo was killed in a hail of forty-one bullets shot by four plain-clothed New York City policemen, who were members of a special Street Crimes Unit, on February 4, 1999. Diallo was unarmed and reaching for his wallet for identification when the killing occurred. A $3 million settlement was reached in January 2004. On March 22, 2003, Ousmane Zongo was shot four times and killed by Bryan Conroy, an undercover New York City police officer, during a warehouse raid on a CD/DVD pirating operation. Among other West African art dealers at the warehouse, Zongo was declared innocent and, two years later, his family was awarded a $3 million settlement.

3. Susan Sachs, "Anger and Protest at Rite for African Killed by the Police," *New York Times*, February 13, 1999, Section A; Herb Boyd, "Homecoming for Amadou Diallo,"

Amsterdam News, February 18, 1999, Section 3:1; Marianne Garvey, "Zongo Is Buried in Africa," *New York Post*, June 10, 2003, News section.

4. Will Herberg, *Protestant, Catholic, Jew; an Essay in American Religious Sociology* (Garden City, NY: Doubleday, 1955).

5. Judith Weisenfeld, "On Jordan's Stormy Banks: Margins, Center, and Bridges in African American Religious History," in *New Directions in American Religious History*, ed. Harry S. Stout and D. G. Hart (New York: Oxford University Press, 1997), 433. For a somewhat dated but excellent treatise on the underclass debate and the correctness of the term itself, see Michael B. Katz, ed., *The "Underclass" Debate: Views from History* (Princeton, NJ: Princeton University Press, 1983).

6. Amy Waldman, "Killing Heightens the Unease"; Janet Allon, "A Little Africa Emerges Along 2 Harlem Blocks," *New York Times*, December 3, 1995, Section 13:6.

7. Western readers may be more familiar with the word *mosque* (many believe it to be of French origin) to indicate the place of worship for Muslims. The Arabic word *masjid*, however, is the proper term in Islam, and it literally means the place where one prostrates, presumably, before God or Allah.

8. Sylviane Diouf-Kamara, "Senegalese in New York: A Model Minority?" *Black Renaissance/Renaissance Noire* 1, no. 2 (1997).

9. I employ the term space or, more precisely, Muslim space to mean the social relations (e.g., Muslim gatherings or ritual performances), cultural productions (e.g., reinvention of old narratives or traditions), and physical objects (e.g., Islamic clothing, Muslim architecture, incense aroma, Islamic bumper stickers) that signify and sustain a Muslim presence or identity. For a more detailed examination, see Barbara Daly Metcalf, ed., *Making Muslim Space in North America and Europe* (Berkeley: University of California Press, 1996).

10. Min Zhou, "Segmented Assimilation: Issues, Controversies, and Recent Research on the New Second Generation," in *The Handbook of International Migration: The American Experience*, ed. Charles Hirschman, Philip Kasinitz, and Josh DeWind (New York: Russell Sage Foundation, 1999), 196–211.

11. The term religious capital, along with other variations like social capital, cultural capital, and human capital, have taken on various usages and meanings over the years and across disciplines. I use the phrase here to mean the ways in which religious practices create a group network leading to the mobilization of resources and some real or imagined form of self-actualization.

12. Paul Stoller, *Money Has No Smell: The Africanization of New York City* (Chicago: University of Chicago Press, 2002), 30–31.

13. Ibid., 34.

14. Most of this African merchant activity was situated along 125th Street between Malcolm X Boulevard (Lenox Avenue) and 7th Avenue.

15. Paul Stoller, "Spaces, Places, and Fields: The Politics of West African Trading in New York City's Informal Economy," *American Anthropologist* 98, no. 4 (1996): 778; see also Stoller, *Money Has No Smell*, 12–14.

16. For an extensive discussion on the relationship between objects and religious identity, see JoAnn D'Alisera, "I ♥ Islam: Popular Religious Commodities, Sites of Inscription, and Transnational Sierra Leonean Identity," *Journal of Material Culture* 6, no. 1 (2001): 91–110; see also her book *An Imagined Geography: Sierra Leonean Muslims in America* (Philadelphia: University of Pennsylvania Press, 2004).

17. See Jonathan P. Hicks, "Vendors Ouster and Boycott Divide Harlem," *New York Times*, October 23, 1994.

18. See Stoller, "Spaces, Places, and Fields," 780.

19. This line was excerpted from Rudyard Kipling's poem "The Ballad of East and West," which is reprinted in many volumes. For a discussion of Rudyard Kipling's work, see Kingsley Amis, *Rudyard Kipling and His World* (London: Thames and Hudson, 1975).

20. For a major proponent of this argument, see the work of this late author, Samuel Huntington, *The Clash of Civilizations and the Remaking of World Order* (New York: Touchstone, 1997).

21. See Sam Roberts, "More Africans Enter US than in Days of Slavery," *New York Times*, February 21, 2005, Section A:1,

22. Stoller, *Money Has No Smell*, 7.

23. Madhulika Khandelwal, *Becoming American, Being Indian: An Immigrant Community in New York City* (Ithaca, NY: Cornell University Press, 2002), 4.

24. John L. Jackson, Jr., *Real Black: Adventures in Racial Sincerity* (Chicago: University of Chicago Press, 2005).

25. Zain Abdullah, "West Africa," in *Encyclopedia of American Immigration*, ed. James Ciment (Armonk, NY: M. E. Sharpe, 2001), 1070–78.

26. For a general religious history of Islam and a history of Islam in West Africa, see Reza Aslan, *No god but God: The Origins, Evolution, and Future of Islam.* New York: Random House, 2005). See also the time-honored work by Peter B. Clarke, *West Africa and Islam: A Study of Religious Development from the 8th to the 20th Century* (London: E. Arnold, 1982). For a concise history of identity formation and Islamization in Black West Africa, see Zain Abdullah, "Negotiating Identities: A History of Islamization in Black West Africa," *Journal of Islamic Law and Culture* 10, no. 1 (2008): 5–18.

27. P. K. Makinwa-Adebusoye, "Emigration Dynamics in West Africa," *International Migration* 33, nos. 3–4 (1995): 459.

28. In contrast to traditional understandings of international migration that view people as uprooted from their home countries and permanently settled in their places of destination, transnational migration or transnationalism recognizes that due to globalization, which refers to the processes that produce a rapid growth of new global technologies, worldwide mass media, and an accelerated transportation system, migrants create a single field of activity that links several nations simultaneously. There are many books on different transnational formations among migrants. For one of the first in anthropology, see Nina Glick Schiller, Linda Basch, and Cristina Blanc-Szanton, *Towards a Transnational Perspective on Migration: Race, Class, Ethnicity, and Nationalism Reconsidered* (New York: New York Academy of Sciences, 1992).

29. Margaret Peil, "Ghanaians Abroad," *African Affairs* 94 (1995): 345–67.

30. Makinwa-Adebusoye, "Emigration Dynamics in West Africa," 435. The concept of "floating population" was borrowed by P. K. Makinwa-Adebusoye from W. M. Freund. See W. M. Freund, "Labour Migration to the Northern Nigerian Tin Mines, 1903–1945," *Journal of African History* 22, no. 1 (1981). Freund uses this term to describe how African laborers, seeking to earn cash income, were forced to migrate in and out of French and English colonial territories—creating a short-term population movement from poor subsistence agricultural areas to more prosperous plantations and mines. Following African independence, the term is applied to a similar labor migration that crosses national borders and stretches as far away as Europe and the United States.

31. Jacques Barou, "In the Aftermath of Colonization: Black African Immigrants in France," in *Migrants in Europe: The Role of Family, Labor and Politics*, ed. Hans C. Buechler and Judith-Maria Buechler (New York: Greenwood, 1987).

32. The percentages allocated for each preference category was quoted in Nancy Kleniewski, *Cities, Change, and Conflict: A Political Economy of Urban Life* (Belmont, CA: Wadsworth,

1997); however, she extracted the data from Gregory Defreitas, "Fear of Foreigners: Immigrants as Scapegoats for Domestic Woes," *Dollars and Sense* 9, no. 33 (1994).

33. Donna L. Perry, "Rural Ideologies and Urban Imaginings: Wolof Immigrants in New York City," *African Today* 44, no. 2 (1997): 229–59.

34. For a discussion on the politics of reception for migrants, see Aristide R. Zolberg, "The Next Waves: Migration Theory for a Changing World," *International Migration Review* 23, no. 3 (1989): 403–30.

35. U.S. Census Bureau, *Characteristics of the Foreign-Born Population in the United States* (Washington, DC: U.S. Government Printing Office, 2000); U.S. Census Bureau, *Characteristics of the Foreign-Born Population in the United States* (Washington, DC: U.S. Government Printing Office, 1990); U.S. Census Bureau, *Current Population Survey* (Washington, DC: U.S. Government Printing Office, 2000).

36. Perry, "Rural Ideologies and Urban Imaginings," 229.

37. Scott L. Malcomson, "West of Eden: The Mouride Ethic and the Spirit of Capitalism," *Transition: An International Review* 6, no. 3 (1996): 30.

38. To protect the identity and privacy of my respondents, I have replaced their actual names and that of stores with pseudonyms. For a more extensive treatment of this ethnographic material, see the forthcoming book by Zain Abdullah, *Black Mecca: The African Muslims of Harlem* (New York: Oxford University Press, in press).

39. My use of the concept is derived from the work of David Mittelberg and Mary Waters. They employ the term to explore similar tensions between American Jews and their Israeli coreligionists and, under different circumstances, African Americans and Haitian immigrants. See David Mittelberg and Mary C. Waters, "The Process of Ethnogenesis among Haitian and Israeli Immigrants in the United States," *Ethnic and Racial Studies* 15, no. 3 (1992): 412–35.

40. Diouf-Kamara, "Senegalese in New York: A Model Minority?"

41. Ibid., 4.

42. Peil, "Ghanaians Abroad," 359.

43. Evans E. Crawford and Thomas H. Troeger, *The Hum: Call and Response in African American Preaching* (Nashville: Abingdon, 1995).

44. Of course, Murids are also Sunnis or orthodox Muslims. I refer to them as "Murid" and their African Muslim counterparts as "Sunni," first, because of the way they self-identify and, second, because it provides a useful categorization. I thank Sylviane Diouf for her question regarding this. The name of this Sufi group is spelled variously: Murid, Mouride, or Mourid.

45. Donal B. Cruise O'Brien, *The Mourides of Senegal: The Political and Economic Organization of an Islamic Brotherhood* (Oxford: Clarendon, 1971), 1.

46. Cheikh Anta Babou, *Fighting the Greater Jihad: Amadu Bamba and the Founding of the Muridiyya of Senegal, 1853–1913* (Athens: Ohio University Press, 2007), 156–57.

47. For a discussion on migrants, religion, and authenticity in the public sphere, see Ruth Mandel, "A Place of Their Own: Contesting Spaces and Defining Places in Berlin's Migrant Community," in Metcalf, *Making Muslim Space*, 147–66.

48. For a discussion on the role Touba plays in the lives of some African Muslims, see Mamadou Diouf, "The Senegalese Murid Trade Diaspora and the Making of a Vernacular Cosmopolitanism," *Public Culture* 12, no. 3 (2000): 679–702.

49. Victoria Ebin, "Making Room Versus Creating Space: The Construction of Spatial Categories by Itinerant Mouride Traders," in Metcalf, *Making Muslim Space*, 92–109.

50. Ibid.

51. Shaykh Seye, "A Little We Know About Shaykh Ahmadou Bamba [*sic*]," *A Special Publication of Murid Islamic Community in America* (2001): 7–8.

52. Hildi Hendrickson, *Clothing and Difference: Embodying Colonial and Post-Colonial Identities* (Durham, NC: Duke University Press, 1996).

53. Joseph Nevadomsky and Aisien Ekhaguosa, "The Clothing of Political Identity: Costume and Scarification in the Benin Kingdom," *African Arts* (1995): 62–100.

54. Roland Barthes, *The Fashion System* (New York: Hill and Wang, 1983).

55. For a full discussion of the Shaykh Ahmadou Bamba Day Parade, see Zain Abdullah, "Sufis on Parade: The Performance of Black, African and Muslim Identities," *Journal of the American Academy of Religion* (forthcoming).

56. Robert C. Smith, *Mexican New York: Transnational Lives of New Immigrants* (Berkeley: University of California Press, 2006), 167.

57. John L. Jackson, Jr., *Harlemworld: Doing Race and Class in Contemporary Black America* (Chicago: University of Chicago Press, 2001), 43.

58. Ibid.

59. Ann Miles, *From Cuenca to Queens: An Anthropological Story of Transnational Migration* (Austin: University of Texas Press, 2004), 32–33.

60. Stoller, *Money Has No Smell*, 153.

61. For a similar discussion on these tensions and how Islam is used as a marker of superiority, see Stoller, *Money Has No Smell*, 153, 165–66.

62. Maxine L. Margolis, *Little Brazil: An Ethnography of Brazilian Immigrants in New York City* (Princeton, NJ: Princeton University Press, 1994), 234.

63. Ibid., 234–35.

64. Stoller, *Money Has No Smell*, 153.

65. Linda Beck, "West African Muslims in America: When Are Muslims Not Muslims?" in *African Immigrant Religions in America*, ed. Jacob K. Olupona and Regina Gemignani (New York: New York University Press, 2007).

66. These sorts of divisions are so endemic to Islam (and all religions and spiritual bodies have similar divisions) that there is a prophetic tradition that predicted Muslims will be divided into seventy-three sects. For a brief and cogent discussion, see Reza Aslan, *No god but God*.

67. Beck, "West African Muslims in America."

68. Bradley Hope, "To Gain Immigrants' Trust, Police Reach Out to African Imams, Revive Dormant Unit," *New York Sun*, January 12, 2007, 3.

69. Michael Moss, with Jenny Nordberg, "A Nation at War: Muslims; Imams Urged to be Alert for Suspicious Visitors," *New York Times*, April 6, 2003, Metropolitan Desk, Late Edition–Final.

70. Besides the little known fact that countless Muslims from West Africa worked at the World Trade Center and died in the attacks, including the cousin of Masjid Aqsa's Ivorian imam, Souleimane Konate, many believe that because West African Harlem business owners were Muslim, customers ostracized them the day following 9/11. See Vinette K. Pryce, "In Harlem, Christians, Muslims, Arabs, African-Americans, Capitalists Confront New American Order," *New York Amsterdam News* 92, no. 38 (September 2001): 4, 2.

71. Malcomson, "West of Eden," 41.

72. Diouf, "The Senegalese Murid Trade Diaspora."

73. For an examination of the problems West Indian immigrant parents face when reunited with their children, see Mary C. Waters, *Black Identities: West Indian Immigrant Dreams and American Realities* (Cambridge: Harvard University Press, 1999).

CHAPTER 15

THE BLACKSTONE LEGACY

ISLAM AND THE RISE OF GHETTO COSMOPOLITANISM

RAMI NASHASHIBI

THIS CHAPTER EXPLORES THE ENCOUNTER OF ISLAM WITH ONE OF AMERICA'S OLDEST black street gangs. The story of the Black P. Stone Nation (also known as the *Blackstones* or the *El Rukns*) is rife with controversy and contradiction. The *New York Times* once dubbed them "the nation's deadliest gang," while the Chicago Tribune described the organization's leader, Jeff Fort, as "the most feared man in Chicago."[1] Yet the Blackstones are also the only American street gang ever to be invited to the White House for a presidential inauguration.[2] Black Entertainment Television (BET) helped fuel the folkloric fascination with this group when it recently broadcast an episode of its *American Gangster* series that focused entirely on the long and complicated history of the Blackstones.[3] This chapter tries to go beyond some of the more sensational and criminological accounts to examine the broader cultural and social aspects of the Blackstones' encounter with Islam in post-1960s ghetto space.

I became interested in the Blackstones first as a community organizer and then as a sociology graduate student at the University of Chicago. While working in different neighborhoods on Chicago's South Side, I became familiar with many individuals who had been associated with the Blackstones. I was curious about the role of Islam in the Blackstones story. Over a ten-year period I conducted multiple interviews, took copious field notes, and spent many hours with people connected to this group's history. I also became increasingly interested in how individuals in this marginalized and highly stigmatized community tried to implement the tenets of Islam on Chicago's gritty postindustrial ghetto streets. Declarations of an Islamic identity by the Blackstones were often dismissed by criminologists and researchers alike as a mere cover-up for the gang's criminal activities.[4] But the Blackstones did make literal, symbolic, and transnational connections to Muslim culture and identity, and, in so doing, engaged in what I call *ghetto cosmopolitanism*, moments of extraordinary and transnational connections made among the most isolated and marginal sectors of our urban periphery.

What is *ghetto cosmopolitanism* and why do I suggest it is important? I begin with a critique of the concept of "cosmopolitanism." In explaining the transatlantic reverberations of the Haitian uprising of 1791 through 1803, literary scholar Ifeoma Nwankwo notes that traditional nineteenth-century understanding of cosmopolitanism were exceedingly elitist: "Cosmopolitanism is reserved for those at the top, and everyone else is viewed as comfortably provincial." Nwankwo sees "black cosmopolitanism" as an alternative terrain for blacks seeking a political subjectivity outside the constraining identities available for most African Americans during this era. "Black cosmopolitanism traces the dialects of a cosmopolitanism from below. It is one that came of age at the same time forces of hegemonic cosmopolitanism in the Atlantic world (cosmopolitanism from above) were forced to reconfigure themselves to deal with the new threats posed by the uprisings in Haiti."[5] Defining cosmopolitanism as "the definition of oneself through the world beyond one's own origins," she concludes that black cosmopolitanism, "does not simply complicate, but also often undercuts traditional understanding of cosmopolitanism."

Following Nwankwo, I view ghetto cosmopolitanism as a form of a "cosmopolitanism from below." The concept is intended to challenge the overwhelmingly constraining parameters of racial and national identities imposed on contemporary ghetto residents by underlining the process of identity construction that gets configured outside the criminalizing conditions of the modern ghetto. Starting in the late 1960s, a combination of forces including deindustrialization, white flight, and the exodus of the black middle class led to the growth of a distinct and devastated urban landscape. Scholars have used various terms to identify and theorize this space: "the global ghetto," "the excluded ghetto," and "the hyperghetto."[6] Depictions of the American ghetto in the media and popular culture have invariably centered on dilapidated housing projects, pervasive violence, and the war on drugs. It is in this isolated and hollowed-out ghetto that the unique intersection of Islam with a legendary black street gang can illustrate what an alternative cosmopolitanism may look like. In this chapter, I provide an overview of the Blackstones' encounter with Islam using primary and secondary sources, including various ethnographic notes, which are in quotations. I examine the Blackstones' history as part of a larger account of how young and criminalized black men developed a *cosmopolitan* narrative, while constrained by the racial and economical constraining borders of the contemporary ghetto.[7]

THE BLACK STONE NATION TO EL RUKNS

It is one of those long and humid Chicago summer days. A group of approximately 250, mostly men with some women and children, gather in what feels like a typical outdoor barbeque in a park on Chicago's South Side. This has become an annual tradition for this multigenerational group of African Americans who had at one point or another had been connected to the almost fifty-year history of the Blackstones. Men in red turbans, bushy gray beards, sunglasses, and traditional Islamic prayer beads dangling from their necks mingle alongside those dressed in the more typical summer urban gear: oversized white T-shirts, do-rags, and baggy blue jeans. Amidst the modest consumption of alcohol and the occasional whiff of marijuana, there is a general deference to Muslim cultural sensibility, particular in the more vigilantly observed Islamic prohibition against the consumption of pork, a point I jokingly allude to as I walk by some of the men flipping the burgers. "Any swine on that grill *akh*?" I ask. "Noo sur! Not on this grill. Never!" a middle aged and balding man emphatically states while nodding his head and smiling. As time for the late afternoon Muslim prayer comes around, one of the men with turbans directs himself toward Mecca and performs the Adhan, call to prayer, in a rich and deeply melodic voice. A space clears and a much smaller group of the men and women gather to prostrate in prayer. I ask one of the younger men more about his life and to reflect on what an outsider may perceive as a contradiction between a gang experience and Islam. He thinks and responds by suggesting that, "Blackstone was his bridge to Islam, and Islam was his ladder to Allah."

Such a gathering speaks in part to the long and complex story of the Blackstones.[8] Under the leadership of Jeff Fort and Eugene Hairston, the Blackstones began in Chicago almost fifty years ago as the Blackstone Rangers and evolved into the late 1960s into the Black P. Stone Nation. The earlier references to *Blackstone* came from the street bearing that name, upon which Fort and Hairston founded the Blackstone Rangers. In the mid-1970s, Fort, then referred to as Chief Malik, became associated with Moorish Americans while in prison. During this era, members of the Blackstone community exemplified more traditional aspects of cosmopolitanism even while subverting many of its key propositions. In the early 1970s, for instance, while still influenced by the Moorish Science teachings, the El Rukns started carrying alternative passports connecting them to a glorious African past. This provided a foundation for alternative notions of citizenship that rejected the degrading racial subjectivities of black urban life. They also immersed themselves in the study of Arabic and Islamic theology as they began to identify with Muslims across the world. Significant numbers of them have traveled to parts of that world for spiritual pilgrimages, marriage, or study. This ongoing engagement with various aspects of Islam would produce a legacy steeped in a tradition beyond the world of their own origins.

Upon release, Fort introduced the first wave of Muslim teachings to the organization. As part of this change, Fort renamed the Black P. Stone Nation into the El Rukns, an Arabic word referring to the Pillars of Islam. Yet the shift to the El Rukns required a number of changes, which many of the Black P. Stone Nation members, including Chief Bull, were not willing to make, and, for some time, both the Black P. Stone Nation and the El Rukns existed simultaneously. Though members of both entities claimed partial or total loyalty to Chief Malik, other gangs as well as the police still considered both groups to be a single organization under his leadership. Yet many who were part of this history remember these events as marking a distinct moment in the evolution of the El Rukns.

The El Rukn's transformation was most evident in the structure, language, and philosophy of the organization. New Arabic concepts were introduced such as *majleek*, in reference to the Islamic governmental term *majlis;* likewise, the terms *amirs and officer muftis* became part of a larger El Rukn lexicon. Strict protocols and codes of conduct were also gradually introduced into the Blackstones. Codes of behavior prohibited cursing and mandated rules about respecting elders, which were enforced by the El Rukn generals through a range of sanctions. Contrary to popular views about the violent imposition of these rules on El Rukn members, one of the worst sanctions was actually a formalized but often temporary exile from the El Rukn community.

During this era, the term "Black Stone" stopped referring to a street on Chicago's South Side and came to stand for the black-colored stone that lay within the Kabbah of Mecca, the center of the Islamic world. This convenient change in reference had great ideological implications for the El Rukns as they grew to identify with a global culture and religious tradition. Such a shift is memorialized in a popular part of the Blackstone literature referred to as the Blackstone Creed:

> Out of the darkness into the light, Blackstone gives us courage, Blackstone gives us sight. Blackstone gives us something that no man should be deprived of, a happiness called "Stone Love." ALL means everything, MIGHTY means great, and that's who we are, Almighty Blackstones. For many are called, but a few are chosen. Many seek, but only a few knows that 9,000 miles across the waters due east in the center of the earth lies a Blackstone called the Kabah, and we are the chosen few to represent it. Who so ever falls upon a stone shall be broken, and whosoever a stone falls upon shall be grounded to dust.

As the El Rukns adopted an overtly Islamic character in the mid-seventies and -eighties, they not only memorized the Blackstone Creed but a longer set of lessons from the Moorish Koran. They also sought to convey an explicitly Islamic affect through dress and other symbols. Turbans, fezzes, and medallions with pyramids on them became part of their Islamic hipster apparel that clearly set them apart from others in the community.

After Fort's release from prison in 1975, he, along with high-ranking members in his organization, formed the El Pyramid Maintenance and Management Corporation. In late 1978, El Pyramid purchased a building at 3947 South Drexel and listed it as the El Rukn Grand Major Temple of America, though it quickly became know as "the Fort." This building became the cultural and spiritual center of all those associated with the Blackstone Nation and would go on to occupy a prominent place in street lore as the ultimate gang headquarters. The Fort explicitly showcased the El Rukn's Islamically informed aesthetics.

"THE HEART OF THE NATION"

The building originally opened on March 3, 1916, as the Oakland Square Theater located at 3947 South Drexel Avenue. Located in the heart of Bronzeville, also known as Chicago's Black Metropolis, it was primarily used as a theater until the 1960s, when it became a black cultural center. Residents still recall how in the 1960s, the theater became a site where representatives of the Black Power movement were invited to speak. Leaders like Dick Gregory, Oscar Brown, Jr., H. Rap Brown, and Stokley Carmichael were all reported to have spoken there during the 1960s and early 1970s.

After the purchase, the El Rukns quickly went about revamping the Fort. A large mural of Mecca was painted outside the building along with pictures of palm trees and pyramids. The rooms were transformed to reflect the group's often contradictory uses of Islamic aesthetics. For instance, while one large room was used for regular Friday and Sunday prayer services, the second floor contained a fully decked disco open to the public every Friday night. Other rooms catered to specific functions of the organization and each one was named (or renamed) to reflect the shifting El Rukn understanding of Islam. In the "Kabah Room," the El Rukns would hold large weekly meetings with many ceremonial procedures, such as a ritual that entailed an El Rukn stylishly sauntering around a circle of his "brothers" while reciting a series of memorized lessons from the Moorish Koran. This process was referred to as "demonstrating" and was the way El Rukns achieved a sense of social and structural status within the organization's hierarchy. In this room, there was also a throne-like structure where Jeff Fort would sit; the throne faced east toward Mecca and was decorated with Islamic adornments. In other rooms members were expected to memorize the "Six El Rukn Requirements," a set of principles that drew from the Moorish Koran and lessons composed by Jeff Fort himself.

The sign hanging in front of the Fort initially declared the building to be the "Moorish Temple of North America" but when the El Rukns incorporated more Arabic, the building was renamed Masjid Al-Malik. This sign was displayed prominently and also stated in bold "Say No to Drugs" and beneath that, "Come All Muslims to *Salah* (prayer)." The calls to prayer at the Fort bespoke the group's shift away from the Moorish Science Temple and to a more orthodox understanding of Sunni Islam. Many suggested that this change took place in 1982, when Fort came upon the Holy Qur'an in prison. *Jummah* prayer would henceforth be held at noon, Ramadan became mandatory and the study of Arabic was taken seriously. The El Rukns also met once a week to study the Qur'an and for Jummah prayers. Many El Rukns, particularly women, also attended what was termed the *Maghreb* meeting on Sunday at sundown. El Rukns learned how to make the five daily prayers, including the communal prayer during the appointed *Jummah* and *Maghreb* times. These gatherings served as a powerful affirmation of the El Rukn culture and provided a great incentive to memorize "the literature." While the literature was memorized and recited dogmatically at highly ritualized meetings, the meaning and application of various laws and rules were constituted through a much more dynamic and interactive process. I recorded an instance of this process through a former El Rukn's comic description of intricate debates over the use of drugs and Islamic protocol:

> I remember how we used to have these long three-hour arguments about when we could wear our fezes. Nobody thought we couldn't wear them when we was smoking weed. There wasn't much of a debate about that, since we all agreed weed was divine. The tricky debates came over when and how we could use cocaine. If we was using cocaine we knew that we shouldn't be wearing our Fezzes. But the real difficult debates came when some dude who thought he was being really philosophical threw out a question about, how about when we smoke weed with cigarettes lined with coke. Now that one jammed all us up for hours.

The Fort became the site where El Rukns would spend countless hours in dialogue negotiating their new spiritual and cultural identity. It was an identity that increasingly reflected an interest in Islam as a means of identifying with a transnational Muslim community.

The El Rukns global outlook and outreach is epitomized by the group's headline-generating contact with the Libyan government in 1986. The contact with Libya was the ideological and cultural culmination of the El Rukns ten-year shift to an internationalist worldview. One former El Rukn recalls the time as a moment of great excitement within the organization:

> We saw Qadaffi giving all this money to Farrakhan and the Nation of Islam thinking that these jokers were Muslim. So Chief sent General Qadaffi this tape of all of us giving him salaams and demonstrating some of our knowledge and when they got the tape they must have been like these guys are the real Muslims, these guys ain't no joke. After that Chief sent three Generals over there to visit with them and everything. When these guys came back they was so blown away by what they saw. They told us they seen each and everyone us over there. I mean for everyone of us here they was someone like that over there that looked and acted like us and they was all Muslim. Man that's when we knew we was part of something larger than we ever imagined. It was deep.

This moment, which, in many ways, captures the peak of the Blackstones' global Islamic consciousness, also precipitated the group's demise. A federal crackdown ensued charging the El Rukns with, among other things, conspiring with the Libyan government to commit acts of terror on American soil and resulted in the dismemberment of the El Rukn infrastructure and yet another indictment against Jeff Fort. This period of growth and spiritual development for the El Rukns came to a symbolic end on a day in the summer of 1989 when city officials had the building razed. On that day, Chicago Mayor Richard Daley addressed the media and a host of local residents. "Today we are here to rid the community of the El Rukn blight once and for all. From this building, they ran their crime syndicate, terrorized the surrounding communities and sold narcotics. People were afraid to walk the streets and afraid to enjoy their community." Police superintendent LeRoy

Martin then stepped to the podium. "This is a symbol of pain and destruction," he said, gesturing toward the doomed building. "It is my vision that with the removal of this building something can be put here to represent life for those young people who have to live in this neighborhood."[9] Twenty years later, where the Fort once stood, a visitor can now see a chain of elegant multimillion dollar single-family homes in what has become one of the most gentrified parts of the historic Bronzeville.

Mo Town and the Reign of Young Prince Keetah

In the wake of the Fort's demolition and the El Rukns's effective dissolution, those remaining loyal to Jeff Fort, whether in prison or in the community, began referring to themselves as *abdullahs* (servants of God), attempting to signify a more refined understanding of Islam. A younger generation of Black P. Stone Nation members claiming Chief Malik as their leader reemerged in the early nineties, scattered throughout Chicago. These Blackstones embraced the amalgam of lessons, symbols, and culture developed throughout the different periods of Blackstone history. In the 1990s, a Blackstone stronghold, known as Mo Town, emerged on Chicago's South Side that indicated the development of a hybrid Blackstone identity.[10]

"Mo Town or No Town" reads the scrawled graffiti adorning one of the many dilapidated buildings in the heart of Chicago's South Side. Sporting different combinations of red and black, with flipped baseball caps cocked unabashedly to the far left, a group of young Black men congregate outside a Palestinian owned liquor and grocery store. Outside, along the wall an arched BPSN (Black P. Stone Nation) is clumsily spray painted on a large five-pointed star and crescent. Inside, overlooking three rows of outrageously priced groceries and other household items, shelves of liquor dwarf the cash register. A verse of the Holy Quran adorns a plastic transparent casing holding cigarettes, condoms, lighters, Phillie blunts and other miscellaneous items. A young Palestinian man stocking the refrigerator with sagging jeans, cap busted to the left and cigarette drooping from the side of his mouth, pauses to engage in an elaborate, sophisticated and clearly well-rehearsed handshake with a Blackstone. The Stones smile as the young and recent immigrant tries his best to shade an unmistakably Arab accent with a Black vernacular. Somewhere between gang nomenclature and a hardcore fellahin (peasant) Palestinian Arabic, the youth manages to rattle off something eliciting great amusement from a group of Stones and second generation Palestinian youth conversing in the corner. Stepping up to the register I offer the Muslim greetings of peace, to which the reply, wa'alaikum salaam (and may peace also be with you) echoes throughout the store.

While Mo Town grew to prominence after the destruction of the Fort, the history of this stronghold should be understood within the larger racial context of thee white working class neighborhoods of Chicago's South Side. Among the small number of black families moving into these neighborhoods during the late

1960s and early 1970s, young black men often joined groups like the Blackstones for protection from violence of white gangs like the Gaylords and the Popes. One resident known as Slick Cat recalled the moment when the El Rukns first stood decisively against the Gay Lords, establishing a balance of power in the gang territory:

> Another time, me and my brother-in-law went over to the park to celebrate his birthday, shit when we were in the car and went over the bridge when I turned over. There was shit 200 maybe 300 of those son of a guns coming across that bridge. They were like, "Go home niggers, go home niggers!" They had bats, rakes and all kinds of stuff. They was trying to kill us. Man it was something else on Bishop, Laflin and Justine. When the El Rukns came over, and we went over to the park, talking about we can't come over there shit. When the Els came we all went over to that Park tat tat tat tat tat (simulating machine-gun fire), you should've seen them motherfuckers running like hell and the next 4th of July, we all was well. Children and kids could run around and play and everything, no more problems, they started moving out. They moved out to Beverly Hills, you know what I mean, I mean when love and unity is together, man that's something you just can't stop! Now we just ain't going just let you hurt us now . . . We ran the first time but we ain't running no more.

For residents like Slick Cat, it was this convergence of "love and unity," two fundamental teachings of the Blackstones, that he remembers fondly.

Many of the El Rukn generals in places like Mo Town were known as practicing Muslims who saw Islam as the Blackstones' official religion and who infused the youth around the community with a respect and deference for the Muslim faith. One such youth was Prince Keetah, Fort's son, who was mentored and raised by older El Rukns, and who grew into a fearsome leader who imposed a remarkable level of discipline and order in Mo Town, helping make it one of the largest and most intimidating stretches of gang territory in Chicago's South Side. Keetah's reputation is still part of Mo Town's lore. Young children swear oaths on his name, and many of the Blackstones who were considered Keetah's right-hand men went so far as to tattoo "YCK" (Young Chief Keetah) on their foreheads and fingertips. Keetah's appreciation for, and emphasis on, Muslim principles and teachings drew on his father's teachings and what he had learned from older El Rukns. Keetah was a figure of contradictions: he underlined the Blackstone commitment to learning and respect for others as aspects of Islam, while peddling drugs and practicing violence to consolidate Blackstone power in Mo Town. Even after his arrest, from behind prison walls, Keetah would repeatedly urge the Blackstones to continue consulting the Qur'an and following the teachings of Islam.

Today, nostalgia for an earlier era among the Blackstones across Chicago has curiously made certain parts of the city more receptive and accommodating of Muslim immigrants than most other communities. Some Blackstones often spoke of Mo Town as if it were an indigenous part of the Muslim world.

"THE PALESTINIAN MOES"

During Keetah's era, a generation of Palestinian children had grown up overseeing their parents' liquor stores or other businesses in the African American community and interacted regularly with the Blackstones. Keetah worked hard at building ties with second generation Palestinian Blackstones while he was running Mo Town. At one point in the mid 1990s, some of these youth created their own gang known as the Arabian Posse—also known as the TAP Boys. The TAP Boys' relationship with a well-entrenched entity like the Blackstones became critical in lending these Palestinian American youth legitimacy in the eyes of opposing gang members. These youth often lived in surrounding inner-city neighborhoods and spent much time cosorting with the Blackstones or other street gangs. Even when their families moved out of the city into the suburbs they came back to places like Mo Town to "represent." Palestinian youth in Mo Town often suggested that they affiliated with the Blackstones because of their mutual respect for Islam as opposed to other gangs who they perceived as "anti-Muslim." The youth were given a sense of importance as they were sought by the Blackstones at meetings to help them learn the Islamic prayers or pronounce different types of greetings. Other Palestinians, especially those who had spent time incarcerated, were also elevated as teachers of the Qur'an, even though, in many cases, second-generation Palestinians growing up on Chicago's South Side knew little about Islam, sometimes knowing less than the average Blackstone. For example, one Palestinian youth told me of his first experience in Cook Country Jail, in which the Blackstones offered him protection after they found out he was Muslim: "It was incredible how nice they treated me, hooked me up with cigarettes, slippers and food. Before we all ate, they asked me to read Fatiha and then we ate together, just like one Muslim family. They would always come to me and ask me how to read this and that from the Quran, I was embarrassed to tell them I didn't know anything about my own religion, what an embarrassment. They're more Muslim then any of these perpetuating fake wanna be Arab out here."

Although this individual's knowledge of Islam was severely limited, the Blackstones valued his ability to help them in Arabic pronunciation of qur'anic verses. One night, the Palestinian youth was accosted and beaten up by a couple of Gangster Disciples. The following morning the Blackstones demanded that the Disciple leadership punish the culprits of the action or threatened that a serious conflict would ensue between the two gangs. The Disciples respected the order and—in the presence of the young Palestinian—carried out the "violation," as punishments carried out by gang members on a fellow member are called. Due to incidents such as these, many Palestinian youth who grew up in the inner city during this era hold places like Mo Town in special regard as one of the historical bases of the Blackstones. (Incidentally, identification with the Blackstones even traveled to Palestine, as members of Chicago's Palestinian American community voyaged back and forth between Ramallah and Chicago in the 1990s. Upon returning to villages and cities of Palestine, these Chicago-raised youth would once again find that familiarity and proximity to the experience with entities like the Blackstones resonated among a younger generation of Palestinians growing up in the West Bank and Gaza.)

Forging an urban masculinity while adhering to an austere Muslim lifestyle is challenging for young black and Palestinian men who situate themselves within the Blackstone legacy. Individuals I interviewed have repeatedly talked about this tension as arising from the contradictory struggles of immigrant Muslims to sustain their spiritual identity, while succeeding economically in the West. The abundant number of liquor stores run by Arab Muslims in cities like Chicago reflects this tension. Many associated with the Blackstone legacy talk about their contradiction of selling drugs and asserting a Muslim identity with a direct reference to their "Arab brothers" selling liquor and other *haram* (prohibited) items. On several occasions, I have been in such spaces and observed debates take place among Blackstones hanging out in the store with an Arab Muslim liquor storeowner over the religious implications of selling pork or alcohol for both of them as Muslims. These conversations show the Blackstones' ability to identify with the presence of middlemen minorities, in this case Arab Muslims, in a way that most other residents do not; the Blackstones, in fact, often themselves acting as middlemen between the Arab grocers and the larger community. This identification is a subtle indicator of a ghetto cosmopolitanism, which emerges from an interpretation of Islam adapted to local urban dynamics.

CONCLUSION

I have presented the idea of ghetto cosmopolitanism as a way to challenge the more exclusive notions of cosmopolitanism associated with the contemporary global era. Not unlike the "cosmopolitanism from below" that served to undercut the racist and elitist notions of nineteenth-century cosmopolitanism and its implications for black subjectivity, the Blackstone legacy's long and complicated relationship with Islam is illustrative of a ghetto cosmopolitanism that undercuts reigning notions of cosmopolitanism. The encounter of Islam with the Blackstone legacy is an instance of cosmopolitanism that takes place in the context of globalization and which challenges us to reconsider the role that Islam has played among the most marginalized and stigmatized actors in this space. While the Blackstones were not the only street organization in Chicago with a relationship to Islam, and many individuals associated with this group did not, and do not, identify themselves as observant Muslims, this group represents a broader urban engagement with Islamic belief and culture.[11] I find that one of the Blackstones' most powerful legacies is a repository of terms and practices that render one of the most stigmatized segments of urban society the most open to accepting and understanding Islamic practices and embracing Muslim immigrants. In the course of my interviews, I came across a significant number of men associated with the Blackstone legacy who have traveled to parts of the Muslim world for the hajj, religious study, or even marriage. Others became intimately familiar with the etiquette, religious symbolism, customs, and transnational networks of first and second-generation South East Asians and Arab Muslim immigrants. Such discursive knowledge gets infused into the everyday ghetto interactions. It is this knowledge and corresponding set of practices that illustrate an alternative cosmopolitanism working in the heart of a marginal ghetto space. Moreover, the Qur'an and the Arabic language become associated with an alternative body of knowledge that is often seen as an important extension of El Rukn and Blackstone identity. Today, many individuals connected to the Blackstone legacy see themselves as part of a larger Muslim community. In many cases, they may be the only segment of similar populations who have managed to connect with the international arena and to think of themselves outside the parameters of one block or any specific Blackstone neighborhood.

The confining, criminalizing, and racializing parameters of the postindustrial ghetto are typically left out of analyses of cosmopolitism. Moreover, the role of Islam in the development of such cosmopolitan movements in these spaces is rarely acknowledged. The Blackstone legacy's complex intersection with Islam is thus particularly instructive. By introducing the concept of "ghetto cosmopolitanism," I have tried to highlight both the economic and political realities of the black urban experience in America as well as the cultural and religious innovation that gangs such as the Blackstones were able to introduce under such conditions.

NOTES

1. "Five Draw Long Sentences for Terrorism Scheme," *New York Times*, December 31, 1987, and the "The Fall of the EL Rukns," Editorial, *Chicago Tribune*, September 16, 1991, 16.

2. This took place in 1969 with President Richard Nixon. Most speculate that the invitation was extended in gratitude for an unusual but critical role that the Blackstones played in supporting the Republicans in 1968 elections among Chicago's black community.

3. The American Gangster series focusing on the Blackstones aired on BET in November 2007. It can be seen at http://www.bet.com/OnTV/BETShows/americangangster/americangangster_gangsterguide_jefffort.htm.

4. See, for instance, criminologist George Knox's well-regarded introductory text on gangs and their usage of religious symbolism: "Fort used religion as a springboard for his first major foray in gang organization. The religious identity . . . Moorish Science Temple of America and Islamic influences was just a front." George W. Knox, *An Introduction to Gangs* (Bristol, IN: Wydom Hall, 1994).

5. Ifeoma Kiddoe Nwankwo. *Black Cosmopolitanism: Racial Consciousness and Transnational Identity in the Nineteenth-Century Americas* (Philadelphia: University of Pennsylvania Press, 2005).

6. The scholars most associated with, "the global ghetto," "the excluded ghetto," and the "hyperghetto" are Carl H. Nightingale, "A Tale of Three Global Ghettos: How Arnold Hirsch Helps Us Internationalize U.S. Urban History," *Journal of Urban History* 29, no. 3 (2003): 257–71; Peter Marcuse "The Ghetto of Exclusion and the Fortified Enclave," *American Behavioral Scientist* 41, no. 3 (1997): 311–26; and Loic Wacquant, *Urban Outcasts: A Comparative Sociology of Advanced Marginality* (Polity, 2008).

7. In piecing together the brief historical narrative of the Blackstone encounter with Islam, I have relied almost exclusively on my interviews and field notes over the years. While there are a few books, like R. T. Sales's *The Blackstone Rangers* (Random Books, 1971), which

provide a look at the early history and formation of the group, there are few academic sources that examine the question of the Blackstones' encounter with Islam.

8. One potential point of confusion in referencing the larger Blackstone legacy is the many names associated with the group: the Blackstone Rangers, Black P. Stone Nation, and El Rukns all refer to different stages in the Blackstones' ideological and cultural development. For simplification purposes, I use the term "Blackstones" in referring to all these stages.

9. Chinta Strausberg, "El Rukn's headquarters smashed," *Chicago Defender*, June 7 1990.

10. Among themselves and others, Blackstones started to use the term "Mo" in the early 1990s, hearkening back to the times when the El Rukns called themselves Moorish Americans. While the genealogy of the term is relatively unknown among many who use it, the ubiquitous greeting of "what's up Mo" became a trademark of the Blackstone parlance. Over the years, Kanye West, Commom, Lupe Fiasco, and other Chicago rappers have lent additional caché to the term by incorporating it into many of their tracks.

11. The Vicelords, the Four Corner Hustlers, and even the Latin Kings have all incorporated some Islamic elements into their structures and literature over the last several decades.

CHAPTER 16

JIHADIS IN THE HOOD

RACE, URBAN ISLAM, AND THE WAR ON TERROR

HISHAAM D. AIDI

IN HIS CLASSIC NOVEL *MUMBO JUMBO*, ISHMAEL REED SATIRIZES WHITE AMERICA'S age-old anxiety about the "infectiousness" of black culture with "Jes Grew," an indefinable, irresistible carrier of "soul" and "blackness" that spreads like a virus contaminating everyone in its wake from New Orleans to New York.[1] Reed suggests that the source of the Jes Grew scourge is a sacred text, which is finally located and destroyed by Abdul Sufi Hamid, "the Brother on the Street." In a turn of events reminiscent of Reed's storyline, commentators are advancing theories warning of a dangerous epidemic spreading through our inner cities today, infecting misguided, disaffected minority youth and turning them into anti-American terrorists. This time, though, the pathogen is Islam—more specifically, an insidious mix of radical Islam and black militancy.

Since the capture of John Walker Lindh, the Marin County "black nationalist"-turned-Taliban,[2] and the arrest of the would-be terrorist José Padilla, a Brooklyn-born Puerto Rican ex-gang member who encountered Islam while in prison, terrorism experts and columnists have been warning of the "Islamic threat" in the American underclass and alerting the public that the ghetto and the prison system could very well supply a fifth column to Osama bin Laden and his ilk. Writing in New York City's *Daily News*, the social critic Stanley Crouch reminded us that in 1986, the powerful Chicago street gang al-Rukn—known in the 1970s as the Blackstone Rangers—was arrested en masse for receiving $2.5 million from Libya's strongman, Muammar Qaddafi, to commit terrorist acts in the United

This chapter originally appeared in *Middle East Report*, Issue 224, Fall 2002.

States. "We have to realize there is another theater in his unprecedented war, one headquartered in our jails and prisons," Crouch cautioned.

Chuck Colson of the evangelical American Christian Mission, which ministers to inmates around the country, penned a widely circulated article for the *Wall Street Journal*, charging that "al-Qaeda training manuals specifically identify America's prisoners as candidates for conversion because they may be 'disenchanted with their country's policies' . . . As U.S. citizens, they will combine a desire for 'payback' with an ability to blend easily into American culture." Moreover, he wrote, "Saudi money has been funneled into the American Muslim Foundation, which supports prison programs," reiterating that America's "alienated, disenfranchised people are prime targets for radical Islamists who preach a religion of violence, of overcoming oppression by jihad."[3]

Since September 11, 2001, more than a few American-born black and Latino jihadis have indeed been discovered behind enemy lines. Before Padilla (Abdallah al-Muhajir), there was Aqil, the troubled Mexican American youth from San Diego found in an Afghan training camp fraternizing with one of the men accused of killing the journalist Daniel Pearl. Aqil, now in custody, is writing a memoir called *My Jihad*. In February, the *New York Times* ran a story about Hiram Torres, a Puerto Rican whose name was found in a bombed-out house in Kabul on a list of recruits to the Pakistani group Harkat al-Mujahedeen, which has ties to al-Qaeda. Torres, also known as Mohamed Salman, graduated first in his New Jersey high school class and briefly attended Yale before dropping out and heading to Pakistan in 1998. He has not been heard from since. A June edition of *U.S. News and World Report* mentions a group of African Americans, their whereabouts currently unknown, who studied at a school closely linked to the Kashmiri militia Lashkar-e Taiba. L'Houssaine Kerchtou, an Algerian government witness, claims to have seen "some black Americans" training at al-Qaeda bases in Sudan and Pakistan.

Earlier this year, the movie *Kandahar* caused an uproar in the American intelligence community because the African American actor who played a doctor was the American fugitive David Belfield. Belfield, who converted to Islam at Howard University in 1970, is wanted for the 1980 murder of the Iranian dissident Ali Akbar Tabatabai in Washington. Belfield has lived in Tehran since 1980 and goes by the name Hassan Tantai.[4] The two most notorious accused terrorists now in U.S. custody are black Europeans, the French Moroccan Zacarias Moussaoui and the English Jamaican shoe bomber, Richard Reid, who were radicalized in the same mosque in the London ghetto of Brixton. Moussaoui's ubiquitous mug shot in orange prison garb, looking like any American inner-city youth with his shaved head and goatee, has intrigued many and unnerved some. "My first thought when I saw his photograph was that I wished he looked more Arabic and less black," wrote Sheryl McCarthy in *Newsday*. "All African-Americans need is for the first guy to be tried on terrorism charges stemming from this tragedy to look like one of our own."

But assessments of an "Islamic threat" in the American ghetto are sensational and ahistorical. As campaigns are introduced to stem the "Islamic tide," there has been little probing of why alienated black and Latino youth might gravitate toward Islamism. There has been no commentary comparable to what the British race theorist Paul Gilroy wrote about Richard Reid and the group of Britons held at Guantánamo Bay: "The story of black European involvement in these geopolitical currents is disturbingly connected to the deeper history of immigration and race politics." Reid, in particular, "manifest[s] the uncomfortable truth that British multiculturalism has failed."[5]

For over a century, African American thinkers—Muslim and non-Muslim—have attempted to harness the black struggle to global Islam, while leaders in the Islamic world have tried to yoke their political causes to African American liberation. Islamism, in the U.S. context, has come to refer to differing ideologies adopted by Muslim groups to galvanize social movements for "Islamic" political ends—the Nation of Islam's "buy back" campaigns and election boycotts or Harlem's Mosque of Islamic Brotherhood lobbying for benefits and cultural and political rights from the state. Much more rarely, it has included the jihadi strain of Islamism, embraced by foreign-based or foreign-funded Islamist groups (such as al-Rukn) attempting to gain American recruits for armed struggles against "infidel" governments at home and abroad. The rise of Islam and Islamism in American inner cities can be explained as a product of immigration and racial politics, deindustrialization and state withdrawal, and the interwoven cultural forces of black nationalism, Islamism, and hip-hop that appeal strongly to disenfranchised black, Latino, Arab, and South Asian youth.

ISLAM IN THE TRANSATLANTIC

The West Indian–born Christian missionary Edward Blyden was the first Afrodiasporan scholar to advocate an alliance between global Islam and Pan-Africanism, the system of thought that is considered his intellectual legacy. After studying Arabic in Syria and living in West Africa, Blyden became convinced that Islam was better suited for people of African descent than Christianity because of what he saw as the lack of racial prejudice, the doctrine of brotherhood, and the value placed on learning in Islam. His seminal tome, *Christianity, Islam and the Negro Race* (1888), laid the groundwork for a Pan-Africanism with a strong Islamic cultural and religious undergirding.

Blyden's counterpart in the Arab world was the Sudanese Egyptian intellectual Duse Muhammad Ali. In 1911, after the First Universal Races Congress held at the University of London, Duse Mohammad launched the *African Times and Orient Review*, a journal championing national liberal struggles and abolitionism "in the four quarters of the earth" and promoting solidarity among "non-whites" around the world. Published in both English and Arabic, the journal was circulated across the Muslim world and African diaspora, running articles by intellectuals from the Middle East to the West Indies (including contributions from

Booker T. Washington). Duse would later become mentor to the American black nationalist Marcus Garvey when he worked at the *Review* in London in 1913 and would leave his indelible stamp on Garvey's Universal Negro Improvement Association, whose mission "to reclaim the fallen of the race, to administer and assist the needy" would become the social-welfare principles animating myriad urban Islamic and African American movements.[6] In 1926, Duse created the Universal Islamic Society in Detroit, which would influence, if not inspire, Noble Drew Ali's Moorish Science Temple and Fard Muhammad's Temple of Islam, both seen as precursors of the modern-day Nation of Islam (NOI).

Blyden's and Duse's ideas, which underlined universal brotherhood, human rights, and "literacy" (i.e., the study of Arabic), had a profound impact on subsequent Pan-Africanist and Islamic movements in the United States, influencing leaders such as Garvey, Elijah Muhammad, and Malcolm X. The latter two inherited an "Arabo-centric" understanding of Islam, viewing the Arabs as God's "chosen people" and Arabic as the language of intellectual jihad—ideas still central to the Nation of Islam today. The NOI's mysterious founder, Fard Muhammad, to whom Elijah Muhammad referred as "God himself," is widely believed to have been an Arab.[7] "Fard was an Arab who loved us so much so as to bring us al-Islam," Minister Louis Farrakhan has said repeatedly. For the past thirty-five years, Farrakhan's top adviser has been the Palestinian American Ali Baghdadi, though the two fell out recently when Farrakhan condemned suicide bombings.[8] In the NOI "typologist" theology, Arabs are seen as a "Sign" of a future people, a people chosen by God to receive the Qur'an but who have strayed, and so God has chosen the American Negro to correct "Arab misunderstanding" of God's ordinances and to spread Islam in the West.[9]

Malcolm X was probably the most prominent African American Muslim leader to place the Civil Rights Movement not just in a Pan-Islamic and Pan-African context but also within the global struggle for Third World independence. In addition to his historic visit to Mecca, where he would witness "Islamic universalism" and eventually renounce the NOI's race theology, Malcolm X would confer with Egyptian President Gamal Abdel Nasser and Algerian President Ahmed Ben Bella, leaders of the Arab League and Organization of African Unity, respectively, and consider taking African American problems to the floor of the United Nations General Assembly.

When Warith Deen Muhammad, who had been educated at Al-Azhar University, took over the Nation of Islam after the death of his father, Elijah, in 1975, he renounced his father's race theology and changed his organization's name to the World Community of al-Islam in the West to emphasize the internationalist ties of Muslim over the nationalistic bonds of African Americans—leading to a split with Farrakhan who then proceeded to rebuild the NOI in its old image. Arab and Islamic states would persistently woo Warith Deen Muhammad, apparently eager to gain influence over U.S. foreign policy. "But," lamented one scholar, "he has rejected any lobbying role for himself, along with an unprecedented opportunity to employ the international pressure of Arab states to improve the social conditions of black Americans."[10]

TARGETING THE DISAFFECTED

Is there any truth to the claim that Muslim states or Islamist groups specifically targeted African Americans to lobby the U.S. government or to recruit them in wars overseas? *U.S. News and World Report* notes that, just in the 1990s, between 1,000 and 2,000 Americans—of whom "a fair number are African-Americans"— volunteered to fight with Muslim armies in Bosnia, Chechnya, Lebanon, and Afghanistan. Many were recruited by radical imams in the United States. According to several reports, in the late 1970s, the Pakistani imam Sheikh Syed Gilani, now on the run for his alleged role in Daniel Pearl's murder, founded a movement called al-Fuqara (The Poor), with branches in Brooklyn and New Jersey, where he preached to a predominantly African American constituency. Using his *Soldiers of Allah* video, Gilani recruited fighters for the anti-Soviet jihad in Afghanistan. Likewise, according to the FBI, working out of his "jihad office" in Brooklyn, the blind cleric Sheikh Omar Abel Rahman raised millions of dollars for the Afghan resistance and sent two hundred volunteers to join the mujahedeen.

According to a recent study, Saudi Arabia has historically exerted the strongest influence over the American Muslim community, particularly since the rise of Organization of Petroleum Exporting Countries in 1973.[11] Through the Islamic Society of North America (ISNA), Muslim Student Associations, the Islamic Circle of North America, and the Saudi-sponsored World Muslim League, the Saudis have financed summer camps for children, institutes for training imams, speakers' series, the distribution of Islamic literature, mosque building, and proselytizing. In addition, the Saudi embassy, through its control of visas, decides who in the American Muslim community goes on the pilgrimage to Mecca. But there is absolutely no evidence suggesting a connection between this influence and terrorism against the United States, as has been alleged by several media outlets.[12]

In the early 1980s, Iran attempted to counter Saudi influence over the American Muslim community and to gain African American converts to Shiism. On November 13, 1979, Ayatollah Khomeini had ordered the release of thirteen African American hostages, stating that they were "oppressed brothers" who were also victims of American injustice. In 1982, a study commissioned by the Iranian government to appraise the potential for Shiite proselytizing in black America attacked the Nation of Islam and Sunni Muslims for their "insincerity" and argued that Saudi proselytizers were in cahoots with the CIA. The report stated, "Besides being dispirited, the African-American Muslims feel that nobody cares about them. [Everyone] only wants to use them for their own personal reasons as they languish . . . The majority of African-Americans really want pure Islam. However, until and unless someone is willing, qualified and able to effectively oppose active Saudi oil money . . . the Islamic movement in America will plod on in a state of abject ineptitude and ineffectiveness."[13] But the Iranian revolution did not have much influence over African American Muslims, with the notable exception of the aforementioned Belfield.

The majority of African Americans, and increasingly Latinos, who embrace Islam do not end up wearing military fatigues in the mountains of Central Asia.

For most, Islam provides order, meaning, and purpose to nihilistic and chaotic lives, but even if most do not gravitate toward radical Islamism, why the attraction to Islam in the first place?

EXITING THE WEST

Many blacks and Latinos in American metropolises live in poverty and feel alienated from the country's liberal political and cultural traditions. Repelled by America's permissive consumerist culture, many search for a faith and culture that provides rules and guidelines for life. Often they are drawn to strands of Christianity that endorse patriarchy, "family values," and abstinence. But many young African Americans, and increasingly Latinos, reject Christianity, which they see as the faith of a guilty and indifferent establishment. Christian America has failed them and stripped them of their "ethnic honor." Estranged from the United States and, in the case of Latinos, from their parents' homelands, many minority youth search for a sense of community and identity in a quest that has increasingly led them to the other side of the Atlantic to the Islamic world. Sunni Islam, the heterodox Nation of Islam, and quasi-Muslim movements such as the Five Percenters and Nuwaubians allow for a cultural and spiritual escape from the American social order that often entails a wholesale rejection of Western culture and civilization.

Family breakdown and family values come up often in conversations and sermons at inner-city mosques as explanations for the younger generation's disenchantment with American society and liberalism. The decline of the two-parent household, which preoccupies discussions of family values, has economic and political roots. In the 1970s and 1980s, the middle classes left for the suburbs, investors relocated, and joblessness in urban areas increased rapidly. As one analyst observed, "The labor market conditions which sustained the 'male breadwinner' family have all but vanished." Matrifocal homes arose in its place. The new urban political economy of the 1980s—state withdrawal and capital flight—led to "the creation of a new set of orientations that places less value on marriage and rejects the dominance of men as a standard for a successful husband-wife family."[14] But in the view of many inner-city Muslim leaders, family breakdown and economic dislocation result from racism, Western decadence, and immorality—they are the effect of straying from the way of God. Raheem Ocasio, imam of New York's Alianza Islamica, contends, "Latinos in the society at large, due to pressures of modern Western culture are fighting a losing battle to maintain their traditional family structure . . . Interestingly, the effects of an Islamic lifestyle seem to mitigate the harmful effects of the Western lifestyle and have helped restore and reinforce traditional family values. Latino culture is at its root patriarchal, so Islam's clearly defined roles for men as responsible leaders and providers and women as equally essential and complementary were assimilated. As a result, divorce among Latino Muslim couples is relatively rare."[15]

By embracing Islam, previously invisible, inaudible, and disaffected individuals gain a sense of identity and belonging to what they perceive as an organized, militant, and glorious civilization that the West takes very seriously. One Chicano ex-convict tried to explain the allure of Islam for Latino inmates and why Mexican Americans sympathize with Palestinians: "The old Latin American revolutionaries converted to atheism, but the new faux revolutionary Latino American prisoner can just as easily convert to Islam . . . There reside in the Latino consciousness at least three historical grudges, three conflicting selves: The Muslim Moor, the Catholic Spanish and the indigenous Indian . . . [For the Mexican inmates] the Palestinians had their homeland stolen and were oppressed in much the same way as Mexicans."[16]

"BRINGING ALLAH TO URBAN RENEWAL"

In the wretched social and economic conditions of the inner city, and in the face of government apathy, Muslim organizations operating in the ghetto and prisons deliver materially. As in much of the Islamic world, where the state fails to provide basic services and security, Muslim organizations appear, funding community centers, patrolling the streets, and organizing people.

As the state withdrew and capital fled from the city in the Reagan-Bush era, social institutions and welfare agencies disappeared, leaving an urban wasteland. Churches have long been the sole institutions in the ghetto, but Islamic institutions have been growing in African American neighborhoods for the past two decades. In Central Harlem, Brownsville, and East New York—areas deprived of job opportunities—dozens of mosques (Sunni, NOI, Five Percenter, and Nuwaubian) have arisen, standing cheek by jowl with dozens of churches that try to provide some order and guidance in these neighborhoods. In the ghettoes of Brooklyn, on Chicago's South Side, and in the barrios of East Harlem and East Los Angeles, where, aside from a heavy police presence, there is little evidence of government, Muslim groups provide basic services. The Alianza Islamica of New York, headquartered in the South Bronx, offers afterschool tutorials, equivalency diploma instruction for high school dropouts, marriage counseling, substance-abuse counseling, AIDS-awareness campaigns, and sensitivity talks on Islam for the New York Police Department. The Alianza has confronted gangs and drug posses, training young men in martial arts to help clean up the streets of the barrios with little reliance on trigger-happy policemen.

One quasi-Islamic group, the United Nation of Islam, which broke away from Farrakhan's NOI in 1993, has adopted the slogan "Bringing Allah to Urban Renewal" and is resurrecting blighted urban neighborhoods across the country, opening up health clinics, employment centers, restaurants, and grocery stores that do not sell red meat, cigarettes, or even soda because they are bad for customers' health.[17] The United Nation of Islam does not accept government funds, fearing that federal money would compromise its mission of "Civilization Development." Similarly, the NOI conducts "manhood training" and mentoring programs in inner cities across the country; earning the praise of numerous scholarly reports, which claim that young men who participate in these programs for an

extended time show "positive self-conception," improved grades, and less involvement in drugs and petty crime.[18]

In addition to delivering basic services, the NOI today tries to provide jobs and
housing. The NOI's Los Angeles branch is currently buying up homes for home-
less young men (calling them "Houses of Knowledge and Discipline"), building AIDS treatment clinics, and starting up a bank specializing in small loans.[19]
In 1997, Farrakhan announced a "three-year economic program" that aimed to
eliminate "unemployment, poor housing and all the other detriments that plague
our community."[20] Farrakhan seems to have reverted to the strategies of economic
nationalism pursued by Elijah Muhammad. One scholar argues that under Elijah,
the NOI was essentially a development organization emphasizing thrift and economic independence among poor black people, with such success that it turned
many followers into affluent entrepreneurs. The organization itself evolved into a
middle-class establishment, allowing Warith Deen Muhammad, after his father's
death, to shed black-nationalist rhetoric and identify with a multiracial *umma*
(community)—moves that resonated with his middle-class constituency.[21] In the
1970s, the NOI had owned thousands of acres of farmland, banks, housing com-
plexes, retail and wholesale businesses, and a university and was described by
C. Eric Lincoln as one of the "most potent economic forces" in black America,
but Warith Deen Muhammad liquidated many of the NOI's assets. When Farrakhan resuscitated the NOI in the 1980s, he revived Elijah Muhammad's message
of black economic empowerment (appealing to many poorer blacks) and began
rebuilding the NOI's business empire. According to *Business Week*, in 1995, the
NOI owned 2,000 acres of farmland in Georgia and Michigan, a produce-trans-
port business, a series of restaurants, and a media-distribution company.

ISLAM BEHIND BARS

Over the past thirty years, Islam has become a powerful force in the American
prison system. Ever since the Attica prison riots in upstate New York in 1971,
when Muslim inmates protected guards from being taken hostage, prison officials
have allowed Muslim inmates to practice and proselytize relatively freely. Prior
to the rise of Islam, the ideologies with the most currency among minorities in
prison were strands of revolutionary Marxism—Maoism and Guevarism—and
varieties of black nationalism. According to one report, nowadays, one third of
the million or more black men in prison are claiming affiliation with the Nation
of Islam, Sunni Islam, or some quasi Muslim group, such as the Moorish Science
Temple.[22] Mike Tyson, during a stint in prison in the mid-1990s, seems to have
combined all three currents, leaving prison as a Muslim convert, Malik Shabazz,
but with Mao and Ché Guevara tattoos. "I'm just a dark guy from the den of
iniquity," the former heavyweight champion explained to journalists.

The presence of Muslim organizations in prisons has increased in the past
decade as the state has cut back on prisoner services. In 1988, legislation made
drug offenders ineligible for Pell grants; in 1992, this was broadened to include

convicts sentenced to death or lifelong imprisonment without parole, and in 1994, the law was extended to all remaining state and federal prisoners. In 1994, Congress passed legislation barring inmates from high education, stating that criminals could not benefit from federal funds, despite overwhelming evidence that prison educational programs not only help maintain order in prison but also prevent recidivism.[23] Legislation also denies welfare payments, veterans' benefits, and food stamps to anyone in detention for more than sixty days.

In 1996, the Clinton administration passed the Work Opportunity and Personal Responsibility Act, preventing most ex-convicts from receiving Medicaid, public housing, and Section Eight vouchers. Clinton forbade inmates in 1998 from receiving Social Security benefits, saying that prisoners "collecting Social Security checks" was "fraud and abuse" perpetrated against "working families" who "play by the rules."[24] All these cutbacks affected minorities disproportionately but African Americans in particular because of the disproportionately high incarceration rates of African American men. Disparate treatment by the criminal justice system—which has a devastating effect on the black family, the inner-city economy, and black political power, since convicts and ex-convicts cannot vote in thirty-nine states—is another powerful factor fueling the resentment of minorities toward the establishment.

In this atmosphere, it is no surprise that Muslim organizations in prisons are gaining popularity. The Nation of Islam provides classes, mentorship programs, study groups, and "manhood training" that teaches inmates respect for women, responsible sexual behavior, drug prevention, and life-management skills. Mainstream American Muslim organizations also provide myriad services to prisoners. At ISNA's First Conference on Islam in American Prisons, Amir Ali of the Institute of Islamic Information and Education described the services and support system that his organization provides to Muslim inmates: regular visits to prisons by evangelists who deliver books and literature, classes in Arabic and Islamic history, correspondence courses in other subjects, twenty-four-hour toll-free phones and collect-calling services for inmates to fall families, mentorship programs for new converts, and "halfway houses" to help reintegrate Muslim inmates into society after release.

Those who study Islam behind bars cast doubt on the assertions of Colson and Crouch. At ISNA's Third Annual Conference on Islam in American Prisons in July 2002, the keynote speaker, David Schwartz, who recently retired as religious services administrator for the Federal Bureau of Prisons, strongly rejected the notion that American prisons were a breeding ground for terrorists and stated that Islam was a positive force in the lives of inmates. Robert Dannin adds, "Why would a sophisticated international terrorist organization bother with inmates— who are fingerprinted and whose data is in the U.S. criminal justice system?"[25]

ISLAM AND HIP-HOP

The street life is the only life I know
I live by the code style it's made PLO

Iranian thoughts and cover like an Arabian
Grab a nigga on the spot and put a 9 to his cranium.
                    —Method Man, "PLO Style" *Tical*. Def Jam 1994

If Rastafarianism and Bob Marley's Third Worldist reggae anthems provided the music and culture of choice for marginalized minority youth two decades ago, in the 1990s, "Islamic hip-hop" emerged as the language of disaffected youth throughout the West.

Arabic, Islamic, or quasi Islamic motifs increasingly thread the colorful fabric that is hip-hop, such that for many inner-city and suburban youth, rap videos, and lyrics provide a regular and intimate exposure to Islam. Many "Old School" fans will recall the video of Eric B and Rakim's "Know the Ledge," which featured images of Khomeini and Muslim congregational prayer, as Rakim flowed, "In control of many, like Ayatollah Khomeini . . . I'm at war a lot, like Anwar Sadat."[26] Self-proclaimed Muslim rap artists proudly announce their faith and include "Islamic" messages of social justice in their lyrics. Followers of Sunni Islam ("al-Islam" in hip-hop parlance), q-Tip (Fareed Kamal), and Mos Def are among the most highly acclaimed hip-hop artists, lauded as representatives of hip-hop's school of "Afro-humanism" and positivity. Mos Def, in an interview with the Web site Beliefnet, described his mission as a Muslim artist: "it's about speaking out against oppression wherever you can. If that's gonna be in Bosnia or Kosovo or Chechnya or places where Muslims are being persecuted; or if it's gonna be in Sierra Leone or Colombia—you know, if people's basic human rights are being abused and violated, then Islam has an interest in speaking out against it, because we're charged to be the leaders of humanity."[27]

The fluidity and variegated nature of Islam in urban America is seen in the different "Islams" represented in hip-hop and most poignantly in the friction between Sunni Muslims and Five Percenters. Today, most "Islamic" references in hip-hop are to the belief system of the Five Percent Nation, a splinter group of the NOI founded in 1964 by Clarence 13X. The Five Percent Nation (or the Nation of Gods and Earths) refashioned the teachings of the NOI, rejecting the notion that Fard was Allah and teaching instead that the black man was God and that his proper name is ALLAH (Arm Leg Leg Arm Head). They taught that 85 percent of the masses are ignorant and will never know the truth. Ten percent of the people know the truth but use it to exploit and manipulate the 85 percent; only 5 percent of humanity know the truth and understand the "true divine nature of the black man who is God or Allah."[28] In Five Percenter theology, Manhattan (particular Harlem) is known as Mecca, Brooklyn is Medina, Queens is the desert, the Bronx is Pelan, and New Jersey is the New Jerusalem. Five Percenter beliefs have exerted a great influence on hip-hop argot and street slang. The expressions "word is bond," "break it down," "peace," "wassup G" (meaning God, not gangsta), and "represent" all come from Five Percenter ideology.

Orthodox Sunni Muslims see Five Percenters as blasphemous heretics who call themselves "Gods." They accuse Five Percenters of *shirk*, the Arabic word meaning polytheism—the diametrical opposite of *tawhid* (unitary nature of God) that defined the Prophet Muhammad's revelation. Since Five Percenters often wear skullcaps and women cover their hair, Sunni Muslims will often greet them with *al-salaam alaykum* (peace be unto you), to which the Five Percenters respond "Peace, God." Five Percenters refer to Sunni Muslims as deluded and "soon to be Muslim." In the "ten percent," Five Percenters include the "white devil," as well as orthodox Muslims "who teach that Allah is a spook."

Busta Rhymes, Wu Tang Clan, and Mobb Deep are among the most visible Five Percenter rappers. Their lyrics—replete with numerology, cryptic "Islamic" allusions, and, at times, pejorative references to women and whites (as "white devils" or "cave dwellers")—have aroused great interest and controversy. The journalist and former rapper Adisa Banjoko strongly reprimands Five Percenter rappers for their materialism and ignorance: "In hip-hop a lot of us talk about knowledge and the importance of holding on to it, yet under the surface of hip-hop's 'success' runs the thread of ignorance [*jahiliyya*, the Arabic term referring to the pagan age in Arabia before Islam]."[29] Like "the original jahiliyya age," hip-hop today is plagued by "jahili territorialism and clan affiliation," a "heavy disrespect of women," and a materialism that "borders on jahili idol worship."[30] The Five Percenter Ibn Dajjal responded angrily to Adisa's criticism: "No amount of *fatwas* or censorship will ever silence the sounds of the NOI and Five Percent *mushrik* (idolator) nations. The group will continue to rise in fame with customers coming from all walks of life: black, white and Bedouin. Far from a masterpiece of style, the book [the Quran] is literally riddled with errors and clumsy style which yield little more than a piece of sacred music . . . Maybe there should be a new hip-hop album titled *Al-Quran al-Karim Freestyle* by Method Man and Ghostface Killa!"[31]

Though it has nothing to do with the jihadi trend, the language of Islam in the culture of hip-hop does often express anger at government indifference and U.S. foreign policy and challenge structures of domination. The outspoken rapper Paris, formerly of the NOI, who galled the establishment with his 1992 single "Bush Killer," has raised eyebrows again with the single "What Would You Do?" (included on his forthcoming LP, *Sonic Jihad*), which excoriates the "war on terror" and the USA PATRIOT Act and implies government involvement in the September 11 attacks. In early 2002, the Brooklyn-based Palestinian American Hammer Brothers, "originally from the Holy Land, living in the Belly of the Beast, trying to rise on feet of Yeast,"[32] released their pro-Intifada cut, "Free Palestine," now regularly blared at pro-Palestinian gatherings in New York. One particularly popular and articulate artist is spoken-word poet Suheir Hammad, the Palestinian-American author of *Born Black, Born Palestinian*, on growing up Arab in Sunset Park, Brooklyn. Hammad appeared on HBO's "Def Poetry Jam" some weeks after September 11 and delivered a stirring rendition—to a standing ovation—of her poem, "First Writing Since," on being an Arab New Yorker with a brother in the U.S. Navy.[33]

"No Real Stake"

Pan-Africanism and Pan-Islam were fused together by African American and Muslim intellectuals over a century ago to fight colonialism, racism, and Western domination. Today, that resistance strategy has been adopted by tens of thousands of urban youth (judging by NOI rallies in the United States and Europe) in the heart of the West. The cultural forces of Islam, black nationalism, and hip-hop have converged to create a brazenly political and oppositional counterculture that has a powerful allure. At root, the attraction for African American, Latino, Arab, south Asian, and West Indian youth to Islam, and movements that espouse different brands of political Islam, is evidence of Western states' failure to integrate minority and immigrant communities and deliver basic life necessities and social welfare benefits—policy failures of which Islamic groups (and right-wing Christian groups) are keenly aware.

Rather than prompt examination of why minority youth in the ghetto and its appendage institution, the prison, would be attracted to Islam—whether in its apolitical Sunni or Sufi, its Five Percenter, or its overtly political Nation of Islam or jihadi variety—the cases of Moussaoui, Reid, and Padilla have led to arguments about how certain cultures are "unassimilable," hysterical warnings of a "black (or Hispanic) fifth column," and aggressive campaigns to counter Islamic influence in the inner city. Evangelical groups are trying to exclude Islamic institutions from George W. Bush's faith-based development initiative. Jerry Falwell has stated that "it is totally inappropriate under any circumstances" to give federal aid to Muslim groups because "the Muslim faith teaches hate, Islam should be out the door before they knock. They should not be allowed to dip into the pork barrel."[34] Another Christian effort, Project Joseph, conducts "Muslim awareness seminars" in inner cities across the country, warning that Muslim leaders are exploiting the weakness of black churches, informing African Americans that conversion to Islam does not imply "recovering their ethnic heritage" and publicly admonishing that "if the conversion rate continues unchanged, Islam could become the dominant religion in black urban areas by the year 2020."[35]

The aspirations of the very poor and disenfranchised in America will continue to overlap with the struggles and hopes of the impoverished masses of the Muslim Third World, who will in turn continue to look toward African Americans for inspiration and help. Minister Farrakhan's recent "solidarity tour" of Iraq and recent meetings between Al Sharpton and Jesse Jackson and Yasser Arafat show that Muslim causes continue to reverberate in the African American community. By and large, African Americans do not seem to share the hostility to Islam that has intensified since September 11. Akbar Muhammad, professor of history at the State University of New York, Binghamton, and son of Elijah Muhammad, wrote in 1985 that because African Americans have "no real political stake in America, political opposition to the Muslim world is unworthy of serious consideration."[36] These words still hold true for many minorities in post-September 11 America.

NOTES

1. Ishmael Reed, *Mumbo Jumbo* (New York: Simon and Schuster, 1996).

2. Many say Lindh was corrupted by reading *The Autobiography of Malcolm X* and by his love of hip-hop. See Shelby Steele, "Radical Sheik," *Wall Street Journal*, December 18, 2001. Lindh often posed as black online, going by the names "Doodoo" and "Prof J." He attacked Zionism, once writing: "Our blackness does not make white people hate us, it is THEIR racism that causes hate. . . . [The "N" word] has, for hundreds of years, been a label put on us by Caucasians . . . and because of the weight it carries with it, I never use it myself." See Clarence Page, "The 'White Negro' Taliban?" *Chicago Tribune*, December 14, 2001.

3. Chuck Colson, "Evangelizing for Evil in Our Prisons," *Wall Street Journal*, June 24, 2002. See also Mark Almond, "Why Terrorists Love Criminals (and Vice Versa): Many a Jihadi Began as a Hood," *Wall Street Journal*, June 19, 2002; Earl Ofari Hutchinson, "Hispanic or African-American Jihad?" *Black World Today*, June 12, 2002; *Christian Science Monitor*, June 14, 2002.

4. *Guardian*, January 10, 2002.

5. Paul Gilroy, "Dividing into the Tunnel: The Politics of Race between the Old and New Worlds," *Open Democracy*, January 31, 2002.

6. Robert A Hill, *Marcus Garvey and the Universal Negro Improvement Association Papers* (Berkeley: University of California Press, 1983), 302.

7. Despite Farrakhan's claim to have renounced race theology, *The Final Call* still prints on its back page that "God appeared in the person of W. Fard Muhammad."

8. Ali Baghdadi, "Farrakhan Plans to Meet Sharon," *Media Monitors Network*, April 14, 2002, available online at http://www.mediamonitors.net (accessed April 20, 2002).

9. See Mattias Gardell, *In the Name of Elijah Muhammad: Louis Farrakhan and the Nation of Islam* (Durham, NC: Duke University Press, 1996), 62, 193.

10. Allen, "Minister Louis Farrakhan and the Continuing Evolution of the Nation of Islam," 73.

11. Dannin, *Black Pilgrimage to Islam*.

12. Gause, "Be Careful What You Wish For."

13. Muhammad Said, "Questions and Answers about Indigenous U.S. Muslims," unpublished manuscript, Tehran, 1982.

14. "In 1993, twenty-seven percent of all children under the age of eighteen were living with a single parent. This figure includes fifty-seven percent of all black children, thirty-two percent of all Hispanic children and twenty-one percent of all white children. Wilson, *When Work Disappears: The World of the New Urban Poor* (New York: Alfred A. Knopf, 1996), 85. Elsewhere, Wilson argues that the sharp increase in black male joblessness since 1970 accounts in large measure for the rise in the number of single-parent families. Since jobless rates are highest in the inner city, rates of single parenthood are also highest there.

15. Rahim Ocasio, "Latinos, the Invisible: Islam's Forgotten Multitude," *Message*, August 1997.

16. *Los Angeles Times*, June 23, 2002.

17. *Christian Science Monitor*, December 1, 1999.

18. Majors and Wiener, *Programs That Serve African-American Youth*.

19. *Los Angeles Times*, February 13, 2002.

20. *Final Call*, February 11, 1997.

21. Mamiya, "Minister Louis Farrakhan and the Final Call."

22. *Newsweek*, October 30, 1995.

23. Josh Page, "Eliminating the Enemy: A Cultural Analysis of the Exclusion of Prisoners from Higher Education" (MA thesis, University of California, Berkeley, 1997).
24. Bill Clinton, radio address, April 25, 1998, transcript available online at http://www.whitehouse.gov (accessed June 10, 2005).
25. Quoted in Hishaam Aidi, "Jihadis in the Cell Block," Africana.com, July 22, 2002.
26. Eric B. and Rakim, "Juice (Know the Ledge)," *Juice*, MCA, 1992.
27. Hishaam Aidi, "Hip-Hop for the Gods," Africana.com, April 30, 2001.
28. Nuruddin, "The Five Percenters."
29. Adissa Banjoko, "Hip-Hop and the New Age of Ignorance," *FNV Newsletter*, June 2001, available online at http://www.daveyd.com/age ofignorance.html (accessed August 22, 2006).
30. Banjoko, "Hip-Hop."
31. Ibid.
32. Hammer Brothers, "Free Palestine," *Free Palestine*, Wax Poetic Productions, 2002.
33. Suheir Hammad, "First Writing Since," appeared in *Middle East Report* 221 (Winter 2001), http://www.teachingforchange.org/News%20Items/first_writing_since.htm (accessed August 22, 2006).
34. *Washington Post*, March 8, 2001.
35. *USA Today*, July 19, 2000.
36. Muhammad, "Interaction between 'Indigenous' and 'Immigrant' Muslims in the United States."

PART IV

CONCLUSION

CHAPTER 17

REDISCOVERING MALCOLM'S LIFE

A HISTORIAN'S ADVENTURES IN LIVING HISTORY

MANNING MARABLE

TO THE MAJORITY OF OLDER WHITE AMERICANS, THE NOTED AFRICAN AMERICAN leaders Malcolm X and Dr. Martin Luther King, Jr., seem as different from each other as night and day. Mainstream culture and many history textbooks still suggest that the moderate Dr. King preached nonviolence and interracial harmony, whereas the militant Malcolm X advocated racial hatred and armed confrontation. Even Malcolm's infamous slogan, "By Any Means Necessary!" still evokes among whites disturbing images of Molotov cocktails, armed shoot-outs, and violent urban insurrection. But to the great majority of black Americans and to millions of whites under thirty, these two black figures are now largely perceived as being fully complementary with each other. Both leaders had favored the building of strong black institutions and healthy communities; both had strongly denounced black-on-black violence and drugs within the urban ghetto; both had vigorously opposed America's war in Vietnam and had embraced the global cause of human rights. In a 1989 "dialogue" between the eldest daughters of these two assassinated black heroes, Yolanda King and Attallah Shabazz, both women emphasized the fundamental common ground and great admiration the two men shared for each other. Shabazz complained that "playwrights always make Martin so passive and Malcolm so aggressive that those men wouldn't have lasted a minute in the same room." King concurred, noting that in one play "my father was this wimp who carried a Bible everywhere he went, including someone's house for dinner." King argued, "That's not the kind of minister Daddy was! All these ridiculous clichés . . . " Both agreed that the two giants were united in the pursuit of black freedom and equality.

As a child of the radical sixties, I was well ahead of the national learning curve on the King vs. Malcolm dialectic. At age seventeen, as a high school senior, I had attended Dr. King's massive funeral at Ebenezer Baptist Church in Atlanta; on April 9, 1968, I had walked behind the rugged mule-driven wagon carrying Dr. King's body, along with tens of thousands of other mourners. The chaotic events of 1968—the Vietnamese Tet offensive in February, President Johnson's surprise decision not to seek reelection, the assassinations of both Dr. King and Bobby Kennedy, the Paris student and worker uprisings that summer, the "police riot" in Chicago at the Democratic National Convention—all were contributing factors in spinning the world upside down.

By the end of that turbulent year, for the generation of African-American students at overwhelmingly white college campuses, it was Malcolm X, not Dr. King, who, overnight, became the symbol for the times we were living through. As leader of my campus black student union, I reread *The Autobiography of Malcolm X* during the winter of 1969. The full relevance and revolutionary meaning of the man suddenly became crystal clear to me. In short, the former "King Man" became almost overnight a confirmed, dedicated "X Man."

Malcolm X was the Black Power generation's greatest prophet who spoke the uncomfortable truths that no one else had the courage or integrity to broach. Especially for young black males, he personified for us *everything* we wanted to become: the embodiment of black masculinist authority and power, uncompromising bravery in the face of racial oppression, the ebony standard for what the African American liberation movement should be about. With Talmudic-like authority, we quoted him in our debates, citing chapter and verse the precise passages from the *Autobiography*, and books like *Malcolm X Speaks, By Any Means Necessary*, and other edited volumes. These collected works represented almost sacred texts of black identity to us. "Saint Malcolm X-the-Martyr" was the ecumenical ebony standard for collective "blackness." We even made feeble attempts to imitate Malcolm's speaking style. Everyone quoted him to justify their own narrow political, cultural, and even religious formulations and activities. His birthday, May 19, was widely celebrated as a national black holiday. Any criticisms, no matter how minor or mild, of Malcolm's stated beliefs or evolving political career were generally perceived as being not merely heretical, but almost treasonous, to the entire black race.

Working-class black people widely loved Brother Malcolm for what they perceived as his clear and uncomplicated style of language and his peerless ability in making every complex issue "plain." Indeed, one of Malcolm's favorite expressions from the podium was his admonition to other speakers to "Make it Plain," a phrase embodying his unshakable conviction that the black masses themselves, "from the grassroots," would ultimately become the makers of their own revolutionary black history. Here again, inside impoverished black urban neighborhoods and especially in the bowels of America's prisons and jails, Malcolm's powerful message had an evocative appeal to young black males. In actor Ossie Davis's memorable words, "Malcolm was our manhood! . . . And, in honoring him, we honor the best in ourselves. And we will know him then for what he was and is—a Prince—our own Black shining Prince!—who didn't hesitate to die, because he loved us so."

The Autobiography of Malcolm X, released into print in November 1965, sold millions of copies within several years. By the late sixties, the *Autobiography* had been adopted in hundreds of college courses across the country. Malcolm X's life story, as outlined by the *Autobiography*, became our quintessential story about the ordeal of being black in America. Nearly every African American at the time was familiar with the story's basic outline. Born in the Midwest, young Malcolm Little became an orphan: his father was brutally murdered by the Ku Klux Klan and his disturbed mother, overwhelmed by caring for seven little children, suffered a mental breakdown and had been institutionalized. Malcolm then relocated east to Roxbury and Harlem. He then became an urban outlaw, the notorious "Detroit Red," a pimp, hustler, burglar, and drug dealer. Pinched by police, "Detroit Red" was sentenced to ten years' hard labor in prison, where he then joined the Black Muslims. Once released, given the new name Malcolm X, he rapidly built the Black Muslims from an inconsequential sect to over one hundred thousand strong. But then Malcolm X grew intellectually and politically well beyond the Muslim. He decided to launch his own black nationalist group, the Organization of Afro-American Unity. He started preaching about human rights and "the ballot or the bullet." Malcolm made a pilgrimage to the holy city of Mecca, converted to orthodox Islam, and became "El-Hajj Malik El-Shabazz." He was then acclaimed by Islamic, African, and Arab leaders as a leading voice for racial justice. Then, at the pinnacle of his worldwide influence and power, Malcolm was brutally struck down by assassins' bullets at Harlem's Audubon Ballroom. This was the basic story nearly every activist in my generation knew by heart.

A number of Malcolm X's associates and others who had known him personally published articles and books in the late sixties, which firmly established the late leader as the true fountainhead of Black Power.[1] Far more influential, however, for popularizing the Malcolm legend was the Black Arts Movement. Poets were particularly fascinated with the magnetic physical figure of Malcolm as a kind of revolutionary black Adonis. In life towering at six feet, three inches tall and weighing a trim 175 to 180 pounds, broad-shouldered Malcolm X was mesmerizingly handsome, always displaying a broad, boyish smile, and always spotlessly well-groomed; in death, he would remain forever young. In photographs, he seemed both strong and sensitive.

After receding somewhat during much of the late 1970s and early 1980s, Malcolm X's cultural reputation among artists, playwrights, and musicians exploded again with the flowering of the hip-hop generation. Malcolm's cultural renaissance began with the 1983 release of Keith LeBlanc's "No Sell-Out," a twelve-inch dance single featuring a Malcolm X speech set to hip-hop beat. Old School group Afrika Bambaata and the Soul Sonic Force followed in 1986 with "Renegades of Funk," declaring that both King and Malcolm X had been bold and bad "renegades of the atomic age." On its classic 1988 hip-hop album, "It Takes a Nation of Millions to Hold Us Back," Public Enemy (PE) generously sampled from a Malcolm X speech, constructing the provocative phrase, "Too Black, Too Strong." On "Party for Your Right to Fight," Public Enemy told the hip-hop nation that "J. Edgar Hoover . . . had King and X set up." PE's massive popularity and its strong identification with Malcolm's image led other hip-hop artists to also incorporate Malcolm X into their own music. In 1989, the Stop the Violence Movement's "Self Destruction" album featured a Malcolm X lecture, and its companion video included beautiful murals of the black leader as the hip-hop background for rappers. The less commercially popular, but enormously talented artist, Paris released "Break the Grip of Shame" in 1990, which prominently featured Malcolm's ringing indictment: "We declare our right on this Earth to be a man, to be a human being, to be respected as a human being to be given the rights of a human being in this society on this Earth in this day, which we intend to bring into existence, by any means necessary!"

As "Thug Life" and "Gangsta Rap" emerged from the West Coast and soon acquired a national commercial appeal, these artists painted Malcolm X in their own cultural contexts of misogynistic and homophobic violence. Ice Cube's 1992 "Predator," for example, sampled a Malcolm address over a beat on one cut; on another, "Wicked," Ice Cube rapped, "People wanna know how come I gotta gat and I'm looking out the window like Malcolm ready to bring that noise. Kinda trigger-happy like the Ghetto Boys." Less provocatively, KRS-One's 1995 "Ah-Yeah" spoke of black reincarnation: "They tried to harm me, I used to be Malcolm X. Now I'm on the planet as the one called KRS." Perhaps the greatest individual hip-hop culture has yet produced, Tupac Shakur, fiercely identified himself with Malcolm X. On Tupac's classic 1996 "Makaveli" album, on the song "Blasphemy," he posed a provocative query: "Why you got these kids minds, thinking that they evil while the preacher being richer. You say honor God's people, should we cry when the Pope die, my request, we should cry if they cried when we buried Malcolm X. Mama tells me am I wrong, is God just another cop waiting to beat my ass if I don't go pop?"

The widespread release and commercial success of Spike Lee's 1992 biofilm "X," combined with hip-hop's celebration of Malcolm as a "homeboy," created the context for what historian Russell Rickford has termed "Malcolmology." Hundreds of thousands of African American households owned and displayed portraits of Malcolm X, either in their homes, places of business, or at black schools. By the 1990s, Malcolm X had become one of the few historical figures to emerge from the black nationalist tradition to be fully accepted and integrated into the pantheon of civil rights legends, an elite of black forefathers, who included Frederick Douglass, W. E. B. Du Bois, and Dr. Martin Luther King, Jr.

As with every mythic figure, the icon of Saint Malcolm accommodated a variety of parochial interpretations. To the bulk of the African American middle class, the Malcolm legend was generally presented in terms of his inextricable trajectory of intellectual and political maturation, culminating with his dramatic break from the Nation of Islam and embrace of interracial harmony. For much of the hip-hop nation, in sharp contrast, the most attractive characteristics of Saint Malcolm emphasized the incendiary and militant elements of his career. Many hip-hop artists made scant distinctions between Malcolm X and his former protégé and later bitter rival, Louis X (Farrakhan). Some even insisted that Malcolm X had never supported any coalitions with whites, despite his numerous public statements in 1964 and 1965 to the contrary. The hip-hop Malcolmologists seized Malcolm as the ultimate black cultural rebel, unblemished and uncomplicated by the pragmatic politics of partisan compromise, which was fully reflected in the public careers of other post-Malcolm black leaders, such as Jesse Jackson and Harold Washington. Despite their black cultural nationalist rhetoric, however, hip-hop Malcolmologists also uncritically accepted the main parameters of the black leader's tragic life story, as presented in *The Autobiography of Malcolm X*. They also glorified Malcolm's early gangster career as the notorious, streetwise "Detroit Red" and tended to use selective quotations by the fallen leader that gave justification for their use of weapons in challenging police brutality.

The widespread sampling of Malcolm's speeches on hip-hop videos and albums, plus the popular acclaim for Lee's biopic, culminated into "Malcolmania" in 1992 and 1993. There were "X" posters, coffee mugs, potato chips, T-shirts, and "X-caps," which newly elected President Bill Clinton wore occasionally when jogging outside the White House in the morning. At the time, CBS News estimated the commercial market for X-related products at $100 million annually. The Malcolmania hype had the effect of transporting the X-man from being merely a black superhero into the exalted status of mainstream American idol.

This new privileged status for Malcolm X was even confirmed officially by the U.S. government. On January 20, 1999, about 1,500 officials, celebrities, and guests crowded into Harlem's Apollo Theater to mark the issuance by the U.S. Postal Service of the Malcolm X postage stamp. Prominently in attendance were actors Ruby Dee, Ossie Davis, and Harry Belafonte. Also on hand was Harlem millionaire entrepreneur, media mogul (and Malcolm's former attorney) Percy Sutton. The Malcolm X stamp was the Postal Service's latest release in its "Black Heritage Stamp Series." Pennsylvania Congressman Chaka Fattah, the ranking Democrat on the House of Representatives Postal Subcommittee, remarked at the festive occasion, "There is no more appropriate honor than this stamp because Malcolm X sent all of us a message through his life and his life's work." To Congressman Fattah, Malcolm X's "thoughts, his ideas, his conviction, and his courage provide an inspiration even now to new generations to come." Few in the audience could ignore the rich irony of this event. One of America's sharpest and most unrelenting critics was now being praised and honored by the same government that had once carried out illegal harassment and surveillance against him. Ossie Davis, who understood the significance of this bittersweet moment better than anyone else, jokingly quipped, "We in this community look upon this commemorative stamp finally as America's stamp of approval."

The Malcolm X postage stamp was the twenty-second release in the "Black Heritage Series," which had previously featured other black heroes such as Frederick

Douglass, Harriet Tubman, Martin Luther King, Jr., Mary McLeod Bethune, and W. E. B. Du Bois. The U.S. Postal Service also released a short biographical statement accompanying the stamp's issuance, noting that the retouched photographic image of Malcolm X had been taken by an Associated Press photographer at a press conference held in New York City on May 21, 1964. The statement explains that soon after this photograph was taken, that Malcolm X "later broke away from the organization," referring to the NOI, and "disavowed his earlier separatist preaching." The most generous thing one could say about his curious statement was that it was the product of poor scholarship. The photograph actually had been taken during an interview in Cairo, Egypt, on July 14, 1964. Malcolm X had publicly broken from the NOI on March 8, 1964, two months earlier than the official statement had suggested. More problematic was the U.S. Postal Service's assertion that Malcolm X had become, before his death, a proponent of "a more integrationist solution to racial problems." But none of these errors of fact and slight distortions disturbed most who had gathered to celebrate. The Malcolm X postage stamp was a final and fitting triumph of his legacy. The full "Americanization of Malcolm X" appeared to be complete.

When, in 1987, I decided to write what was to have been a modest "political biography" of Malcolm X, there was already a substantial body of literature about him. By 2002, those published works had grown to roughly 930 books, 360 films and Internet educational resources, and 350 sound recordings. As I plowed through dozens, then hundreds, of books and articles, I was dismayed to discover that almost none of the scholarly literature or books about him had relied on serious research that would include a complete archival investigation of Malcolm's letters, personal documents, wills, diaries, transcripts of speeches and sermons, his actual criminal record, FBI files, and legal court proceedings. Some informative articles had appeared written by individuals who had either worked closely with Malcolm X or who described a specific event in which they had been brought into direct contact with the black leader. But these reminiscences lacked analytical rigor and critical insight. What staggered the mind, however, was the literal mountain of badly written articles, the turgid prose, and various academicstyled ruminations about Malcolm X's life and thought, nearly all based on the same, limited collection of secondary sources.

There was remarkably little Malcolm X literature that employed the traditional tools of historical investigation. Few writers had conducted fresh interviews with Malcolm X's widow, Dr. Betty Shabazz, any of his closest coworkers, or the extended Little family. Writers made few efforts to investigate the actual criminal record of Malcolm X at the time of his 1946 incarceration. Not even the best previous scholarly studies of Malcolm X—a small group of books including Peter Goldman's *The Death and Life of Malcolm X* (1973), Karl Evanzz's *The Judas Factor: The Plot to Kill Malcolm X* (1992), and Louis DeCaro's *On the Side of My People: A Religious Life of Malcolm X* (1966)—had amassed a genuine "archival" or substantive database of documentation in order to form a true picture of Malcolm-X-the-man rather than the pristine icon.[2] One problem in this was Malcolm

X's inescapable identification as the quintessential model of black masculinity, which served as a kind of gendered barricade to any really objective appraisal of him. Cultural critic Philip Brian Harper has observed that Malcolm X and the Black Powerites who later imitated him constructed themselves as virile, potent, and hypermasculinist, giving weight to the false impression that racial integrationists like King were weak and impotent.[3]

Nearly everyone writing about Malcolm X largely, with remarkably few exceptions, accepted *as fact* most, if not all, of the chronology of events and personal experiences depicted in the *Autobiography*'s narrative. Few authors checked the edited, published "transcripts" of Malcolm X's speeches as presented in *Malcolm X Speaks* and *By Any Means Necessary* against the actual tape recordings of those speeches or the transcribed excerpts of the same talks recorded by the FBI. Every historian worth her or his salt knows that "memoirs" like the *Autobiography* are inherently biased. They present a representation of the subject that privileges certain facts, while self-censuring others. There are deliberate omissions, the chronological reordering of events, and name changes. Consequently, there existed no comprehensive biography of this man who arguably had come to personify modern, urban black America in the past half-century.

There continued to be, for me, so many unanswered basic questions about this dynamic yet ultimately elusive man that neither the *Autobiography*, nor the other nine hundred-plus books written about him, had answered satisfactorily. The most obvious queries concerned his murder. Substantial evidence had been compiled both by Goldman and attorney William Kuntsler that indicated that two of the men convicted in 1966 for gunning down Malcolm at the Audubon Ballroom, Thomas 15X Johnson and Norman 3X Butler were completely innocent. In 1977, the only assassin who had been wounded and captured at the crime scene, Talmadge Hayer, had confessed to his prison clergyman that both Johnson and Butler had played absolutely no roles in the murder, confirming that, in fact, they had not even been present at the Audubon that afternoon.

There had always been whispers, for years, that Louis Farrakhan had been responsible for the assassination; he had been Malcolm X's closest protégé, and then following his vitriolic renunciation of Malcolm, inherited the leadership of Harlem's Mosque Number Seven following the murder. Then I had to explain the inexplicable behavior of the New York Police Department (NYPD) on the day of the assassination. Usually one to two dozen cops blanketed any event where Malcolm X was speaking. Normally at the Audubon rallies, a police captain or lieutenant was stationed in a command center above the Audubon's main entrance, on the second floor. Fifteen to twenty uniformed officers, at least, would be milling at the periphery of the crowd, a few always located at a small park directly across the street from the building. On February 21, 1965, however, the cops almost disappeared. There were no uniformed officers in the ballroom, at the main entrance, or even in the park at the time of the shooting. Only two NYPD patrolmen were inside the Audubon but at the opposite end of the building. When the NYPD investigation team arrived, forensic evidence was not properly collected and significant eyewitnesses still at the scene were not interviewed for days and, in several instances, weeks later. The crime scene itself was preserved for only a couple of hours. By 6 p.m., only three hours after Malcolm X's killing, a housekeeper with detergent and a bucket of water mopped up the floor, eliminating the bloody evidence. A dance was held in the same ballroom at 7 p.m. that night, as originally scheduled.

Perhaps I could never completely answer the greatest question about Malcolm X: if he *had* lived, or somehow had survived the assassination attempt, what could he have become? How would have another three or four decades of life altered how we imagine him and the ways we interpret his legacy? The legion of books that he inspired presented widely different, and even diametrically opposing, theories on the subject. Virtually every group—the orthodox, Sunni Islamic community, black cultural nationalists, Trotskyists, prisoners and former prisoners, mainstream integrationists, and hip-hop artists—had manipulated the "black shining prince" to promote their own agendas, or to justify their causes. The enormous elasticity of Malcolm's visual image could be universally appropriated, stretching from Ice Cube's 1992 apocalyptic "predator" to being used as the template for the film character "Magneto" in the 1999 block buster hit, "The X-Men," illustrated the great difficulty I now confronted. Malcolm X was being constantly *reinvented* within American society and popular culture.

But the first, most original, and most talented revisionist of Malcolm X was Malcolm X himself. I slowly began to realize that Malcolm X continuously and astutely refashioned his outward image, artfully redesigning his public style and even language to facilitate overtures to different people in varied contexts and yet beneath the multiple layers of reinvention, *who was he?* Was the powerful impact of his short thirty-nine years of existence actually grounded in what he had really accomplished, or based on the unfulfilled promise of what he might have become? Malcolm X is memorialized by millions of Americans largely because of the *Autobiography*, which is today a standard text of American literature. But was Malcolm's *hajj* to Mecca in April 1964, the dramatic turning point of the *Autobiography*, the glorious epiphany Malcolm claimed it was at the time, and that virtually all other interpreters of him have uncritically accepted? Was this spiritual metamorphosis, the embracing of color blindness, and the public denouncing of Elijah Muhammad's sexual misconduct all just part of the political price he was now prepared to pay to gain entry into the Civil Rights Movement's national leadership? Was this not the final "reincarnation" the necessary role change for El-Hajj Malik El-Shabazz to reach inside the court of the Saudi royal family and to gain access to the corridors of governmental power throughout the newly independent nations of Africa and Asia?

With so many unanswered questions to explore, there seemed to me to exist paradoxically a *collective conspiracy of silence surrounding Malcolm X*, an unwitting, or perhaps witting, attempt not to examine things too closely, to stick to the accepted narrative offered by the *Autobiography* and Lee's biopic. By not peering below the surface, there would be no need to adjust the crafted image we have learned to adore, frozen in time. We could simply all find enduring comfort in the safe, masculinist gaze of our "black shining prince."

Historians are trained in graduate school to state only what we can actually *prove*, based primarily on archival or secondary source evidence. Information we collect from oral interviews can only be used from informed subjects, who have an opportunity to review what they have said for the record. Thus, the discipline itself provides certain safeguards to interviewees and informants. Most historians, in other words, do not see themselves as investigative reporters or would-be "cold case" investigators. Yet the skills of both seemed to me necessary in order to crack open the Malcolm X collective conspiracy of silence. Malcolm's actual legacy was dogmatically preserved and fiercely guarded by nearly everyone privy to important information pertaining to him. This was a highly unusual situation for a researcher to confront, especially considering that Malcolm X lived a very "public" existence, appearing on numerous television shows, and speaking at literally thousands of venues across the country.

When I started my biography of Malcolm X in 1987, I was then Chair of Black Studies at Ohio State University. Working with several graduate students, we began compiling photocopies of articles about Malcolm X that appeared in academic journals. We began a newspaper-clipping file of more recent media coverage related to our subject (remember, these were the days before the Internet and World Wide Web). I knew that I would need to penetrate four principal, core areas of investigation in order to present a really balanced and fair portrait of the man. These four broad areas were (1) the black organizations in which Malcolm X played a significant leadership role—the Nation of Islam, the Muslim Mosque, Inc., and the Organization of Afro-American Unity; (2) the surveillance of Malcolm X by the FBI and other governmental agencies; (3) the materials of Alex Haley, coauthor of the *Autobiography*, used in preparing the book; and, of course; (4) the family of Malcolm X, especially his widow, Dr. Betty Shabazz, and their access to any manuscripts, correspondence, texts, or transcripts of speeches and sermons, legal documents, and odd paraphernalia. All four of these areas, for different reasons, proved to be intractable. In 1989, I accepted a professorship in ethnic studies at the University of Colorado in Boulder, and in the following academic year, I organized a research team of six to ten graduate students and work-study assistants, who were dedicated to reconstructing Malcolm's life. After three hard years, we had made, at best, marginal headway. I then accepted my current appointment at Columbia. I had no idea at that time that another decade would elapse before I could really successfully infiltrate these four core areas of Malcolm-related investigation.

The first nearly overwhelming difficulty was the lack of a comprehensive, well-organized archive on Malcolm X. Primary source materials, such as correspondence and personal manuscripts, were literally scattered and fragmented. For some inexplicable reason, the Shabazz family had never authorized a group of historians or archivists to compile these rare documents into a central, publicly accessible repository. By my own count, as of 2003, chunks of Malcolm's core memorabilia were located at seventy-three different U.S. archives and libraries, including the Library of Congress, New York University's Tamiment Library, the Schomburg Center of the New York Public Library, Cornell University Library, Wayne State University Library, the State Historical Society of Wisconsin, Emory University Library, Howard University Library, and Columbia University's Oral History Research Center. I contrasted this chaotic situation to the professionally archived life records of Dr. Martin Luther King, Jr., then at Atlanta's King Center, that would serve as the core database for historian Clayborne Carson's magnificent, thirty-year-long effort, the King Papers Project. Booker T. Washington's papers, carefully archived and preserved, fill exactly 1,077 linear feet of archival boxes at the Library of Congress. Most dedicated Malcolmologists also knew that Dr. Shabazz still retained hundreds of documents and manuscripts by her late husband in her Mount Vernon, New York, home. But no one really had a clue how much primary source material there was and whether any efforts had been made to preserve it.

I had more than a nostalgic desire to preserve memorabilia. As a historian, I also knew that all artifacts made by human beings inevitably disintegrate. Paper, left unprotected, without a climate-controlled environment and acid-free folders, "lives" only about seventy-five years. Audiotape recordings based on magnetic recording technology survive about forty years. People who had worked closely with Malcolm X and who had known him intimately would nearly all be dead in another two decades. Only the Shabazz family had the moral authority to initiate such an undertaking, to secure Malcolm X's place in history. It simply did not make sense. Much later, in 2002, when the near-public auction by Butterfield's of a major cache of Malcolm memorabilia fetching offers of $600,000 and more came to light, I discovered that the Shabazz family had squirreled away several hundred pounds of Malcolm X-related documents and material. Maliakah Shabazz, the youngest daughter, had, without the rest of the family's knowledge, managed to pack and transport her father's materials to a Florida storage facility. Her failure to pay the storage facility's monthly fee led to the seizure and disposition of the bin's priceless contents. The new purchaser, in turn, had contacted eBay and Butterfield's to sell what he believed to be his property. Only a legal technicality voided the sale, returning the memorabilia to the Shabazzes.

After the international publicity and outcry surrounding the Butterfield's abortive auction, however, the Shabazzes decided to deposit their materials at the Schomburg Center in Harlem. In January 2003, the Schomburg publicly announced to the media its acquisition on the basis of a seventy-five-year loan. I have previously written in detail about the Butterfield's abortive auction fiasco, and I had been extensively involved in the financial negotiations with the auction house and the Shabazzes on behalf of Columbia University. But what *none* of the principals, including myself, could bear to ask ourselves and the Shabazz Estate in public, is *why* hundreds of pounds of documents, speeches, manuscripts, Malcolm's Holy Qur'an, and so on, been left deteriorating in storage in their basement *for thirty-five years*?

The intransigence of the Shabazzes forced me to contemplate negotiations with the Nation of Islam. I had written extensively, and quite critically, about Louis Farrakhan and his philosophy over a number of years. Yet during my research, I had learned that Muslim ministers like Malcolm X, under the strict authoritarian supervision of Elijah Muhammad, had been required to submit weekly reports about their mosques' activities. All sermons they delivered were audiotaped, with the tapes mailed to the national headquarters in Chicago. When I broached the possibility of examining their archives through third parties, the NOI curtly refused, explaining that they wished to "protect Dr. Betty Shabazz and her family."

Another potential avenue of biographical inquiry existed among Malcolm X's friends and associates in Muslim Mosque, Inc., and the Organization of Afro-American Unity (OAAU), two groups formed in 1964 and that had disintegrated in the months after Malcolm's assassination. Here I had better luck. Prior friendships with several prominent individuals, such as actor Ossie Davis, provided valuable oral histories of their relationships with Malcolm X. Most key individuals I wanted to interview, however, were either reclusive or elusive. Some were literally "underground" and living in exile in either South America, the Caribbean, or in Africa. A few, such as writer Sylvester Leaks, cordially agreed to converse off the record, then angrily refused to be formally interviewed. Some pivotal figures such as Malcolm X's personal bodyguard, Reuben Francis, had literally disappeared months following the murder. I subsequently learned that Francis had somehow been relocated to Mexico sometime in 1966 and, from then, fell into complete obscurity. Lynn Shiflett, OAAU secretary and a trusted personal assistant to Malcolm X, had refused all interviews and even written contacts since 1966.

The FBI avenue of inquiry proved to be even more daunting. Despite the passage of the Freedom of Information Act, which required the bureau to declassify its internal memoranda that required secrecy for the sake of national security, by 1994, only about 2,300 pages of an estimated *50,000 pages* of surveillance on Malcolm X was made public. Much of this information was heavily redacted or blacked out by FBI censors, supposedly to protect its informants or to preserve "national security." For several years, a group of my student research assistants helped me to make sense of this maze of FBI bureaucratic mumbo jumbo. Eventually, I learned that whatever the FBI's original motives, they fairly accurately tracked Malcolm X's precise movements, public addresses, and dozens of telephone calls, all without legal warrants, of course. In 1995, Farrakhan had proposed announcing a national campaign to pressure the bureau to open up its archives about Malcolm X and especially to release any relevant information concerning his assassination. According to Farrakhan, the Shabazz family insisted that there be no effort to force the bureau to divulge what it knew. Friends close to the family subsequently explained to me that the memories were still too painful, even after thirty long years. A public inquiry would be too traumatic for all concerned.

Without my knowledge, historian Clayborne Carson at Stanford University was, in the early 1990s, working independently along parallel lines. He successfully annotated the FBI memoranda available at that time, publishing an invaluable reference work, *Malcolm X: The FBI File* (New York: Ballantine Books, 1995). Prior to the book's publication, however, attorneys for Dr. Shabazz expressed concerns that Carson should severely *limit* the amount of original material lifted by the FBI from Malcolm X's orations, writings, and wiretapped conversations. Thus, the book that was comprised of letters and transcribed tape recordings already heavily censored by the FBI was, in effect, *censored a second time* for the purposes of not violating copyright infringement. When black studies scholar Abdul Alkalimat prepared a primer text, *Malcolm X for Beginners*, Dr. Shabazz threatened a lawsuit based on the unusual legal claim that *anything* ever uttered by her late husband was her "intellectual property." Alkalimat finally consented to surrender any and all claims to royalties from the book to the Shabazz estate.

I then confronted the enigma of Alex Haley. Haley was the highest-selling author of black nonfiction in U.S. history. His greatest achievement had been the 1976 *Roots*, which, like the coauthored *Autobiography*, had become a celebrated, iconic text of black identity and culture. Yet statements about Malcolm X made by Haley shortly before his death seemed, to me, strangely negative. Haley had even asserted that both Malcolm X and Dr. King were going "downhill" before their deaths.[4] Haley had placed his papers at the University of Tennessee's archives in January 1991. Yet there remained unusual restrictions on scholarly access to his personal records. I personally visited his archives in Knoxville twice. No photocopying of any document in the Haley files is permitted without the prior written approval of his attorney, Paul Coleman of Knoxville, Tennessee. My letters to Coleman were unanswered. When in Knoxville during my second visit, I persuaded the archive's curator to phone Coleman directly on my behalf. Attorney Coleman then explained to me over the telephone that he needed to know the *precise pages* or documents to be photocopied, *in advance*! In practical terms, scholars are forced to copy passages in pencil, by hand, from Haley's archives. This laborious model of information transferral worked well for monks in the Middle Ages but seems inappropriate for the age of digital technology.

As luck would have it, several years before Haley's death, he had named researcher Anne Romaine as his "official biographer." Romaine was a white folksinger, trained neither as a historian nor as a biographer. Yet she was apparently diligent and serious about her work. Between the late 1980s until her death in 1995, Romaine had conducted audiotaped interviews with over fifty individuals, some of whom covered the background to Haley's role in producing the *Autobiography*. The great bulk of Romaine's papers and research materials pertaining to the *Autobiography* were also donated to the University of Tennessee's archives. To my delight, there were absolutely no restrictions on Romaine's papers—everything can be photocopied and reproduced. One folder in Romaine's papers includes the "raw materials" used to construct Chapter 16 of the *Autobiography*. Here, I found the actual mechanics of the Haley-Malcolm X collaboration. Malcolm X apparently would speak to Haley in "free style"; it was left to Haley to take hundreds of sentences into paragraphs and then appropriate subject areas. Malcolm also had a habit of scribbling notes to himself as he spoke. Haley learned to pocket these sketchy notes and later reassemble them, integrating the conscious with subconscious reflections into a workable narrative. Although Malcolm X retained final approval of their hybrid text, he was not privy to the actual editorial processes superimposed from Haley's side. Chapters the two men had prepared were sometimes split and restructured into other chapters. These details may appear mundane and insignificant. But considering that Malcolm's final "metamorphosis" took place in 1963 through 1965, the exact timing of when individual chapters were produced takes on enormous importance.

These new revelations made me realize that I also needed to learn much more about *Haley*. Born in Ithaca, New York, in 1921, Alex Haley was the oldest of three sons of Simon Alexander Haley, a professor of agriculture, and Bertha George Palmer, a grade school teacher. Haley had been raised as a child in Henning, Tennessee. As a teenager, in 1939, Haley enlisted in the U.S. Coast Guard as a mess boy. During World War II, he had come to the attention of white officers for his flair as a talented writer. During long assignments at sea, Haley had ghostwritten hundreds of love letters for sailors' wives and sweethearts back at home. While Haley's repeated efforts to gain print publication for his unsolicited manuscripts failed for eight long years, his extracurricular activities gained the approval and admiration of his white superiors. By the late 1940s, Haley was advanced into a desk job; by the mid-1950s, he was granted the post of "chief journalist" in the Coast Guard. After putting in twenty years' service, Haley started a career as a professional freelance writer. Politically, Haley was both a Republican and a committed advocate of racial integration. He was not, unlike C. Eric Lincoln or other African American scholars who had studied the NOI's activities during the late 1950s, even mildly sympathetic with the black group's aims and racial philosophy.

To Haley, the separatist Nation of Islam was an object lesson in America's failure to achieve interracial justice and fairness. As Mike Wallace's controversial 1959 television series on the Black Muslims had proclaimed, they represented "The Hate That Hate Produced." Haley completely concurred with Wallace's thesis. He, too, was convinced that the NOI was potentially a dangerous, racist cult, completely out of step with the lofty goals and integrationist aspirations of the Civil Rights Movement. Haley was personally fascinated with Malcolm's charisma and angry rhetoric but strongly disagreed with many of his ideas. Consequently, when Haley started work on the *Autobiography*, he held a very different set of objectives than those of Malcolm X. The Romaine papers also revealed that one of Haley's early articles about the NOI, coauthored with white writer Alfred Balk, had been written *in collaboration with the FBI*. The FBI had supplied its information about the NOI to Balk and Haley, which formed much of the basis for their *Saturday Evening Post* article that appeared on January 26, 1963, with the threatening title, "Black Merchants of Hate."[5]

I then began to wonder, as I poured through Romaine's papers, what Malcolm X really had known about the final text that would become his ultimate "testament." Could I not discover a way to find out what was going on inside Haley's head, or at Doubleday, which had paid a hefty $20,000 advance for the *Autobiography* in June 1963? And why, only three weeks following Malcolm X's killing, had Doubleday canceled the contract for the completed book? The *Autobiography* would be eventually published by Grove in late 1965. Doubleday's hasty decision would cost the publisher millions of dollars.

The Library of Congress held the answers. Doubleday's corporate papers are now housed there. This collection includes the papers of Doubleday's then-executive editor, Kenneth McCormick, who had worked closely with Haley for several years as the *Autobiography* had been constructed. As in the Romaine papers, I found more evidence of Haley's sometimes-weekly private commentary with McCormick about the laborious process of composing the book. These Haley letters of marginalia contained some crucial, never previously published intimate details about Malcolm's personal life. They also revealed how several attorneys retained by Doubleday closely monitored and vetted entire sections of the controversial text in 1964, demanding numerous name changes, the reworking and deletion of blocks of paragraphs, and so forth. In late 1963, Haley was particularly worried about what he viewed as Malcolm X's anti-Semitism. He therefore rewrote material to eliminate a number of negative statements about Jews in the book manuscript, with the explicit covert goal of "getting them past Malcolm X," without his coauthor's knowledge or consent. Thus, the censorship of Malcolm X had begun well *prior* to his assassination.

A cardinal responsibility of the historian is to relate the full truth, however unpleasant. In the early 1960s, the Nation of Islam had been directly involved with the American Nazi Party and white supremacist organizations—all while Malcolm X had been its "national representative." This regrettable dimension of Malcolm's career had to be thoroughly investigated yet few scholars, black or white, had been willing to do so. In 1998, in my book *Black Leadership* (New York: Columbia University Press, 1998), I had described Farrakhan's anti-Semitic, conservative, black nationalism as an odious brand of "Black Fundamentalism." Farrakhan had been Malcolm's prime protégé, and the question must now be posed whether Malcolm X was partially responsible for the bankrupt political legacy of black anti-Semitism and Black Fundamentalism. The whole truth, not packaged icons, can only advance our complete understanding of the real man and his times.

The Romaine papers also had provided clear evidence that the lack of a clear political program or plan of action in the *Autobiography* was no accident. Something was indeed "missing" from the final version of the book as it appeared in print in late 1965. In Haley's own correspondence to editor Kenneth McCormick, dated January 19, 1964, Haley had even described these chapters as having "the most impact material of the book, some of it rater lava-like."[6] Now my quest shifted to finding out what the contents of this "impact material" were. The trail now led me to Detroit attorney Gregory Reed. In late 1992, Reed had purchased the original manuscripts of the *Autobiography* at the sale of the Haley Estate for $100,000. Reed has, in his possession in his office safe, the three "missing chapters" from the *Autobiography*, which still have never been published. I contacted Reed and, after several lengthy telephone conversations, he agreed to show me the missing *Autobiography* chapters. With great enthusiasm, I flew to Detroit and telephoned Reed at our agreed-upon time. Reed then curiously rejected meeting me at his law office. He instead insisted that we meet at a downtown restaurant. I arrived at our meeting place on time and a half hour later, Reed showed up, carrying a briefcase.

After exchanging a few pleasantries, Reed informed me that he had not brought the entire original manuscript with him. However, he would permit me to read, at the restaurant table, small selections from the manuscript. I was deeply disappointed but readily accepted Reed's new terms. For roughly fifteen minutes, I quickly read parts from the illusive "missing chapters." That was enough time for me to ascertain, without doubt, that these text fragments had been dictated and written sometime between October 1963 and January 1964. This coincides with the final months of Malcolm's NOI membership. More critically, in these missing chapters, Malcolm X proposed the construction of an unprecedented, African American united front of black political and civic organizations, including both the NOI and civil rights groups. He perhaps envisioned something similar in style to Farrakhan's Million Man March of 1995. Apparently, Malcolm X was aggressively pushing the NOI beyond Black Fundamentalism into open, common dialogue and political collaboration with the civil rights community. Was this the prime reason that elements inside both the NOI and the FBI may have wanted to silence him? Since Reed owns the physical property, but the Shabazz estate retains the intellectual property rights of its contents, we may never know.

With each successive stumbling block, I became more intrigued. The complicated web of this man's life, the swirling world around him, his friends, family, and intimate associates, became ever more tangled and provocative. The tensions between these at-times feuding factions, the innuendos, the missed opportunities, the angry refusals to speak on the record, the suppression of archival evidence, the broken loyalties, and constant betrayals all seemed too great. It required of me a difficult journey of many years, even to possess the knowledge of how to untangle the web, to make sense of it all. What I acquired, however, by 2003 and 2004, was a true depth of understanding and insight that was surprising and much more revelatory than I had ever imagined. I finally learned that the answer to the question—why was this information about Malcolm X so fiercely protected—because the life, and the man, had the potential to become much more dangerous to white America than any single individual had ever been.

Malcolm X, the *real* Malcolm X, was infinitely more remarkable than the personality presented in the *Autobiography*. The man who had been born Malcolm Little, and who had perished as El-Hajj Malik El-Shabazz, was no saint. He made many serious errors of judgment, several of which directly contributed to his

murder. Yet despite these serious contradictions and personal failings, Malcolm X also possessed the unique potential for uniting black America in any unprecedented coalition with African, Asian, and Caribbean nations. He alone could have established unity between Negro integrationists and black nationalists inside the United States. He possessed the personal charisma, the rhetorical genius, and the moral courage to inspire and motivate millions of blacks into unified action. Neither the *Autobiography* nor Spike Lee's 1992 movie revealed this powerful legacy of the man or explained what he could have accomplished. What continues to be suppressed and censored also tells us something so huge about America itself, about where we were then and where we, as a people, are now. Malcolm X was potentially a new type of world leader, personally drawn up from the "wretched of the earth" into a political stratosphere of international power. Telling that remarkable, true story is the purpose of my biography.

NOTES

1. Immediately following Malcolm X's assassination, several individuals who had worked closely with the fallen leader sought to document his meaning in the larger black freedom struggle. These early texts include Leslie Alexander Lacy, "Malcolm X in Ghana," in *Malcolm X: The Man and His Times*, ed. John Henrik Clarke (New York: Macmillan, 1969), 217–55; Ossie Davis, "Why I Eulogized Malcolm X," *Negro Digest* 15, no. 4 (February 1966): 64–66; Wyatt Tee Walker, "On Malcolm X: Nothing But A Man," *Negro Digest* 14, no. 10 (August 1965): 29–32; and Albert B. Cleage, Jr., "Brother Malcolm," in *The Black Messiah*, ed. Albert B. Cleage, Jr. (New York: Sheed and Ward, 1968), 186–200. The advocates of Black Power subsequently placed Malcolm X firmly within the black nationalist tradition of Martin R. Delany and Marcus Garvey, emphasizing his dedication to the use of armed self defense by blacks. Amiri Baraka's essay, "The Legacy of Malcolm X, and the Coming of the Black Nation," in LeRoi Jones, *Home: Social Essays* (New York: William Morrow, 19966), 238–50, became the template for this line of interpretation. Following Baraka's black nationalist thesis were Eldridge Cleaver, "Initial Reactions on the Assassination of Malcolm X," in *Soul on Ice*, ed. Eldridge Cleaver (New York: Ramparts, 1968), 50–61; James Boggs, "King Malcolm, and the Future of the Black Revolution," in *Racism and the Class Struggle: Further Pages from a Black Worker's Notebook*, ed. James Boggs (New York: Monthly Review Press, 1970), 104–29; Cedrick Robinson, "Malcolm Little as a Charismatic Leader," *Afro-American Studies* 2, no. 1 (September 1972): 81–96; and Robert Allen, *Black Awakening in Capitalist America* (Garden City, NY: Anchor/Doubleday, 1970), esp. 30–40.

2. The best available studies of Malcolm X merit some consideration here. Although originally written more than three decades ago, *Newsweek* editor-journalist Peter Goldman's *The Death and Life of Malcolm X* (New York: Harper and Row, 1973) still remains an excellent introduction to the man and his times. Well-written and researched, Goldman based the text on his own interviews with the subject. Karl Evanzz's *The Judas Factor: The Plot to Kill Malcolm X* (New York: Thunder's Mouth Press, 1992) presents a persuasive argument explaining the FBI's near-blanket surveillance of the subject. Evanzz was the first author to suggest that NOI National Secretary John Ali may have been an FBI informant. Louis A. DeCaro has written two thoughtful studies on Malcolm X's spiritual growth and religious orientation, *On the Side of My People: A Religious Life of Malcolm X* (New York: New York University Press, 1996) and *Malcolm and the Cross: The Nation of*

Islam, Malcolm X and Christianity (New York: New York University Press, 1998). DeCaro graciously agreed to be interviewed in 2001 for the Malcolm X Project at Columbia. The field of religious studies has also produced other informative interpretations of Malcolm X. These works include Lewis V. Baldwin, *Between Cross and Crescent: Christian and Muslim Perspectives on Malcolm and Martin* (Gainesville: University of Florida, 2002); a sound recording by Harnam Cross, Donna Scott, and Eugene Scals, "What's Up with Malcolm? The Real Future of Islam" (Southfield, MI: Readings for the Blind, 2001); Peter J. Paris, *Black Religious Leaders: Conflict in Unity* (Westminster: John Knox Press, 1991).

3. Philip Brian Harper, in his book, *Are We Not Men? Masculine Anxiety and the Problem of African-American Identity* (New York: Oxford University Press, 1996), argues that the simplistic stereotypes of King and his courageous followers as being "non-masculine" and "effeminate" and leaders such as Malcolm X and Stokely Carmichael as "super-masculine, Black males" became widely promulgated. "The Black Power movement," Harper observes, was "conceived in terms of accession to a masculine identity, the problematic quality of those terms notwithstanding" (68).

4. In an extraordinary interview with writer Thomas Hauser, Alex Haley stated that he had "worked closely with Malcolm X, and I also did a *Playboy* interview with Martin Luther King during the same period, so I knew one very closely and the other a little." Based on his knowledge of both men, he had concluded that they had "both died tragically at about the right time in terms of posterity. Both men were . . . beginning to decline. They were under attack." In Haley's opinion, Malcolm, in particular, "was having a rough time trying to keep things going. Both of them were killed just before it went really downhill for them, and as of their death, they were practically sainted." See Thomas Hauser, *Muhammad Ali: His Life and Times* (New York: Touchstone/Simon and Schuster, 191), 508.

5. Alfred Balk had contacted the FBI in October 1962, seeking the bureau's assistance in collecting information about the Nation of Islam for the proposed article he and Haley would write for the *Saturday Evening Post*. The bureau gave Balk and Haley the data they requested, with the strict stipulation that the FBI's assistance not be mentioned. The bureau was later quite pleased with the published article. See M. A. Jones to Mr. DeLoach, FBI Memorandum, October 9, 1963, in the Anne Romaine Papers, Series 1, Box 2, folder 16, University of Tennessee Library Special Collections. See also Alfred Balk and Alex Haley, "Black Merchants of Hate," *Saturday Evening Post*, January 26, 1963.

6. On January 9, 1964, Haley wrote to Doubleday Executive Editor Kenneth McCormick and his agent, Paul Reynolds, that "the most impact material of the book, some of it rather lava-like, is what I have from Malcolm for the three essay chapters, 'The Negro,' 'The End of Christianity,' and "Twenty Million Black Muslims.'" See Alex Haley to Kenneth McCormick, Wolcott Gibbs, Jr., and Paul Reynolds, January 19, 1964, in Annie Romaine Collection, the University of Tennessee Library Special Collection.

ABOUT THE CONTRIBUTORS

Hishaam D. Aidi is a lecturer at Columbia University's School of International and Public Affairs and a contributing editor of *Souls: A Critical Journal of Black Culture, Politics and Society*. A 2008–9 Carnegie Scholar, he is the author of *Redeploying the State* (Palgrave Macmillan, 2009) and has written for *The New African*, *Africana*, *Colorlines*, and *Middle East Report*.

Zain Abdullah is an assistant professor in the Religion Department at Temple University where he also teaches in the Department of Geography and Urban Studies. His research interests include Islam-American culture and religion, African diaspora studies, film production and photography, globalization, and transnationalism. His forthcoming book *Black Mecca: The African Muslims of Harlem* (Oxford University Press, 2009) explores how West Africans navigate their presence in New York City by negotiating their black, African, and Muslim identities. He served as an advisor for the documentary Dollars & Dreams: West Africans in New York (2007), which screened at the FESPACO Pan-African Film Festival in Burkina Faso, West Africa.

Vivek Bald is a documentary filmmaker and an assistant professor of writing and digital media at the Massachusetts Institute of Technology. His films include *Taxivala/Auto-biography* (1994), about South Asian immigrant taxi drivers in New York City, and *Mutiny: Asians Storm British Music* (2003), about South Asian youth, music, and antiracist politics in 1970s through 1990s Britain. His current work, which focuses on the desertion and settlement of Indian Muslim merchant sailors in U.S. port cities in the late nineteenth and early twentieth centuries, is the basis for a forthcoming book, *Bengali Harlem and the Hidden Histories of South Asian New York*, and a documentary film, *In Search of Bengali Harlem*.

Moustafa Bayoumi is an associate professor of English at Brooklyn College, City University of New York (CUNY). He is the coeditor of *The Edward Said Reader* (Vintage) and the author of *How Does It Feel to Be a Problem? Being Young and Arab in America* (Penguin, 2008). His essays have appeared in *The Best Music Writing 2006*, *The Nation*, *The London Review of Books*, *The Village Voice*, *The Yale Journal of Criticism*, *Transition*, *Interventions*, *Arab Studies Quarterly*, and other places.

Edward E. Curtis IV is Millennium Chair of the Liberal Arts and professor of religious studies at Indiana University–Purdue University Indianapolis (IUPUI). He is the author or editor of several books on Muslim American and African American history, including *Muslims in America: A Short History* and *The Columbia Sourcebook of Muslims in the United States*, which was selected by Choice as an outstanding academic title of 2008. A former NEH Fellow at the National Humanities Center, he has also been awarded Carnegie, Fulbright, and Mellon fellowships. Most recently, Professor Curtis has completed work as general editor of the two-volume *Encyclopedia of Muslim-American History*.

Sohail Daulatzai is an assistant professor in African American Studies and Film and Media Studies at the University of California, Irvine. His current book projects include *Born to Use Mics* (with Michael Eric Dyson), which is a meditation on Nas's 1994 album *Illmatic*; and *Return of the Mecca*, a project that explores the connections between Muslim diasporas and black radicalism through social movements, cinema, sports, literature, and hip-hop culture. He is also currently in production on a graphic novel on the 1965 film *Battle of Algiers*, and is an executive producer and creator of the forthcoming hip-hop album titled *Free Rap*, a benefit album for imprisoned Muslim leader Imam Jamil Al-Amin (formerly known as H. Rap Brown).

Keith P. Feldman is an assistant professor of Comparative Ethnic Studies at the University of California, Berkeley. His current book project, *Racing the Question: Israel/Palestine and U.S. Imperial Culture*, takes a comparative racialization approach to reading a range of cultural production circulating between the United States, the Middle East, and North Africa. He has published in *CR: New Centennial Review, MELUS, postmodern culture*, and in the edited collection *Arab Women's Lives Retold: Exploring Identity through Writing* (Syracuse University Press, 2007), as well as in encyclopedias of ethnic, African American, and postcolonial literatures.

Zareena Grewal is an Ethnicity, Race, and Migration Scholar at Yale's Center for International and Area Studies and a lecturer in American Studies at Yale. She received her PhD in anthropology and history from the University of Michigan, has taught at the University of Michigan and Vassar College, and was awarded the Fulbright's prestigious Islamic Civilization Grant. She also directed and produced a documentary on patriotism, racism, and Muslims in the United States, titled *By Dawn's Early Light: Chris Jackson's Journey to Islam* (2004), which is distributed by Cinema Guild and was recently featured on ABC News Now and the Documentary Channel.

Alex Lubin is an associate professor and chair in the Department of American Studies at the University of New Mexico. He is the author of *Romance and Rights: The Politics of Interracial Intimacy, 1945–1954* (University Press of Mississippi, 2005) and the editor of *Revising the Blueprint: Ann Petry and the Literary Left* (University Press of Mississippi, 2007). He is currently working on a manuscript on African American intellectual and political history concerning the question of Palestine.

Sherman A. Jackson is presently a professor of Arabic and Islamic Studies, a visiting professor of law, and a professor of Afro-American Studies at the University of Michigan. He is author of *Islamic Law and the State: The Constitutional Jurisprudence of Shihâb al-Dîn al-Qarâfî* (Brill, 1996), *On the Boundaries of Theological Tolerance in Islam: Abû Hâmid al-Ghazâlî's Faysal al-Tafriqa* (Oxford University Press, 2002), *Islam and the Blackamerican: Looking Towards the Third Resurrection* (Oxford University Press, 2005), and *Islam and Black Theodicy: Classical Islamic Theology in Modern America* (Oxford University Press, forthcoming).

Jamillah Karim is an assistant professor in religious studies at Spelman College. She obtained her PhD in Islamic studies at Duke University. She specializes in Islam in America, women and Islam, and race and immigration. She is author of the book *American Muslim Women: Negotiating Race, Class, and Gender within the Ummah*, which explores relations between African American and South Asian immigrant Muslims in the United States.

Su'ad Abdul Khabeer is a doctoral candidate in the Department of Anthropology at Princeton University. Su'ad's dissertation research explores the ways American Muslim youth negotiate their religious, racial, and cultural identities through hip-hop. Su'ad is also a research affiliate with the Center for Arts and Cultural Policy Studies at Princeton University and Senior Project Advisor for the documentary New Muslim Cool. Her publications include "Rep that Islam: The Rhyme and Reason of American Muslim Hip Hop" in the January 2007 issue of *The Muslim World* and "A Day in the Life," poetry that appeared in the anthology *Living Islam Out Loud: American Muslim Women Speak.*

Samir Meghelli is a Richard Hofstadter Fellow and PhD candidate in history at Columbia University. He is coauthor of *Tha Global Cipha: Hip Hop Culture and Consciousness* (Black History Museum Press, 2006) and has organized public history programs throughout the New York and Philadelphia areas. He received his BA (magna cum laude) from the University of Pennsylvania and his MA and MPhil from Columbia University. Meghelli's research interests include U.S. and African American cultural and political history, histories and theories of citizenship, and African diasporic cultures in the Atlantic and Francophone worlds.

Rami Nashashibi is a doctoral candidate in sociology at the University of Chicago and is working on a dissertation exploring the encounter of Islam with hip-hop and black street gangs in the postindustrial ghetto. He is a recipient of the Robert Park Fellowship, for which he developed and taught the course "Theorizing the Global Ghetto" at the University of Chicago. He is the author of "Ghetto Cosmopolitanism: Making Theory at the Margins," in a book edited by Saskia Sassen titled *Deciphering the Global: Its Scales, Spaces, and Subjects* (Routledge, 2007). Rami is also a full-time community organizer and Executive Director of the Inner-City Muslim Action Network (IMAN), a nonprofit organization that provides direct services while organizing around social justice issues affecting inner-city communities. Rami and IMAN were featured in a BBC Radio documentary, on the cover of the Chicago Tribune, and in a recent PBS documentary.

Richard Brent Turner, PhD, Princeton University, is the coordinator of the African American Studies Program and an associate professor in the Department of Religious Studies at the University of Iowa. He specializes in African American religious history and has been an associate at the W. E. B. Du Bois Institute at Harvard University. His books include *Islam in the African-American Experience* (2nd ed., Indiana University Press, 2003) and *Jazz Religion, the Second Line, and Black New Orleans*, which will be published by Indiana University Press in 2009. Dr. Turner is also the author of *Islam and Black Americans* (a volume in the Schomburg Studies on the Black Experience) and numerous journal articles and book chapters on African American and New World African religions.

INDEX

CPSIA information can be obtained at www.ICGtesting.com
Printed in the USA
BVOW08s0919250215

389244BV00005B/35/P

9 781403 977816